Islamic Ethics and the Genome Question

Studies in Islamic Ethics

Editorial Board

Emad Shahin (*Hamad Bin Khalifa University* (*HBKU*), *Doha, Qatar*)
Mutaz al-Khatib (*Research Center for Islamic Legislation & Ethics, HBKU*)
Ray Jureidini (*Research Center for Islamic Legislation & Ethics, HBKU*)

Managing Editor
P.S. van Koningsveld

VOLUME 1

The titles published in this series are listed at *brill.com/sie*

Islamic Ethics and the Genome Question

Edited by

Mohammed Ghaly

This publication is sponsored by the Research Center of Islamic Legislation and Ethics in Doha (Qatar), which is affiliated to the Faculty of Islamic Studies, Hamad Bin Khalifa University.

The Library of Congress Cataloging-in-Publication Data is available online at http://catalog.loc.gov
Library of Congress Control Number: 2018962012

Typeface for the Latin, Greek, and Cyrillic scripts: “Brill”. See and download: brill.com/brill-typeface.

ISSN 2589-3947
ISBN 978 9004 39212 0 (hardback)
ISBN 978 9004 39213 7 (e-book)

To my children:
Khadija, Maryam, Mustapha, Aisha, Hamza
and their generation,
born in the age of genomics

Hoping that such studies will help them
navigate their life, full of challenges ahead,
with due ethical commitment

Contents

Preface IX
About the Authors XI
Introduction 1
Mohammed Ghaly

PART 1
Collective Ijtihād and Genomics

1 Sharia Scholars and Modern Biomedical Advancements: What Role for Religious Ethics in the Genomic Era? 17
Mohammed Ghaly

2 Islamic Ethics and Genomics: Mapping the Collective Deliberations of Muslim Religious Scholars and Biomedical Scientists 47
Mohammed Ghaly

3 Transformation of the Concept of the Family in the Wake of Genomic Sequencing: An Islamic Perspective 80
Ayman Shabana

PART 2
Genomics and Rethinking Human Nature

4 Conceptualizing the Human Being: Insights from the Genethics Discourse and Implications for Islamic Bioethics 113
Aasim I. Padela

5 Islamic Perspectives on the Genome and the Human Person: Why the Soul Matters 139
Arzoo Ahmed and Mehrunisha Suleman

6 The Ethical Limits of Genetic Intervention: Genethics in Philosophical and Fiqhi Discourses 169
Mutaz al-Khatib

PART 3
Widening the Scope of Ethical Deliberations

7 In the Beginning Was the Genome: Genomics and the Bi-Textuality of Human Existence 203
Hub Zwart

8 Creation, Kinds and Destiny: A Christian View of Genome Editing 224
Trevor Stammers

9 *Living with the Genome,* by Angus Clark and Flo Ticehurst, within the Muslim Context 241
Ayman Shabana

PART 4
Contributions in Arabic

10 الجينوم والطبيعة البشرية: مقاربة تحليلية في ضوء الفلسفة والعلم التجريبي والأخلاق الإسلامية 253
سعدية بن دنيا

11 سؤال الجينوم بين الخِلْقة والأخلاق: مقاربة دلالية معرفية في أخلاقيات علم الجينوم من منظور إسلامي 284
عباس أمير

12 الجينوم والحياة: تمديد الحياة وأثره الأخلاقي على المجتمعات الإسلامية 304
عمارة الناصر

فهرس 331

Index 335

Preface

This volume originated from the proceedings of the three-day international seminar "Islamic Ethics and the Genome Question", organized by the Research Center for Islamic Legislation & Ethics (CILE) in Doha, Qatar, between 3rd and 5th April 2017.[1] This seminar makes part of the CILE series of interdisciplinary seminars which solicit contributions from researchers, scholars and experts in various fields in order to address key ethical questions from an Islamic perspective. Like other seminars in the series, this seminar was preceded by a Call-For-Papers (CFP), espoused with a background paper explaining its main themes and key questions.[2] All submissions were reviewed by an internal committee and a limited number of the submissions was selected. Besides the submissions coming from the CFP, direct invitations were sent to some participants, whose published research shows their ability to cover topics that were not addressed by the CFP submissions. Throughout the three days of the seminar, the two groups, coming from the CFP and direct invitations, presented their papers and feedback on each other's papers and exchanged ideas and insights on many issues related to the main themes of the seminar. Benefiting from the intensive discussions during the seminar, the authors worked on revising their papers. Finally, a few new papers were written after the seminar in order to cover some lacunas revealed by the discussions during the seminar. A post-seminar internal review was made inside CILE then the selected material went through the peer-review process managed by Brill. The papers which successfully went through these various layers of review are included in this volume.

Throughout the long journey, which started by mere proposals of rough ideas about the seminar up until this publication, I have received invaluable help and support from a great number of people whose list is too long to be included here. Every researcher is well aware that working on refining the lan-

1 The research-related activities conducted before, during and after the seminar, which resulted in this publication, were made possible by the NPRP grant "Indigenizing Genomics in the Gulf Region (IGGR): The Missing Islamic Bioethical Discourse", no. NPRP8-1620-6-057 from the Qatar National Research Fund (QNRF), a member of The Qatar Foundation. The statements made herein are solely the responsibility of the authors. In my capacity as the Lead Principal Investigator (LPI) of the IGGR project, I submit my due thanks to the QNRF for their generous and continuous support.

2 Both the call-for-papers and the Background Paper were published on the CILE website (www.cilecenter.org) in both English and Arabic. The English version was also advertised via the *Times Higher Education*.

guage and unifying the referencing style of pieces written by different authors is time-consuming and laborious. Most of this work was accomplished by two of our brilliant students of "Islamic Thought and Applied Ethics" specialization in the College of Islamic Studies, namely Mariam Taher and Reem Al-Sahlawi. Furthermore, a splendid job was achieved by the two competent and hard-working Research Assistants, Noha Abdel Ghany and Shaimaa Moustafa, who assisted me in my current two research projects funded by the Qatar National Research Fund (QNRF). Additionally, all colleagues working in CILE were very supportive throughout all the elevations and depressions of this extensive journey.[3] Finally, my due thanks go to Dr. P.S. Van Koningsveld, who worked temporarily as the Managing Editor of this series. His advice and wisdom were crucial for bringing this publication to light.

I keep the last word here for my dear family, to whom I remain indebted my whole life. My wife, Karima, has always been far and beyond the ideal woman I could have ever imagined in my dreams. Personally, my children, the twin Maryam and Khadija, Mustapha, Aisha and Hamza have practically demonstrated that "pure innocence" exists and I hope they will be up to the challenges ahead in their life, while keeping their precious innocence intact. My mother, Fawiza, is an example of the simple villager whose strength lies in her ability to selflessly give without waiting for a reward in return. My late father, Mustafa, is the great personality in my life whom I miss dearly. I say to these great figures in my life: "I am related to you all not only through a shared genome, but also through unbated love and mutual care".

Mohammed Ghaly
Doha, Qatar
August 2018

3 For the full list of the CILE team, please check https://www.cilecenter.org/en/staff/

About the Authors

Aasim I. Padela

Dr. Aasim Padela is a clinician-researcher and bioethicist whose scholarship lies at the intersection of community health and religion. He utilizes diverse methodologies from health services research, religious studies, and comparative ethics to examine the encounter of Islam with contemporary biomedicine through the lives of Muslim patients and clinicians, and in the scholarly writings of Islamic authorities. Through systematic research and strategic interventions, he seeks (1) to improve American Muslim health outcomes and healthcare experiences, and (2) to construct a multidisciplinary field of Islamic bioethics. Dr. Padela holds an MD from Weill Cornell Medical College, completed residency in emergency medicine at the University of Rochester, and received an MSc in Healthcare Research from the University of Michigan. His Islamic studies expertise comes via a BS in Classical Arabic from the University of Rochester, seminary studies during his secondary school, and continued tutorials with traditionally trained Islamic authorities.

Abbas Ameir (عباس أمير)

Born in Iraq in 1969, Dr. Abbas Ameir Muariz ʿAbbūd al-Shammarī is Professor and head of the Quranic Sciences Department in the College of Education, University of Al-Qadisiyah in Iraq. He is the president of the "Quran for All Foundation for Social and Humanity Researches and Studies". Dr. Ameir is a member of both the Iraqi and Arab Writers' Unions. He is a member of the expert-committee on cultural affairs at the Ministry of Culture in Iraq. He is also a member of the supreme committee at the Ministry of Education in Iraq concerned with preserving Arabic language, and of the central committee for improving knowledge resources at the Iraqi universities. Dr. Ameir is the author of five published books and fifteen research papers and he received four Arab and Iraqi awards. He participated in twenty-five Arab and Iraqi academic conferences and symposiums.

Amara Naceur (عمارة الناصر)

Dr. ʿAmāra al-Nāṣir is Professor of philosophy at the University of Mostaganem in Algeria. He is the head of the research unit "Hermeneutic Research: Texts and Translations". His research interests include areas like philosophy of hermeneutics and argumentation, contemporary Western philosophy and Bioethics. Besides participating in many international conferences and seminars, Dr. al-Nāṣir wrote and translated numerous books and he is also the author of articles published in peer-reviewed journals. His published works address themes like Western hermeneutics and Arab-Islamic interpretation, research methods in Philosophy, phenomenology of intuition and expression, hermeneutical and ethical approaches to the concept of disease.

Arzoo Ahmed

Arzoo Ahmed is Director at the Centre for Islam and Medicine (CIM), which conducts research on themes at the intersection of healthcare, religion and ethics. The centre also produces educational and training resources that facilitate religious literacy within the healthcare setting in the UK. Arzoo read Physics at the University of Oxford, graduating with a BA, after which she completed an MPhil in Medieval Arabic Thought at the Oriental Institute, with a research thesis titled 'Reason and Revelation in the Works of Raghib al-Isfahani.' She has an Alimiyyah degree in traditional Islamic studies, under the supervision of Shaykh Akram Nadwi and is currently studying on the philosophy MA programme at King's College London.

Ayman Shabana

Dr. Ayman Shabana is Associate Research Professor at Georgetown University's School of Foreign Service in Qatar (SFS-Q). Prior to joining SFS-Q, he taught at several institutions including the University of California, Los Angeles, the University of Tennessee, Knoxville, and Florida International University. He received his Ph.D. from the University of California, Los Angeles, his MA from Leiden University in the Netherlands, and his BA from al-Azhar University in Egypt. His teaching and research interests include Islamic legal history, Islamic law and ethics, human rights, and bioethics. He is the director of the Islamic Bioethics Project, which has been supported by three consecutive grants from Qatar National Research Fund's National Priorities Research Program. In 2012, he received the Research Excellence Award at the Qatar Annual Research Forum and during the academic year 2013-2014 he was a visiting research fellow at the Islamic Legal Studies Program at Harvard Law School.

Hub (H.A.E) Zwart

Dr. Hub Zwart is Professor of Philosophy at the Faculty of Science, Radboud University Nijmegen, the Netherlands, and the Director of the Institute for Science in Society. His research focuses on philosophical and ethical issues in emerging life sciences (genomics, synthetic biology, neuroscience, transplantation medicine) from a 'continental' perspective (dialectics, phenomenology, psychoanalysis). He is Editor-in-Chief of the journal *Life Sciences, Society and Policy*, together with Ruth Chadwick. Special attention is given to the use of genres of the imagination (cinema, novels, plays, poetry) in research and education. Publications available at: https://radboud.academia.edu/HubZwart.

Mehrunisha Suleman

Dr. Mehrunisha Suleman is a post-doctoral researcher at the Centre of Islamic Studies, University of Cambridge. Her research involves an ethical analysis of the experiences of end of life care services (EOLC) in the UK from Muslim perspectives. Dr. Suleman holds a DPhil in Population Health from the University of Oxford and a BA in Biomedical Sciences Tripos from the University of Cambridge. She also holds a medical degree and an MSc in Global Health Sciences from the University of Oxford. She has worked with Sir Muir Gray on the Department of Health's QIPP Right Care Programme. She is an expert for UNESCO's Ethics Teacher Training Programme and was awarded the 2017 National Ibn Sina Muslim News Award for health. She has an 'Alimiyyah degree in traditional Islamic studies, which she was given under the supervision of Shaykh Akram Nadwi at Al Salam Institute in 2013.

Mohammed Ghaly

Dr. Mohammed Ghaly is professor of Islam and Biomedical Ethics at the Research Center for Islamic Legislation & Ethics (CILE), College of Islamic Studies at Hamad Bin Khalifa University in Doha, Qatar. He has B.A. degree in Islamic Studies from Al-Azhar University (Egypt) and M.A. and PhD degrees in the same specialization from Leiden University (the Netherlands). During the period 2007-2013, Ghaly was a faculty member at Leiden University. The intersection of Islamic Ethics and biomedical sciences is Ghaly's main specialization. He is the editor-in-chief of the *Journal of Islamic Ethics* (published by Brill). Since 2011, Ghaly has been a faculty member at the Erasmus Mundus Program; the European Master of Bioethics, jointly organized by a number of European universities. During the academic year 2014-2015, he was Visiting Researcher of the Kennedy Institute of Ethics at Georgetown University, USA. In 2017-2018, he was Visiting Scholar of the School of Anthropology and Museum Ethnology at the University of Oxford. Ghaly is also the Lead Principal Investigator (LPI) and research consultant of a number of funded research projects. His publications can be accessed via https://cilecenter.academia.edu/MohammedGhaly

Mutaz al-Khatib

Dr. Mutaz al-Khatib is Assistant Professor of Methodologies and History of Islamic Ethics at the Research Center for Islamic Legislation & Ethics (CILE), College of Islamic Studies at Hamad Bin Khalifa University in Doha, Qatar. He did Islamic Studies in Damascus (BA, 1997) and studies Arabic Literature at al-Azhar University in Cairo. Al-Khatib was a founding member of the Intellectual Forum for Innovation (1999), and the anchor of al-Sharia and Life program on Al Jazeera Channel, (2004 -2013). He acted as Editor-in-Chief of the section "Islam and Contemporary Affairs" on IslamOnline.net (2003-2008). He was a visiting fellow at Zentrum Moderner Orien (ZMO) in Berlin (2006), and a visiting scholar at the Forum Transregionale Studien, Berlin (2012-2013). Al-Khatib was a Visiting Lecturer at both the Islamic University of Beirut and Qatar University. He is a reviewer for a number of journals including *Islamization of Knowledge* (International Institute of Islamic Thought), *Journal of al-Tajdīd* (International Islamic University, Malaysia) and the Arab Center For Research & Policy Studies (Qatar). Al-Khatib authored and edited several books and over 20 academic articles.

Saadia Bendenia (سعدية بن دنيا)

Dr. Saadia Bendenia is professor of sociology, Faculty of Social Sciences, University of Mostaganem in Algeria. She conducted research on "Perceiving the Platonic Text within the Christian Culture: Augustine as a Model." Her doctoral dissertation was on "Neoplatonism and Its Influence on Islamic Philosophy: Al-Farabi's Approaches to the Platonic Code." Dr. Bendenia is a research team leader at the Laboratory of Philosophy and Humanities, University of Mostaganem. One of her several published works is "Al-Mafāhīm al-Falsafiyya ka Qiyam Dīniyya: Aflūṭīn namūdhajan" (Philosophical Concepts as Religious Values: Plato as a Model)', published in 2007 in *Kitābāt Mua'aṣira*, issue no. 63. Another published article is entitled "Al-Umma al-Islāmiyya wa mafhūm ṣidām al-ḥaḍārāt" (The Muslim Nation and the Concept of the Clash of Civilizations), published in 2009 in *al-Kalima*, issue no. 64. She also wrote "Al-Ḥadāthah wa ma ba'd al-ḥadāthah wa al-qiyam al-siyāsiyya al-akhlāqiyya al-Islāmiyya" (Modernism, Postmodernism, and Islamic Ethical and Political Values), published in 2011 by *al-Kalima*, issue no. 73.

Trevor Stammers

Dr. Trevor Stammers is Reader in bioethics at St Mary's University, Twickenham, London and Director of its Centre for Bioethics and Emerging Technologies. He was previously a clinician for over 30 years and is a past chair of the Christian Medical Fellowship. He is Editor-in-Chief of *The New Bioethics,* a multidisciplinary journal of biotechnology and the body and is the author of over 50 papers covering a wide spectrum of medical and bioethical topics.

Introduction

Mohammed Ghaly[1]

Addressing the deep ontological and ethical questions raised by the field of genomics, named here "the Genome Question" (GQ) represents one of the key challenges that both religious and non-religious ethical traditions face in the modern time. This holds true to the Islamic tradition; one of the main and fastest-growing world religions. The international library is now replete with academic publications which address the GQ from secular bioethical perspectives. When it comes to the religious perspectives, the list of available publications considerably declines. When it comes to genomics and Islamic ethics in particular, one can hardly come across any distinct publication. This volume is meant to fill in this gap, without claiming to be all-inclusive, and to open up new venues for future studies and publications in this field.

Before reviewing the various chapters included in this volume, it is pertinent to explain what we mean by the "Genome Question" and how it should be addressed from an Islamic bioethical perspective. The GQ widely includes, and certainly not limited to, a set of ethical questions raised by the cutting-edge technologies of genomics, which the Islamic tradition would ordinarily respond to by providing immediate and short-term answers through judging specific applications, like genomic testing, DNA paternity and selective abortion, through the lens of ethics. This usually happens by employing tools from the discipline of Islamic Jurisprudence (*fiqh*). The authors who contributed to this volume could strongly argue that the GQ goes much further and deeper than exploring how far certain technologies in particular situations are (in) compatible with specific ethical traditions. The GQ is much broader in scope than these direct ethical questions which appear on the surface. The field of genomics itself needs to be critically examined, because the very birth and further progress of genomics are, implicitly or explicitly, indicative of certain perceptions we hold about ourselves as human beings (including individuals, families and societies) and the ways through which we answer questions like:

1 Professor of Islam and Biomedical Ethics, Research Center for Islamic Legislation & Ethics (CILE), College of Islamic Studies, Hamad Bin Khalifa University, Doha, Qatar, mghaly@hbku.edu.qa

What makes us distinctively human? Are genome and soul related? If yes, in what way? Is our human identity fixed and we must keep it intact, or is it evolving in nature and we can/should always try to improve it, this time through "fixing" or "enhancing" our genome? What kind of individuals, families and societies would result, dominate or die out because of the genetic/genomic technologies which (will) allow choosing the sex of the children in addition to certain physical and cognitive characteristics?

The breadth and depth of the GQ, we argue, should be coupled with a parallel breadth and depth in the Islamic ethical discourse for two main reasons. The first reason deals with the complex and multidimensional nature of the GQ itself. The larger and deeper questions of genomics, even some of the direct and specific questions, cannot be properly addressed by depending exclusively on the discipline of *fiqh*. The second reason attends to the nature and scope of the field of Islamic Bioethics, as we envisage it. As it is the case with the field of mainstream bioethics, Islamic Bioethics is to be characterised with interdisciplinarity. Besides the discipline of *fiqh*, Islamic bioethical discourse should truly reflect the richness of the Islamic tradition by incorporating insights from a broad spectrum of other disciplines including philosophy, theology, Sufism, Qurʾān exegesis, Hadith commentaries, belles-lettres (*adab*), … etc. Alongside these disciplines, usually imprecisely called "religious" sciences, Islamic Bioethics should also benefit from the critical perspectives developed by social sciences and their interaction with biomedical sciences, like medical anthropology and medical sociology. By incorporating social sciences in the Islamic bioethical discourse, people can reach more comprehensive and informed conclusions in which not only the "ought" aspect will be examined, but also the "is" aspect, sometimes called in Islamic literature as people's realities (*aḥwāl al-nās*). Exploring *aḥwāl al-nās* and incorporating them into the contemporary Islamic bioethical discourse, in our view, cannot be made through conducting surveys only. What is needed is a much more sophisticated analysis, which takes into consideration the nuances of lived experiences, power imbalances, and the particularities of certain (sub-)communities, … etc. The more difficult question to study here is how the perspectives and insights coming from social sciences will be integrated in the Islamic religious discourse on making moral judgement and determining the ethically acceptable and objectionable choices.[2] There is, however, another layer of interdisciplinarity that needs to be considered for Islamic Bioethics, viz. engaging with bioethical

2 For critical remarks on the interplay of social science and bioethics, see Callahan 1999, 275-294.

deliberations from outside Islamic tradition, including both the religious (e.g. Jewish and Christian) and secular discussions.

Striking a well-justified balance between these different layers of interdisciplinarity will remain a real challenge for those who want to engage in seminal contributions in the field of Islamic Bioethics. It is not so difficult to uncritically adopt/reject bioethical perspectives developed outside the Islamic tradition and justify their supposed (in)compatibility with Islam just by quoting passages from the foundational scriptures of Islam, namely the Qur'ān and Sunna.[3] On the other hand, it can hardly be methodologically justified to approach ethical questions triggered by biomedical technologies, which were produced outside the Islamic tradition, by focusing exclusively on one discipline within the Islamic tradition, namely *fiqh*. This position, however, does not mean that *all* these levels and dimensions of interdisciplinarity should *always* be integrated in the Islamic discourse on *any* bioethical issue. Undoubtedly, this is not feasible as various factors (e.g. the nature of the issues at hand, the (un)availability of experts, and the different settings of the discussions) should be considered before deciding to what extent the interdisciplinary the Islamic bioethical discourse should be. While preparing for the seminar whose proceedings are published in this volume, we tried to solicit contributions which guarantee as much interdisciplinarity as possible. This explains the inclusion of chapters which analyse insights from ethical traditions outside Islam. During the seminar, these contributions were fully integrated in the deliberations. Over and above exchanging ideas and critical remarks throughout the seminar, the authors of these chapters were asked to prepare written responses on other chapters, which address genomics from an Islamic perspective. Additionally, the other participants were asked to prepare written responses on these chapters and to present them during the seminar. These procedures did improve the cross-fertilization of ideas and insights among all participants, which will hopefully be reflected throughout this volume. As for the chapters which approached the GQ from an Islamic perspective, interdisciplinarity was also underscored. Although we completely defend the centrality of Islamic Jurisprudence (*fiqh*) in developing an authentic Islamic ethical discourse, we equally problematize the proposition which reduces Islamic ethics to *fiqh* only. As explained above, many other disciplines, as well as *fiqh*, should be employed and operationalized in order to produce a rigorous and productive Islamic ethical discourse on genomics. The contributions included in this volume explored how insights from disciplines like philosophy, theology and Qur'ān exegesis

3 For some illustrative examples of this approach when principlism was examined from an Islamic ethical perspective, see Ghaly 2016, 6-27.

can be of added value in this regard. Unfortunately, some disciplines, such as Sufism, are not covered in this volume. This is due to the fact that we could not find experts in Sufism with interest in exploring genomics-related issues. Moreover, the volume does not include contributions from social scientists. This is partially because the field of genomics has just recently found its way into the Muslim world and thus there is still no concrete "social reality" of genomics research to be explored.[4] We anticipate that these and other missing aspects will be attended to in future studies.

Part 1: Collective *Ijtihād* and Genomics

In order to put the discussions on the interplay of Islamic ethics and genomics in their proper context, the first part of this volume fathoms out the contemporary Islamic bioethical discourse by highlighting some of its distinctive features. In order to develop an ethical position rooted in the Islamic tradition, one needs to consult its two main scriptural sources, namely the Qurʾān and Sunna. Like many other topics, one cannot expect finding direct answers to the questions raised by the field of genomics by surveying the content of these two Scriptures. Thus, developing an Islamic ethical position necessitates exerting extra intellectual and scholarly efforts guided by a set of methods and principles developed throughout the Islamic history. The whole process, commencing from the point of understanding the question or the issue at hand leading to deducing the religious ruling (*ḥukm Sharʿī*) or developing the ethical position, is known in the Islamic tradition as *ijtihād*, which literally means exerting one's utmost effort (Weiss 1978). For various reasons, some of which are explained in the chapters included in this part, Muslim religious scholars collaborated with biomedical scientists and thus the process of *ijtihād* became collective in nature (Ghaly 2015).[5]

The first chapter in this part, "Sharia Scholars and Modern Biomedical Advancements: What Role for Religious Ethics in the Genomic Era", presents a historical review of contemporary Islamic Bioethics, which goes back to the beginning of the twentieth century. Unlike the field of mainstream bioethics where religion gradually lost its central role, in this chapter, Mohammed Ghaly explains how Muslim religious scholars continued to play a central role in constructing and shaping the Islamic bioethical discourse. Building upon the the-

4 For the status quo of genomics in the Muslim world and more particularly in the Gulf region, see Ghaly 2016a.

5 For more information on the concept of *ijtihād*, see Weiss 1978.

sis "Sharia is valid for all times and places", shared by Muslim religious scholars, they acquired the challenge of demonstrating how the religio-ethical system of Islam (Sharia) is still viable enough to address the vexing questions raised by modern biomedical technologies. Ghaly also highlights the difficulties that these religious scholars encountered in this regard because of their educational background. This usually focused exclusively on mastering the Arabic language and the disciplines of knowledge that help these scholars understand the Islamic Scriptures, namely the Qur'ān and Sunna. Because of this genre of "religious" education, Muslim religious scholars had no access to updated biomedical information, and most of them could not read first-hand sources. This situation necessitated employing an interpretive mechanism through which Islamic Scriptures will be approached collectively, rather than individually, through a group of people who will collaborate to deduce religious rulings and ethical judgments compliant with Sharia. The chapter explains how this mechanism, known as collective *ijtihād*, functioned within the field of Islamic Bioethics, and what developments this mechanism went through from the beginning of the twentieth century up to the current genomic era. The chapter also raises some critical remarks about how the very term "Sharia" is to be defined.

As a follow up for the first chapter, the second chapter, "Islamic Ethics and Genomics: Mapping the Collective Deliberations of Muslim Religious Scholars and Biomedical Scientists", analyzes how the mechanism of collective *ijtihād* was employed to address the Genome Question (GQ). In this chapter, Mohammed Ghaly presents a comprehensive overview of the key conferences and expert meetings which facilitated the interdisciplinary discussions among Muslim religious scholars and biomedical scientists, from the beginning of the 1990s onwards. Ghaly identified two main approaches in these discussions, namely the "precaution-inclined approach" and the "embracement-inclined approach". Within the precaution-inclined approach, genomics is perceived as something almost alien to the Islamic tradition and thus should be approached with great caution. For instance, the advocates of this approach argue that genomics-associated technologies are in principle forbidden unless proved otherwise, and that Muslims should rather wait for concrete research results before joining the genomic revolution. On the other hand, the advocates of the embracement-inclined approach contextualize genomics within the call of Islam to search for beneficial knowledge (*ʿilm nāfiʿ*), which God made accessible to all humans who work intensly. Within this approach, genomics is not only something permissible, but it is seen as a collective duty (*farḍ kifāya*), which means that Muslims are collectively required to engage in and contribute to. Recent developments in the Muslim world, especially in the Gulf region, showed that the second approach proved to be more appealing. However, the advocates of both approaches agreed that Sharia-based determinants (*ḍawābiṭ Sharʿiyya*) should be developed in order to guide the research in the field of genomics and its resulting applications and technologies.

The final chapter in this part, "Transformation of the Concept of the Family in the Wake of Genomic Sequencing: An Islamic Perspective" by Ayman Shabana, explores the details of some of the Sharia-based determinants (*ḍawābiṭ Sharʿiyya*) developed through the mechanism of the collective *ijtihād*. These deal with the family institution, which has a pivotal place in the ethical edifice of Islamic tradition. Within this edifice, each family has a certain ideal, with its own structure, characteristics and regulations governing the relationship among the members belonging to this institution. Genomics and associated technologies have not only challenged some aspects of this family ideal, but also created new possibilities for reshaping some constituents of this ideal. This chapter examines how the collective deliberations among religious scholars and biomedical scientists addressed these challenges, in addition to the opinions of individual scholars, and whether the newly created possibilities should be seen as ethically defensible or objectionable options. These questions are addressed at the hand of three applied examples, namely premarital genetic testing, fetal sex selection and germline genetic modification.

Part 2: Genomics and Rethinking Human Nature

After setting the scene and examining the status quo of Islamic bioethical deliberations on genomics, the second part of this volume tries to impose new frontiers, explore new dimensions, and raise some of the deep questions which were not (fully) covered in the discussions facilitated by the mechanism of collective *ijtihād*. The main thread which connects the three chapters included in this part is exploring how human nature can/should be comprehended, revisited or even reshaped in the light of genomics and the new possibilities it has created. In the first chapter "Conceptualizing the Human Being: Insights from the Genethics Discourse and Implications for Islamic Bioethics", Aasim Padela argues that formulating an ethical position towards complex issues in the field of genomics are usually premised on specific ontological perceptions about the nature of the human being, although these ontologies sometimes remain implicit and unspoken. Based on an extensive literature review, Padela holds that Western bioethical deliberations on issues related to genetics and genomics are indicative of three main ontological perspectives, which perceive the human being as (a) a data store that houses information, (b) a reproductive organism, or (c) an evolving biological entity. Each of these ontological perceptions, Padela argues, impacts the ethical conclusions adopted by Western bioethicists. Padela's proposal for improving the religious, and particularly Islamic, bioethical discourse on genomics is not only to evaluate the technological applications by counting the strengths and weaknesses or the direct benefits and harms of each application but also by questioning and critically examining the underlying ontologies.

The second chapter, "Islamic Perspectives on the Genome and the Human Person: Why the Soul Matters" by Mehrunisha Suleman and Arzoo Ahmed, can be seen as a natural extension of the ontological discussions outlined by Padela, but with more focus on the Islamic tradition. The two authors start from the premise that the information unlocked by human genetics and genomics greatly influenced how we perceive our human nature. Against this backdrop, the chapter examines the relationship between the genome and the human person. A substantial part of the chapter is dedicated to studying the concept "soul/spirit (*nafs/rūḥ*)", and associated terms. The authors surveyed the references to these terms in the Qurʾān and how they were analysed by Qurʾān exegetes, philosophers and theologians. The aim here is to explore how such narratives rooted in the Islamic tradition can provide novel perspectives to the understanding of the human person and the ethical considerations surrounding genomics. The chapter adopts an interdisciplinary approach by engaging insights from different disciplines within the Islamic tradition, especially Quranic exegesis, Islamic theology and philosophy.

The last chapter in this part, "The Ethical limits of Genetic Intervention: Genethics in Philosophical and *Fiqhi* Discourses" by Mutaz al-Khatib, again provides an interdisciplinary investigation of how the new fields of genomics and genetics can influence our understanding of human nature and how far we can subject the human person to these cutting-edge technological interventions. According to al-Khatib, the various ways through which people address such questions are usually determined by one's stance towards other deeper questions, namely on how we understand the nature of genetic/genomic technologies (e.g. are they neutral and value-free, or do they imply certain value judgements, pre-assumptions and convictions?) and how we understand the nature of the human being, especially during the early pre-implantation stage and the following phases of embryonic development. Throughout the chapter, al-Khatib employs the analytical tools of Western philosophy and Islamic jurisprudence (*fiqh*) to see how such questions are (to be) addressed and what kind of similarities and differences exist between these two disciplines.

Part 3: Widening the Scope of Ethical Deliberations

The third part of this volume purposefully aims to enrich the abovementioned levels of interdisciplinarity of Islamic bioethical discourse by incorporating insights from outside the Islamic tradition. In the first chapter, "In the Beginning Was the Genome: Genomics and the Bi-Textuality of Human Existence", Hub Zwart continues investigating the possible impact of genomics on our understanding of human nature. In agreement with almost all authors who contributed to this volume, Zwart argues that focusing on specific applied issues like selective abortion, artificial reproduction and paternity testing does not do justice to the complexity of the religion-science relationship in the context of genomics. According to him, such an approach would typically present science as the progressive and liberating power and religion as the conservative and restrictive one. Instead of this reductionist approach, Zwart proposes perceiving human existence as the result of a reciprocal interaction between two types of texts, namely the text written in the language of molecular biology consisting of the alphabet of nucleotides, and the text recorded in the religious Scriptures like the Bible and the Qurʾān. Within this proposed framing, Zwart develops what he calls an "occidental perspective" which builds upon the works of prominent philosophers like Hegel, Teilhard and Lacan. Through this perspective, he tries to revisit the relationship between science, represented here in genomics, and religion, particularly world religions like Islam and Christianity.

Just to advance venues for possible cross-fertilization of insights, we refer to the fact that the very idea of having two interrelated texts is not alien to the Islamic tradition. The prominent Muslim religious scholar Ibn Taymiyya (d. 1328) spoke about a similar idea when he divided God's words into two types, namely "religious words (*kalimāt dīniyya*)" and "universal words (*kalimāt kawniyya*)". The former is communicated through revealed Books like the Bible and the Qurʾān, while the latter is communicated through the universe (Ibn Taymiyya n.d., 5/8-17). The idea of Ibn Taymiyya received commentaries from contemporary religious scholars. One can see its clear impact on the contemporary school of "Islamization of Knowledge", especially their ideas about the two readings (*al-qirāʾtān*); reading the written Book, i.e., the Quran, and reading the observable book, namely the universe (Malkāwī 1981, 43-57). Thus, our aspirations are for this chapter to stimulate researchers in the field of Islamic Studies in approaching the Genome Question through this lens.

The second chapter, "Creation, Kinds and Destiny: A Christian View of Genome Editing" by Trevor Stammers, presents a perspective on the Genome Question rooted in the Christian tradition. The reader will notice a number of common themes and parallels between this chapter and the second chapter in Part II, written by Mehrunisha Suleman and Arzoo Ahmed, although each chapter addresses the Genome Question through the lens of a different religious tradition. This became already clear during the seminar, where the three authors benefited from sharing thoughts and critical remarks in improving the earlier drafts of these two chapters. A great deal of this chapter is dedicated to explaining the accounts of creation and Fall and related concepts like *imago dei* (image of God), *sicut deus* (like God), embodiment, dominion and co-creation. Stammers is keen to expose the internal diversity within Christianity on explaining these accounts and concepts. He does so by granting space for different opinions expressed by authoritative voices in Christianity like Irenaeus (130-202 AD), Augustine of Hippo (AD 354-430), and Dietrich Bonhoeffer (d. 1945). He also shows how this diversity continued when contemporary Christian ethicists, like John Wyatt and Ronald Cole-Turner, tried to interpret these accounts and concepts within the context of genomics-related technologies and genomic/genetic engineering. The author also touches upon the question whether genomics could problematize the distinction used to be made by ethicists between therapy and enhancement. Finally, the idea of perceiving the genome as the secular alternative to the religious soul is critically examined from a Christian perspective, with a focus on the question of *telos* or end purpose, which makes the Christian vision sometimes quite different from the secular vision.

The third item in this part is a review essay in which Ayman Shabana presents the book *Living with the Genome: Ethical and Social Aspects of Human Genetics*, edited by Angus Clark and Flo Ticehurst. The material included in the reviewed volume is based on selections from a voluminous work, namely the five-volume and three-million word *Encyclopedia of the Human Genome*. The forty-two chapters included in this volume were meant to provide a collection of concise and accessibly written articles on the social and ethical aspects of human genetics and genomics. Although published in 2006, Shabana argues that the book provides a useful introduction to the range of ethical, legal and social implications of genetics and genomics, most of which remain relevant today. Thus, this is an ideal book to make specialists, and those interested, in Islamic Bioethics aware of the types of ethical questions and modes of reasoning which can be interpreted in international, mainly Western, ethical deliberations. Shabana gives a concise overview of the six main parts of the volume, namely the Human Genome Project; Genetic Disease; Disability, Genetics and Eugenics; Genetics and Society; Genetic Explanations; and Reproduction, Cloning and the Future. In conjunction with presenting the key issues discussed in each part, Shabana also suggests how these issues are (to be) approached from an Islamic bioethical perspective. In addition to the agreements between what is presented in the reviewed volume and the parallel discussions in Islamic Bioethics, Shabana highlights issues which may create potential tension between the mainstream (Western) bioethical discourse and the Islamic tradition, e.g. the emphasis on individual freedom and autonomy in the former against more inclination towards communitarian ethics in the latter.

Part 4: Contributions in Arabic

The material included in the fourth and last part of this volume was subsumed together purely for linguistic considerations. The three chapters which compose this part are all written in Arabic. From the very beginning when the pre-publication seminar was still a vague idea, we were keen to have contributions in both English and Arabic; that is the reason why the Call-For-Papers and the Background Paper of the seminar were published in both English and Arabic. In order to facilitate the communication among the participants in the seminar, all papers were translated into English or Arabic and there was a simultaneous translation throughout the three days of the seminar. In our view, having contributions from different languages is much more than just linguistic diversity. Bilingual authors are well acquainted that writing in a specific language predominantly determines the content as well, e.g. which sources should be consulted, what kind of questions should be prioritized to address the concerns of the targeted audience, how these questions should be approached, ... etc. Consequently, we hope that the inclusion of this part will be of added value, especially for those who are curious about the Arabic writings on Islamic Bioethics.[6] In order to make this material accessible to the readers who do not master the Arabic language, an English translation of the material included in this chapter will be available afterwards on the CILE website (www.cilecenter.org).

The first chapter, by Saadia Bendenia (Saʿdiyya Bin Dunyā), "Al-Jīnūm wa al-ṭabīʿa al-bashariyya: Muqāraba taḥlīliyya fī ḍawʾ al-falsafa wa al-ʿilm al-tajrībī wa al-akhlāq al-Islāmiyya" (Genome and Human Nature: Analytical Approach in the Light of Philosophy, Experimental Sciences and Islamic Ethics) addresses the interplay of human nature and genome. The author starts her study by recognizing the complexity and elusiveness of the human nature and

6 It is to be noted that abovementioned CILE seminar included more Arabic contributions than those included in this chapter. Some of these contributions will compose a thematic issue in the Arabic journal *Tabayyun* (https://tabayyun.dohainstitute.org), to be published in 2019.

thus proposes an interdisciplinary approach to understand (certain aspects of) this nature. She emphasizes that the breathtaking advancements in fields like genetics and genomics, despite their significance, cannot alone explain what human nature is and that insights from other fields must be consulted. In this chapter, Bendenia tries to enrich her multidimensional analysis of the human nature through insights from different disciplines, including philosophy, biology (especially genetics and genomics) and Islamic ethics. Concerning philosophy, the chapter provides an extensive overview of perspectives, which spans many centuries of thinking about what the human nature exactly is. References are made to well-known philosophers like Socrates, Plato, Aristotle, René Descartes, John Locke, David Hume, Karl Marx, Sigmund Freud, Jürgen Habermas and others. As for the biological perspectives on human nature, Bendenia makes reference to influential names like Darwin, Gregor Mendel, Francis Crick, James Watson and also to the Human Genome Project. She explains how the big discoveries about the biological nature of humans led to the prominence of genetic determinism which was also countered by social determinism. Another consequence of these discoveries is that they opened up the possibility of understanding human nature as well as (re)shaping and modifying it. The last section of the chapter is dedicated to exploring human nature through the lens of Islamic ethics. According to Bendenia, human nature, from an Islamic ethical perspective, is a mixture of material (*mādda*) and soul (*rūḥ*) and that striking a good balance between the needs of these two aspects is the ideal way to remain healthy, in both the biological and moral sense. That is why the author believes that none of the two polarizing positions, viz. genetic determinism and social determinism, could capture the true character of human nature, which is actually a mix. The author interprets some Quranic verses, e.g. 25:54, in a way to support this idea. Using technologies like genomic editing in order to facilitate moral enhancement seems to be welcomed by the author because, she argues, it will generally improve human nature. Similarly, improving the material aspect of the human nature by making people taller or stronger does not seem to be problematic for the author. She bases herself on historical reports in the Qur'ān speaking about earlier generations of humans whose physical make-up was much stronger and bigger than ours. Thus, the author argues, the physical capacities that humans have now do not represent a fixed part of the human nature, but an evolving and improvable one.

The second chapter, by Abbas Ameir ('Abbās Amīr), is entitled "Su'āl al-jinūm bayna al-khilqa wa al-akhlāq: Muqāraba dilāliyya maʿrifiyya fī akhlāqiyyāt ʿilm al-jīnūm min manẓūr Islāmī" (The Genome Question between physical make-up and ethics: A Semantic and Epistemological Approach to Genomic Ethics from an Islamic Perspective). The chapter investigates the links between parallel concepts like genotype and phenotype, and between *khilqa*, an Arabic word which means physical make-up, and *akhlāq* or *khuluq*, which means ethics or morality. The overall content of the chapter indicates that the author here touches upon the famous nature vs. nurture controversy. The main thesis of this chapter is that there is as a strong link between *khilqa* or one's physical, including molecular, structure and one's *khuluq* or ethics as the link between the genotype and phenotype. This means that any intervention in someone's molecular structure, e.g. through genomic editing, can have an impact on one's moral character. For Ameir, this does not necessarily mean that all types of genetic intervention are ethically objectionable, but that they should always be approached very cautiously. He distinguishes between a fixed part in our humanness that should never be touched, and a changeable part that can always be improved through human intervention. As an illustration, he refers to the freedom of human individuals to make choices about their own lives and the diversity of people's identities in life as components of the fixed part that should not be touched by the genomic editing or similar technologies. As a specialist in Quranic Studies, the author elaborates on this thesis throughout the chapter by depending heavily on references from the foundational Islamic Scripture, viz. the Qur'ān. He uses various hermeneutical and exegetical tools in interpreting about fifteen Quranic verses to show their possible relevance to the abovementioned thesis and associated ideas.

In the third chapter "Al-Jīnūm wa al-ḥayāh: Tamdīd al-ḥayāh wa atharuh al-akhlāqī ʿalā al-mujtamaʿāt al-Islāmiyya" (Genome and Life: Extending Lifespan and its Moral Impact on Muslim Societies), Amara Naceur (ʿAmāra al-Nāṣir) is somehow elaborating on the thesis outlined by Abbas Ameir in the previous chapter, but by focusing on one concrete example, namely modifying the genome for the sake of extending life and how far this would impact the world of ethics. According to the author, the attempts to postpone ageing and to have an extended life in principle do not go against human nature (*fiṭra*), as created by God. He makes use of prophetic traditions and references in the Qur'ān including historical reports about persons with extremely long lives, like Prophet Nūḥ (Noah), to argue that such attempts do not fall outside the borders of normalcy and natural course of life. On the other hand, the author argues that modifying one's genome for the sake of extending his/her lifespan poses complex ethical and philosophical questions related to our longstanding understanding of what "life" itself means. The social structure that people have in life, the author explains, and the associated values built throughout centuries are all linked to the average or "normal" lifespan that people used to live. Extending these "normal" lifespans will eventually mean that both the social structure of life and the associated values cannot continue without radical changes. Nāṣir makes use of the works of Western authors, like Francis Fukuyama, to show some of the concrete problems that can be created by extending the lifespan through biotechnological means, like the one described below.

The social structure of our current life largely depends on age-graded hierarchies, which usually assume a pyramidal structure. Besides artificial constraints such as fixing a (mandatory) retirement age, death remains one of the main factors which recede old generations from the pool of competitors for the top ranks in society. By extending the current lifespan to the extent that people will be expected to live and work until the age of 90 or even later, various generations will simultaneously exist and compete. Within this scenario, the elderly who are already at the top of the social hierarchies will not easily make space for the younger generations but will usually use their considerable influence to protect their positions, despite the likely declines in physical and professional capabilities because of age-related complications. This means that generational succession, which is a major stimulant of progress and change, will possibly be hindered. Eventually, we will have to think of other possible social structures, together with their fitting moral values (Fukuyama 2002, 76-79). Again, this shows that modifying the genome is not only a biological issue but a moral one as well.

References

Callahan, Daniel.1999. The Social Sciences and the Task of Bioethics. *Daedalus* 128 (4): 275-294.

Fukuyama, Francis. 2002. Our Posthuman Future: Consequences of the Biotechnology Revolution. New York: Farrar, Straus and Giroux.

Ghaly, Mohammed. 2015. "Biomedical Scientists as Co-Muftis: Their Contribution to Contemporary Islamic Bioethics". *Die Welt des Islams* 55: 286-311.

Ghaly, Mohammed. 2016. Deliberations within the Islamic Tradition on Principle-Based Bioethics: An Enduring Task. In Ghaly, Mohammed (ed.). *Islamic Perspectives on the Principles of Biomedical Ethics*. pp. 3-39. London: World Scientific Publishing & Imperial College Press.

Ghaly, Mohammed. 2016a. *Genomics in the Gulf Region and Islamic Ethics: The Ethical Management of Incidental Findings*. Doha, Qatar. based World Innovative Summit for Health (WISH). Available at http://www.wish.org.qa/wp-content/uploads/2018/01/Islamic-Ethics-Report-EnglishFINAL.pdf, retrieved 28 May 2018.

Ibn Taymiyya, Taqī al-Dīn. N. D. *Majmūʿat al-rasāʾil wa al-masāʾil*. Cairo: Lajnat al-Turāth al-ʿArabī.

Malkāwī, Fatḥī. 1981. *Manhajiyyat al-takāmul al-maʿrifī: Muqaddimāt fī al-manhajiyya al-Islāmiyya*. Herndon, Virginia: International institute of Islamic Thought.

Weiss, Bernard. 1978. Interpretation in Islamic Law: The Theory of *Ijtihād*. *The American Journal of Comparative Law* 26 (2): 199-212.

PART 1

Collective Ijtihād and Genomics

CHAPTER 1

Sharia Scholars and Modern Biomedical Advancements: What Role for Religious Ethics in the Genomic Era?

Mohammed Ghaly[1]

Historical Context of Medical and Biomedical Breakthroughs: What Role Would Religion Have?[2]

By the beginning of the twentieth century, it became clear that the ramifications of the breathtaking biomedical advancements and associated technologies will not remain within the confines of scientific and clinical practices. The complex questions and challenges raised by these advancements and technologies also necessitated profound ethical considerations. Various religions and philosophies addressed these questions and challenges as part of their historical role in responding to peoples' concerns and curiosities, in addition to demonstrating that they still hold influential roles in the age of modernity, with all its new challenges. By the middle of the twentieth century, the role of religious thought in the field of biomedical ethics in Western scholarship, particularly in the United States and Western Europe, started to wane. Furthermore, several scholars who specialized in religion and theology in their academic studies brushed aside religious discourse and instead, adopted a sec-

---

1 Professor of Islam and Biomedical Ethics, Research Center for Islamic Legislation & Ethics (CILE), College of Islamic Studies, Hamad Bin Khalifa University, Doha, Qatar, mghaly@hbku.edu.qa

2 This research was made possible by the NPRP grant "Indigenizing Genomics in the Gulf Region (IGGR): The Missing Islamic Bioethical Discourse", no. NPRP8-1620-6-057 from the Qatar National Research Fund (a member of The Qatar Foundation). The statements made herein are solely the responsibility of the author. An earlier draft of this chapter was presented at the 6th Annual International Conference of Social and Human Sciences, organized by the Arab Center for Research and Policy (March 18-20, 2017), Doha, Qatar, whose Arabic version is scheduled for publication in the Arabic journal *Tabayyun*.

DOI:10.1163/9789004392137_003

ular one when they embarked on the field of biomedical ethics. In his study on the history of the relationship between religion and bioethics, Albert Jonsen (Professor of the History of Medical Ethics at the University of Washington) drew a comparison between what the Italian missionary Matteo Ricci did in 1582 when he crossed the Western borders traveling to the then "forbidden empire", viz. China, and what a large number of theologians did 400 years later when they decided to specialize in the field of bioethics, in the sense that they "doffed the intellectual garb of religious ethics and donned, if not the white coats of doctors, the distinctly secular mentality of modern medicine" (Jonsen 2006, 23).

In this regard, secular discourse does not necessarily clash with religion in its essence: although, it excludes any central role it may play in the common area of ethics by distinguishing between two types of morality. On the one hand, there is the "common morality", which is universal in nature and orientation and through which the public at large can be addressed. This type of morality is developed and communicated by a secular ethical discourse. On the other hand however, there is the "particular morality", which addresses specific groups of people. The ethical discourse which is premised in religious beliefs and religiously-tented terminologies falls within the category of particular morality (Beauchamp and Childress, 2013, 2-6). It should be emphasized here that religious bioethical discourse did not completely disappear from the field of bioethics in Western scholarship. For instance, various contributions coming from the three monotheistic religions, viz. Judaism, Christianity and Islam, could always find their way to the public in addition to specialized journals like *The National Catholic Bioethics Quarterly*. However, these contributions remained confined to the scope of ethics as practiced by certain groups of people, such as Muslims living as a religious minority in the West or the Christian Jehovah's Witnesses. This situation continued to make the impact of such contributions marginal and narrow in comparison with those inspired by secular thought.

Islamic discourse, particularly the discipline of Islamic jurisprudence (*fiqh*), was not detached from these historical developments. It is true that the majority of the achievements in the field of biomedical sciences occurred outside the Muslim-majority world and in a social, political and cultural environment not familiar to many contemporary Muslim jurists. However, these jurists, together with all those who believed that Islam is a religion which remains relevant for our contemporary world and not just an ancient religious tradition[3],

3 It seems that the concern of proving the contemporaneity of Islam as a religious tradition was not exclusive to Sharia scholars. Some Muslim physicians also attempted to tackle these

felt the necessity of addressing the questions and challenges triggered by the modern biomedical advancements. Unsurprisingly, there is a link between this belief and the ongoing debates at that time about the validity of Islam and its religious ethical system (Sharia) in guiding various aspects of modern life. The position of Sharia scholars and like-minded thinkers at this time was couched in the famous phrase "Sharia is valid for all times and places."[4] Clearly, these scholars felt the risk of the marginalization of Sharia in the emerging field of biomedical sciences, seen as it is one of the marvelous achievements of modernity. Hence, addressing these modern questions and challenges, which subsequently came to be known as the field of "Islamic Bioethics" became an integral part of the quest to prove the contemporaneity of Islam and its possible active and impactful role in the age of modernity.

Accommodating Contemporary Challenges: The Evolving Role of Physicians

Some early signs of Sharia scholars' concerns that some may believe that Sharia and modern biomedical sciences would conflict, were expressed in shaykh Rashīd Riḍā's fatwa (d. 1935). The fatwa was issued in response to a question about the perspectives of early Muslim jurists regarding the possible maximum duration of pregnancy, as outlined in classical *fiqh* manuals, and their incompatibility with modern facts established by credible sciences such as medicine and anatomy. The Tunisian questioner raised the point made by "Frankish doctors (*al-aṭibbā' al-ifrinj*)" working in his country regarding the impossibility of the continuation of pregnancy for such long periods that could extend into years, as claimed by early jurists. The questioner added: "They excused the position of Muslim scholars in this regard [by assaying] that the science of medicine did not disclose its secrets in past times the way it does in our present time". In his response to this question, Riḍā premised his thesis

issues, as exemplified by the book written by the famous Egyptian physician 'Abd al-'Azīz Ismā'īl. The first edition of this book appeared in 1939, followed by two editions in 1954 and 1959. The book's preface was written by the then Shaykh of al-Azhar Muḥammad Muṣṭafā al-Marāghī (d. 1945). See, Ismā'īl (1959).

4 The objective here is not to provide a thorough investigation of the debates about this thesis and the (counter-) arguments of each party, but just to highlight the fact that there was relationship between these debates on one hand and the supposed role of Sharia in the to-be-born field of Islamic bioethics. For further information about the literature which addressed this topic in general, see for example, Ḥusayn (1999), Tirmanīnī (1977), Qaraḍāwī (1993), Ibrāhīm (2004), Zakariyyā (1986), Jābirī (1996), and 'Ashmāwī 2004.

that the contentions of early jurists cannot be accepted at face value on the idea that adopting these classical opinions would entail dismissing what "has been established by physicians of our time, who hail from all kinds of religions and creed, despite their vast knowledge of medicine, anatomy and physiology, and their reliance on their research and trials on instruments, sensors, probes, and the X-rays which pierce through the skin and flesh, making the body transparent and exposing its interior to the naked eye, in addition to basing their knowledge on experimentation and induction, and their collaborative work, despite living in different countries, thanks to the ease in postal and telegraphic communication". At the end of the fatwa, Riḍā argued that upholding such outdated contentions and neglecting the achievements of modern science could eventually result in a wide range of harms including, the attempts of non-Muslims to "defame and discredit our Sharia based on science and experimentation, not on prejudice and fanaticism, which would eventually preclude them from converting to our religion and prevent revealing its truth to those who do not know the origin of these claims within our tradition. This also entails spreading doubt among many Muslims about the truthfulness of our Sharia and its divine nature. I mean by 'many' all those who are learning medicine and are aware of, and satisfied with, the opinions adopted by today's physician and scientists about the duration of pregnancy despite knowing that these [opinions] are incompatible with what they think as the established Sharia, attested by the Qurʾān and the Sunna" (Riḍā, 1910).

It seems that advocating a discourse which accommodates modern biomedical sciences and thus demonstrates the contemporaneity of Sharia and its relevance to the modern age was not exclusive to those who had a reformist agenda of the so-called reformists. In fact, the advocates of such a discourse comprised a wide range of scholars coming from different backgrounds including those who had conservative inclinations. Just as an illustrative example, we refer to the fatwa issued by the prominent Najdi scholar, Shaykh ʿAbd al-Raḥmān al-Saʿdī (d. 1955) on human organ transplant, which was seen by him as a groundbreaking medical intervention at this time: "Many questions are raised these days about what recently happened in modern medicine regarding the removal of part of a human body and transplanting it into another person who direly needs it". After presenting the views of the proponents and opponents on the adoption of this modern technological advancement, al-Saʿdī expressed his support for the position of the proponents. He concluded his fatwa by emphasizing the benefits that accrue to the Islamic religion as a result of adopting this position. In this regard he said that "It is also to be noted that people should know that Islam does not stand as a barrier against genuine and preponderant benefits (*maṣāliḥ*). On the contrary, it adjusts to the times and conditions by keeping track of the comprehensive and partial benefits and interests. Atheists delude ignorant people that Islam cannot keep up with the modern developments, and this is a calumny on their part, for the Islamic religion is the quintessence of absolute good in its comprehensive and partial facets. It provides solutions for each and every problem, specific or general, and all other systems are inherently fallible".[5]

However, it was not easy for Sharia scholars to deal with such modern questions, since providing answers concerning biomedical matters requires understanding the precise nature of these questions and the relevant subject matters or, using jurisprudential language, "the correct perception" (*taṣawwur ṣaḥīḥ*) of the issue at hand, which plays a pivotal role in the process of "verifying of the effective rationale" (*taḥqīq al-manāṭ*). In this context, we come across early and important references to the significant role of physicians assisting religious scholars in developing the right perception of modern medical issues. Because of their educational background which almost exclusively consisted of religious and Sharia-related sciences, contemporary Muslim jurists could not have direct access to the right perception of these modern biomedical advancements. In the abovementioned fatwa of Rashīd Riḍā on the maximum possible duration of pregnancy, the questioner himself was aware of this complexity. He realized that the question was two-fold where juristic and medical aspects intersect with one another. He hinted that the medical aspects would be undertaken by the physician Muḥammad Tawfīq Ṣidqī, a cherished friend of Riḍā and one of the contributors to *al-Manār* journal (Riḍā 1910, 900). Although Riḍā did not reveal whether he actually consulted Ṣidqī while drafting the fatwa, the published text suggests that part of the information comes from a physician rather than a jurist. In this context, it is also worth mentioning the contacts between Shaykh Muṣṭafā al-Maraghı (d. 1945), the former Shaykh of al-Azhar, and they physician ʿAbd al-ʿAzīz Ismāʿīl whom Muḥammad Farīd Wajdī described as "A notable authority in medicine in the Orient, and his position today in this honorable field resembles that of Ibn Sīnā and Abū Bakr al-Rāzī during the golden Arab age of science." (Ismāʿīl 1952, 8). Shaykh al-Marāghī wrote a preface to Ismāʿīl's book *Al-Islām wa al-ṭibb al-ḥadīth* (Islam and modern medicine) (Ismāʿīl 1952, 5-7). The book originally appeared in the form of articles published in *Al-Balāgh Newspaper* and subsequently in *Al-Azhar Journal* (Ismāʿīl 1952, 9). Al-Saʿdī also referred to this issue in his

5 Saʿdī (2011, 95-100). Also, in his book on human organ transplant, Yūsuf al-Qaraḍāwī reported and endorsed the full text of the fatwa, hailing its author as a "prominent Saudi scholar from Najd who conforms to the Ḥanbali jurisprudential doctrine, but who enjoys broad horizons and an innovative propensity in his interpretations and fatwas." See Qaraḍāwī (2010, 61-67).

aforementioned fatwa by saying "All issues occurring at all times, whether their overall genus or identical cases took place before, should be conceptualized in the first place, and if their essence has become known, their characteristics have been diagnosed and one has fully conceptualized them, in their essence, their premises and results, they should then be applied to scriptural texts and their overall fundamentals." Moreover, al-Saʿdī explained that experts in medicine have a major role to play in the conceptualization process or developing the right perception, "Whenever the highly skilled physicians unanimously agree that the organ donor will not be subjected to harm, and we realized the interest gained by others from this, [organ transplantation] becomes a genuine and pure benefit." (Saʿdī 2011, 95-97).

Such examples show that preserving Sharia's role in the modern age of biomedical breakthroughs would not have been possible, at least from the perspective of contemporary jurists, without resorting to physicians whose expertise fall outside the scope of Sharia specialists. It was also clear that modern medicine, whose religious and ethical ramifications occupied the minds of Sharia scholars, was not different from the medicine that their predecessors dealt with. Though having Greek roots, pre-modern medicine gradually became an integral part of the Arabic-Islamic civilization. Thus, this old medicine was not unfamiliar to early jurists, at least at the level of the Arabic language which was its *lingua Franca*. Modern medicine, however, comprises an integral part of the Western civilization, and its scientific and technical aspects cannot be grasped without studying the output of Western academies and institutions which do not use Arabic as a research language.

With time, the need to rely on physicians becoming more and more demanding, for various reasons. For instance, one can refer to the rapid and complex evolution of biomedical sciences and the rise of a large number of techniques, which posed new ethical dilemmas, such as organ transplantation, resuscitation, Assisted Reproductive Techniques (IVF), stem cell research, gene therapy, etcetera. As described by some contemporary Muslim religious scholars, these techniques used to be part of supernatural miracles in the past (Qaraḍāwī 1996, 104). On the other hand, the modern educational system fragmented these sciences into various disciplines,[6] and created multiple special-

6 It seems that the complaint about the presence of a rupture between science and various specializations was not confined to Sharia scholars in the Muslim world only. This issue was addressed by a number of Western intellectuals, philosophers and also physicians, who talked about the need for the interconnectivity of sciences and humanities, or of the "two cultures", as named by the English writer C.P. Snow in his famous lecture delivered in 1955, see Ten Have 2012.

izations and subdivisions within each discipline to the extent that it became impossible for contemporary Sharia scholars to keep track of them, not to mention to grasp their new techniques and subtleties. The inaccessibility to this type of information for Muslim religious scholars has to do, among other things, with the fact that relevant information is usually available in English only or in other languages that these scholars do not master.[7] These developments necessitated further expansion of the role played by physicians in these bioethical discussions so that they can provide religious scholars with the right perception. The increasing need for a more intensive and systematic integration of the contributions made by physicians in the process of religio-ethical reasoning (*ijtihād*) eventually led to activation of the mechanism of collective *ijtihād* in the field of Islamic Bioethics by the beginning of the 1980s. Through this mechanism, the collaboration between Sharia scholars and biomedical scientists will reach new heights later.[8]

The mechanism of the collective *ijtihād* was institutionalized through the establishment of a number of religio-scientific institutions, three of which are to be singled out here because of their seminal contribution to the discourse

---

7 Under the title "Conditions of *ijtihād*," Yūsuf al-Qaraḍāwī cited the requirement of "knowledge of people and life," stating that it is a new condition which was not mentioned by scholars of the Islamic science of fundamentals in the past. He added that this is not a prerequisite for attaining the rank of *ijtihād*, but it should enable *ijtihād* to be accurate and appropriate. Speaking of this requirement, he said that the *mujtahid* should acquire as much as possible scientific knowledge, such as biology, physics, chemistry, mathematics, and other similar subjects, because they constitute the cultural foundation necessary for every contemporary person. He also praised the experience of al-Azhar which had introduced "these sciences in the curriculum a long time ago" (Qaradawi 1996, 104). In reality, the introduction of these sciences in al-Azhar's curricula did not bring about the required integration and interconnectivity, for the student who enrolled in the Faculty of Sharia or in other legal specializations should have spent his secondary school education in the literary studies section, where the list of required subjects excludes almost any subject related to modern medical sciences. Rather, these subjects are studied by those who join the Scientific Section so that at the university the student has the right to specialize in medicine, engineering or other modern sciences. This is what we experienced as students at al-Azhar, be it at school or university level, in the 1980s and 1990s, which is the period during which this book was written and published.

8 Islamic history has witnessed interactions between Islamic jurists, especially those who assumed positions in the judiciary, and those who were known back then as "experts" in different areas of knowledge, including medicine. However, these interactions were only confined to individual cases and matters that were dealt with on a case-by-case basis, and this cooperation and interaction were not institutionalized the way they are nowadays in a number of jurisprudential institutions and academies. (See Shaham 2010, and Wiryāshī 2016).

on Islamic Bioethics. The Islamic Organization for Medical Sciences (*Al-Munaẓẓama al-Islāmiyya li-al-ʿUlūm al-Ṭibbiyya*), which was established in Kuwait in 1984, is one of the most prominent institutions in this regard whose activities are exclusive to addressing bioethical issues from an Islamic perspective. The IOMS collaborates with two other institutions, namely the Islamic Fiqh Academy (*Al-Majmaʿ al-Fiqhī al-Islāmī*), founded in 1977 and operating under the umbrella of the Muslim World League in Mecca, as well as the Jeddah-based International Islamic Fiqh Academy (*Majmaʿ al-Fiqh al-Islāmī al-Dawlī*) founded in 1981 and operating under the umbrella of the Organization of Islamic Cooperation.

The thesis that Islamic Sharia is compatible with and still relevant to contemporary life and is also valid for all times and places, as mentioned earlier, lies at the heart of the work of these institutions, especially when they tackle such emerging bioethical issues. Explicit reference to the question of Sharia has been made in some of the resolutions adopted by these institutions, including the resolution of the IIFA adopted during its fifth session held in Kuwait on 10-15 December, 1988. The introductory text of the resolution entitled *Taṭbīq aḥkām al-Sharīʿa* (The Implementation of the Rulings of the Islamic Sharia) reads as following: "Bearing in mind that the International Islamic Fiqh Academy, which emerged from the good will of the Third Islamic Conference Summit held in Mecca, in order to seek Sharia-based solutions to the problems of the Muslim nation and the organization of the lives of Muslims in conformity with the guidelines of the Islamic Sharia, as well as the removal of all obstacles to the application of God's Sharia and the establishment of all the necessary means for its implementation" (IIFA 1988, 3471).

Despite the widespread use of the mechanism of collective *ijtihād* since the 1980s and its output, which is generally characterized by good quality, individual *ijtihād* continued to play a role in addressing these issues. Furthermore, through the papers and studies presented to the conferences organized by these *fiqh* academies and institutions, individual *ijtihād* represented the foundation of the positions adopted collectively by these institutions. Some contemporary jurists also disseminated the resolutions and recommendations of these academies by incorporating them into their own writings, which sometimes included studies presented within the framework of collective *ijtihād* (Quradāghī and Muḥammadī 2006). However, these collective resolutions and recommendations, in spite of their importance and earnestness, remain within the non-binding jurisprudential opinions which do not qualify for consensus (*ijmāʿ*) according to the majority of scholars, though they are less prone to error in comparison with individual *ijtihād*. Therefore, the area of *ijtihād* remains open even after the issuance of such resolutions (Sharafī 2013)[9]. There are some instances of individual *ijtihād* submitted to these academies which despite not finding their way to the collective resolution, become disseminated in the form of individual opinions which may find acceptance in other contexts.

For example, the opinion of Yūsuf al-Qaraḍāwī on human milk banks was not endorsed when it was debated during the second session of the International Islamic Fiqh Academy in 1985. However, Qaraḍāwī published his opinion on this matter as part of the collection of his own fatwas, which was subsequently adopted by the European Council for Fatwa and Research during its twelfth session held in 2004 (Ghaly 2012). Finally, there are also some cases of individual *ijtihād* which go contrary to the resolutions and recommendations adopted by the collective *ijtihād* institutions, and even call for the review of the latter. For instance, the Jordanian religious scholar, ʿAbd al-Nāṣir Abū al-Baṣal expressed his critical comments on the resolution of the International Islamic Fiqh Academy on cloning (Abū al-Baṣal 2004, 16).

The Role of Sharia in the Era of Genomics[10]

The early years of the twenty-first century witnessed the completion of the famous "Human Genome Project," with the United States playing an avant-garde role in it, along with several other countries. With the completion of this project, for the first time in history, the contemporary man has become able to identify himself almost entirely at the genetic level. Besides the physical structure of the human being, his organs and tissues (phenotype), human genetic structure (genotype) has been identified, and the latter constitutes the basis for the physical structure and how it functions. The human body contains

9 For information on some critical opinions about collective *ijtihād* and its contributors, see Raysūnī 2013, 64-72. Yūsuf al-Qaraḍāwī spoke in an idealistic manner, when advocating a refined collective *ijtihad* to be undertaken by an international Islamic scholarly academy with specific qualifications. He also stated that the agreement of this academy on a given issue which necessitates *ijtihād* represents "the consensus of the *mujtahids* of the era, claiming its own authority and becoming binding in fatwa and legislation." It is clear, however, that al-Qaraḍāwī does not mean just any of the established Islamic academies. See Qaraḍāwī 1996, 184.

10 Some call it the "era of genomics" given that the study of genomes has become of interest to researchers in various scientific specializations, while others opt for the term "post-genomic era" as the decoding of the genome actually paved the way for current research in the field of medical and biomedical sciences.

nearly thirty-seven trillion cells, inside each of which—excluding red blood cells—lies DNA in the form of tightly wrapped and packed threads. Hypothetically, if the DNA threads were unwrapped and stretched out, they would span the return distance between the earth and the sun by nearly 200 times. The genome is the complete set of DNA, where DNA represents the main structure of the genome. Such a simplified image helps present the extent of the achievement of decoding the human genome, which is the sum-total of the genetic composition and includes about 30,000 genes. It is to be noted that mutations in one single gene may cause 4,000 diseases (Collins 2006, 1-2; DePamphilis and Bell 2011, 20; Lewis 2014, 1-12).

Since its inception in the last decade of the twentieth century, the Human Genome Project (HGP) has drawn significant global attention, which increased after the completion of the project and the publication of its findings in leading scientific journals. This event was compared to other major achievements in the history of science, such as the exploration of space and the discovery of nuclear fission. The HGP has equally led to major changes in the philosophy of modern medicine and its technical applications, some of which have already been used, while others are still expected. The human DNA structure has generated significant interest among biomedical scientists and physicians, and it now plays a pivotal role in various aspects of healthcare, such as in determining the appropriate diets and lifestyles for each individual, and the predictability of potential diseases even in the absence of physical symptoms. Because genetic composition varies from one individual to another, the tendency now is to embrace "personalized medicine" or "precision medicine," which stipulates that the incidence of a particular disease does not necessarily require prescribing the same type and dose of medication for all patients. Instead, these should be determined on the basis of the genetic makeup of each person. In addition, the information inscribed in the genome reveals the biological kinship of the genome bearer and his distant ancestors, which could extend to hundreds of years, and the connection to his (future) progeny. Accordingly, we often hear the phrase "book of life" in reference to the huge amount of information inherent in the human genome, which does not only relate to health and sickness, but also to human life in general (Ibn ʿAbd al-ʿAzīz 2000).

Once again, the debate about the role of Sharia in dealing with these new scientific developments resurfaced. This time, however, the historical context of the twenty-first century differed from the one that prevailed in the debate about Sharia and modern medicine by the beginning of the twentieth century. With the development of communication technologies, global distances shrunk, knowledge of scientific ventures, like the (HGP), became much more accessible, even in early stages. On the other hand, the mechanism of collective *ijtihād* which brings together Muslim religious scholars and biomedical scientists became institutionalized by the beginning of the 1980s. The abovementioned difficulties created by the type of education received by contemporary religious scholars, however, have not changed. There are even indications that these difficulties have been exacerbated, particularly with respect to the human genome. This is explicitly stated by Muḥammad ʿAlī al-Bār, one of the physicians known for their regular participation in, and influential contribution to, the collective *ijtihād* discussions on bioethical issues. He explained these difficulties in his comments on the discussions that took place during the Symposium held by the Islamic Organization for Medical Sciences (IOMS) held in 1998, which will be outlined below.[11] Thus, these developments in the genomic era have magnified the role of biomedical sciences and strengthened the need of religious scholars for biomedical scientists to explore and demonstrate the role of Sharia in this new era.

In 1993, just a few years after the official declaration of the start of the Human Genome Project, discussions took off in the Muslim world. The Faculty of Science at the University of Qatar, in cooperation with the Islamic Educational, Scientific and Cultural Organization (ISESCO) and the World Islamic Call Society, organized a symposium under the title "Ethical Implications of Modern Researches in Genetics," which brought together religious scholars and biomedical scientists. The symposium was held on February 13-15, 1993, and issued twelve recommendations, the fifth of which was on the Human Genome Project, describing it as "the largest scientific project in the history of humanity." (Īsiskū 1993, 360). These issues were also addressed by the conference on "Genetic Engineering between Shariah and Law", convened by the Faculty of Sharia and Law at the United Arab Emirates University on May 5-7, 2002. The first session of the conference was devoted to the theme "Human Genome: Its Essence and Future." The second edition was held on November 20-22, 2007 and revolved around the Sharia-based determinants (*ḍawābiṭ Sharʿiyya*) for

11 Al-Bār cites the book *The Ethics of the Human Genome*, in which the author Hānī Rizq reveals that he spent one hour explaining to jurists at this symposium the scientific aspects of the genome, but the jurists did not understand his jargon and asked for an interpretation of what he said, "because we did not understand anything at all." Al-Bār adds that although other specialists, including the late Ḥassān Ḥatḥūt and Al-Bār himself, tried to explain these scientific aspects more than once throughout the symposium, some of the participating jurists did not fully grasp them. (Tawṣiyyāt 1998, 1112). It is worth mentioning that the published proceedings of this symposium do not include Hānī Rizq's paper, but he did present the papers written by Ṣāliḥ ʿAbd al-ʿAzīz Karīm and Muḥammad al-Yashuwī, because neither author could attend the symposium. (al-ʿAwaḍī and al-Jundī 2000, 5-10, 103, 107, 133).

genomic research and genetic testing. Also, the Dubai-based Pan Arab Human Genetics Conference is periodically organized by the Center for Arab Genomic Studies. Moreover, in collaboration with other institutions based in Qatar, the Research Center for Islamic Legislation and Ethics (CILE) organized two activities, which focused on genomics and Islamic ethics. On October 2, 2014, a public symposium entitled "Islamic Ethics in the Era of Genomics" was organized in collaboration with the Supreme Council of Health in Qatar. Furthermore, in its session held in 2015, the World Innovation Summit for Health collaborated with CILE to organize a symposium on "Healthcare and Ethics: Genomics". A number of Sharia scholars and biomedical scientists from Qatar and abroad participated in these two symposia.[12]

As for the contributions of the abovementioned three key institutions, which adopted the mechanism of collective *ijtihād* in its institutional form, the Islamic Organization for Medical Sciences (IOMS) hosted a symposium on "Genetics, Genetic Engineering, Human Genome and Gene Therapy: An Islamic Perspective" on October 13-15, 1998 ('Awaḍī and Jundī 2000). The final recommendations of this symposium still represent an authoritative source for most of the subsequent debates between religious scholars and biomedical scientists. In its eleventh session held on November 14-19, 1998, the International Islamic Fiqh Academy (IIFA) discussed the recommendations of this symposium, but the relevant resolution was deferred to a future session. The Islamic Fiqh Academy (IFA) also held its sixteenth session on January 5-10, 2002, which debated several issues, including the potential areas for the use of genetic fingerprinting, and issued a number of resolutions. The seventh resolution refers in passing to the human genome, emphasizing that it should not be commodified: "The human genome may not be sold to a race, to a people, or to an individual, for whatever purpose, and it may not be donated to any party, given the unethical consequences that result from these types of transactions." (IFA 2002, 360).[13] Furthermore, the IOMS organized a symposium on February 6-9, 2006 entitled "Human Genetic and Reproductive Technologies: Comparing Religions and Secular Perspectives," whose recommendations included a section entitled "Declaration of Principles," citing ad verbatim segments from the recommendations adopted during the IOMS symposium held

12 In 2015, the Center was awarded a prestigious grant from the Qatar National Research Fund to undertake the scientific research project "Indigenizing Genomics in the Gulf Region (IGGR): The Missing Islamic Bioethical Discourse," which was officially launched by the beginning of September 2016 and would run for a period of three years.

13 *Proceedings of the Sixteenth Session of the Islamic Fiqh Academy in Makkah* (Makkah: The Islamic Fiqh Academy/The Muslim World League, 2002), p. 360.

in 1998. Apparently, the aim here was to garner support for those principles from religious and secular voices coming from outside the Islamic discourse. (al-ʿAwaḍī and al-Jundī 2008, 1173-1175). Several years later, during its twentieth session held on September 13-18, 2012, the International Islamic Fiqh Academy (IIFA) discussed anew the recommendations of the IOMS symposium held in 1998. Once again, the resolution was deferred to a future session, yet, the participants recommended the organization of a specialized symposium to discuss these recommendations. This symposium was held in Jeddah on February 23-25, 2013 and was jointly organized by the IIFA and IOMS. Finally, during its twenty-first session held on November 18-22, 2013, the IIFA approved the resolution of the IOMS recommendations, which had been made fifteen years before, with some modifications.[14] The table below provides an outline of the most important symposia and conferences which adopted the mechanism of collective *ijtihād* to address the ethical issues related to genomics and associated technologies.

14 The resolution of the Academy's is entitled "Resolution on Genetics, Genetic Engineering and the Human Genome" http://www.iifa-aifi.org/2416.html (Retrieved August 4, 2017).

Chronology of Collective *Ijtihād* Debates on Genomics (1993-2015)

Symposium/Conference	Venue	Date	Organizer
Ethical Implications of Modern Researches in Genetics	Doha, Qatar	February 13-15, 1993	The Islamic Educational, Scientific and Cultural Organization/World Islamic Call Society/Faculty of Science, University of Qatar
Genetics, Genetic Engineering, Human Genome and Genetic Therapy: An Islamic Perspective	Kuwait	October 13-15, 1998	The Islamic Organization for Medical Sciences
11th Session	Manama, Bahrain	November 14-19, 1998	The International Islamic Fiqh Academy
16th Session	Makkah, Saudi Arabia	January 5-10, 2002	The Islamic Fiqh Academy
Genetic Engineering between Shariah and Law	Al Ain, United Arab Emirates	May 5-7, 2002	Faculty of Sharia and Law, United Arab Emirates University
Human Genetic and Reproductive Technologies: Comparing Religions and Secular Perspectives	Cairo, Egypt	February 6-9, 2006	The Islamic Organization for Medical Sciences
The Ethical Perspectives of Human Genetic Applications in the Arab World	Dubai, United Arab Emirates	November 20, 2007	Center for Arab Genomic Studies
20th Session	Oran, Algeria	September 13-18, 2012	The International Islamic Fiqh Academy
Genetics, Genetic Engineering and the Human Genome	Jeddah, Saudi Arabia	February 23-25, 2013	The International Islamic Fiqh Academy/ The Islamic Organization for Medical Sciences
21st Session	Riyadh, Saudi Arabia	November 18-22, 2013	The International Islamic Fiqh Academy
Islamic Ethics in the Era of Genomics	Doha, Qatar	October 2, 2014	Research Center for Islamic Legislation and Ethics/ Supreme Council of Health, Qatar
Healthcare and Ethics: Genomics	Doha, Qatar	February 17, 2015	World Innovation Summit for Health/ Research Center for Islamic Legislation and Ethics

It is worth noting here that approaching ethical issues from an Islamic perspective has not been limited to the institutionalized activities which adopted the mechanism of collective *ijtihād*. Some Muslim jurists have provided their own individual *ijtihād*, as reflected for instance in the writings of Nūr al-Dīn al-Khādimī, a Sharia scholar who participated in the collective *ijtihād* activities referred to previously, e.g. the conference organized by the Faculty of Sharia and Law in 2007 and the symposium organized by the Center for Arab Genomic Studies in 2007, both held in the United Arab Emirates. Besides these participations, in his capacity as a religious scholar, Al-Khādimī, published some works that convey his individual *ijtihād* (Khādimī 2003 and 2004). There is also the work of the late Muḥammad Ra'fat 'Uthmān (d. 2016), who participated in the conference organized by the Faculty of Sharia and Law in the UAE in 2002. Additionally, several research studies which address some of the issues related to genomics and its applications have been published, even though, to my knowledge, their authors did not participate in the discussions of collective *ijtihād* referred to earlier. Such studies, however, remain limited in quantity (Kan'ān 2003; Idrīs 2003; Āl Shāfi' 2007; Yashū 2015-2016; 'Ubaydī 2017).

The Role of Sharia in the Genomic Era: Three Main Characteristics

In the perception of many Muslim religious scholars and biomedical scientists, the Human Genome Project represents an important historical landmark in the development of modern science. They argued that it has provided humans with an amazing power to know themselves, especially at the molecular level, in an unprecedentedly precise and profound way. It is also, to manage diseases which used to be seen as incurable, and even to improve human physical and mental capacities. Considering the perception of religious scholars and biomedical scientists who participated in the deliberations on genomics and Islamic ethics, one can argue that the role of Sharia in the genomic era is characterized by three main points:

The first point involves demonstrating the contemporaneity of Sharia. This issue remained a recurrent point in any discussion about the role of Sharia in the discussions on genomics. The old emphasis on the ability of Sharia to deal with emerging issues of any kind is repeated while stressing that genomic technologies make no exception in this regard. In line with the aforementioned statements of Rashīd Riḍā and 'Abd al-Raḥmān al-Sa'dī, Ḥasan Yashū (College of Sharia and Islamic Studies at Qatar University) in the introduction to his research on genomics, sums up the mainstream position on genomics among contemporary Muslim jurists: "Since Sharia is characterized by its transcendental and everlasting nature, it has managed through its general texts, governing principles and consistent rules, as well as by enabling flexible *ijtihād,* to keep abreast of all the developments and to contribute profusely to the solution of various problems, especially those related to contemporary medicine" (Yashū 2015-2016, 18).

The second point which characterized the role of Sharia in the genomic era is the adoption of a position that goes beyond what Riḍā, al-Saʿdī and their colleagues in the twentieth century were trying to defend. Riḍā and his like-minded religious scholars were arguing that there is no contradiction between Sharia and modern science. However, the participants in the discussions of genomics and ethics, be it at the level of collective or individual *ijtihād*, were not just concerned about the (im)permissibility of conducting genomic research, but mainly about how Muslims (should) contribute to this emerging field. For instance, the aforementioned symposium organized by the Faculty of Science at the University of Qatar stated "At a time when concerted efforts are being deployed by many countries to achieve the largest scientific project in the history of mankind, i.e. the comprehensive study of the complete genetic information of the entire human race and its genetic makeup known as the Human Genome Project, Muslims should not remain mere spectators who do not contribute to the study of the biological heritage of mankind and to the study of man's future. Therefore, the participants at the meeting call upon the Islamic countries that are capable of providing strong financial support commensurate with the magnitude of the project to share in this earnest human project so that we can benefit from its significant outcomes" (Īsiskū 1993, 360-361). In later discussions, involvement in genomic research was seen as an act of collective duty (*farḍ kifāya*) as advocated in the final communiqué of the IOMS symposium held in 1998: "Given that genomic sequencing is the means to identify some genetic diseases or the susceptibility to them, it is therefore of added value to health studies and medical sciences in their endeavor to prevent or cure diseases. This makes it fall within the category of collective duties in society." (Awaḍī and Jundī 2000, 2, 1047-1048). The communiqué also invited the Islamic countries to embrace the field of genetic engineering by establishing research centers in this area, whose *raison d'être* should be in conformity with Sharia. The same position was adopted again by the International Islamic Fiqh Academy in its session held in 2013, where Muslim countries were called to join the genomic revolution.[15] This drive to encourage Muslim countries to participate in genomic research was perhaps among the factors that paved the

15 See http://www.iifa-aifi.org/2416.html

way for launching national genomic projects with huge financial allocations in a number of countries led by Qatar and Saudi Arabia. The projects of these two countries were publicly announced by the end of 2013.[16]

The third point is the concern that genomic research and its associated technologies can result in devastating consequences especially if control is lost. Almost all religious scholars and biomedical scientists who contributed to the discussions on the interplay of genomics and Islamic ethics shared this concern, despite their differences on many other questions. They agreed that developing Shari-based determinants (*ḍawābiṭ Sharʿiyya*) is indispensable to make sure that genomic research and its applications will not end up deviating from the principles, ethics and provisions of Sharia. It is to be noted here that fear of ethical slippage in genomic research is not exclusive to Muslim jurists. This fear has to do with the enormous potential of genomics, which also raised concern among those in charge of the Human Genome Project (HGP) since its inception, starting from the renowned scientist and Nobel Prize laureate James Watson, who oversaw the management of the project in its early phases. Given these concerns, part of the budget of the HGP was allocated to the study of the ethical aspects of the project in the framework of the Ethical, Legal and Social Implications (ELSI) program. The ELSI program started concurrently with the scientific research in the project. This was contrary to the standard course of action in the field of biomedical ethics, where ethical issues are usually addressed after the scientific research has already taken great strides and the ethical dilemmas and challenges would emerge thereafter. In the case of genomics, however, it was clear from the onset that these dilemmas and challenges are inevitable (Green and Collins 2015). Strikingly enough, the word "religion" did not appear in this work and religious discourse was marginalized in the research output. In the program evaluation report, which assessed the first ten years of the ELSI work, this point was mentioned as part of the criticism directed to it (ELSI Research Planning and Evaluation Group 2000). This criticism caused people responsible for ELSI to become increasingly aware of the need to consider ethics-based research studies from a religious perspective. This type of research, nevertheless, remained marginal in comparison with the literature which approaches the ethical issues from a secular perspective. Undoubtedly, the marginalized position of religious discourse in this program,

16 For further elaboration on projects concerning genomic research in the Gulf region, especially in Qatar and Saudi Arabia, see the report published in the proceedings of the World Innovation Summit for Health (WISH) held in Doha on November 29-30, 2016. See also Ghaly et al. (2016), 7-15.

reflects the marginalized status of religious discourse in general in the field of biomedical ethics in the West, as noted at the beginning of this chapter.

Against this background, it was natural for the participants in the ethical deliberations in the Muslim world to have concerns about the relationship between genomics and religion and to feel the need to emphasize that genomic research should be conducted in conformity with religious principles and particularly the provisions of Sharia. Signs of this apprehensive concern were reflected in the frequent and recurrent reference to Sharia in the communiqués and recommendations of the aforementioned symposia and conferences. Just as illustrative examples, few quotations are given below from the final communiqué of the symposium organized by the IOMS in 1998 and the recommendations of IIFA in 2013. These two documents represent the essence of what collective *ijtihād* produced about genomics:

> But the findings of this research should not automatically move to the level of practical applications before they go through the filter of Sharia-based determinants (*ḍawābiṭ Sharʿiyya*). Whatever proved to be compatible with Sharia should be approved, and whatever is incompatible should not be permitted
>
> No research shall be undertaken or treatment or diagnosis carried out in connection with a person's gene or genome unless a rigorous assessment is undertaken beforehand to gauge the potential risks and benefits associated with these activities, ensuring in the process adherence to the provisions of Sharia
>
> No research in the human genome or any of its applications, particularly in the fields of biology, genetics and medicine, should prevail over the provisions of Sharia
>
> Islamic countries should engage in the world of genetic engineering by establishing the relevant research centers whose raison d'être should be in conformity with the Sharia[17]
>
> The genome shall not be used in a harmful way or in any form which is contrary to Sharia

---

17 Abdul Rahman al-Awadhi and Ahmed Rajai al-Jundi (eds.), *Genetics, Genetic Engineering, Human Genome and Genetic Therapy: An Islamic Perspective*, Part II, pp. 1045-1052.

> Underscoring the Sharia-based determinants (*ḍawābiṭ* guidelines related to the human genome, as outlined in the recommendations of the symposium on Genetics, Genetic Engineering, Human Genome and Gene Therapy: An Islamic Perspective, organized by the Islamic Organization for Medical Sciences in cooperation with the International Islamic Fiqh Academy in 1998
>
> No clinical research (clinical trials) on the human genome or any of its applications shall be conducted, particularly in the areas of biology, genetics and medicine, as long as it violates the provisions of Sharia or the human rights which is recognized by Islam (ʿAwaḍī and Jundī 2000, 2, 1045-1052).

The aforementioned three points, especially the third one, represent serious challenges to demarcate a role for Sharia in the age of genomics, to both Muslim religious scholars and biomedical scientists. To the same extent that genomic techniques determine the priorities and aspirations of medical and biomedical sciences in the near future, the ethical issues resulting from these techniques will likewise determine the research agendas of biomedical ethics in general. The bulk of the ethical questions posed by modern techniques, such as Assisted Reproduction Technologies (ARTs), artificial insemination, and genetic engineering, fall within the purview of the major questions raised by the field of genomics. Yet, the genomic context usually adds new dimensions and complexities to these questions. As far as the third point is concerned, a couple of studies tried to spell out the Sharia-based determinants (*ḍawābiṭ Sharʿiyya*) tailored for specific issues with relevance to genomics like medical treatment, DNA paternity, privacy and abortion (Kanʿan 2003; Khādimī 2004; Yashū 2015-2016; ʿUbaydī 2017). These studies filled certain gaps because the abovementioned expressions regarding the necessity of taking Sharia provisions seriously remain too general and in need of more detailed studies. This problem was highlighted by a number of Sharia scholars who argued that genomics is too complex to be governed by general and sweeping rules and standards. Al-Khādimī's statement below is an illustrative example:

> The human genome is not just one thing so that it can be governed by these Sharia-based determinants (*ḍawābiṭ Sharʿiyya*) in a homogenous way. Rather, it is a renewed phenomenon with its own scientific identity, characteristics, uses, outcomes, overlaps and ramifications. It also moves from one stage to another, thus increasing in complexity, fragmentation, and is multifaceted. Additionally, it is a sensitive and thorny phenome-

> non as far as it relates to human dignity and sanctity, the edifice of ethics and values, as well as the security of peoples, countries and individuals. It is first and foremost premised on certain backgrounds and intersects with specific motivations and purposes, where not only the scientific and commercial aspects overlap. But, possibly also, the doctrinal, intellectual, political, cultural and religious aspects (Khādimī 2007).

In their bid to explore the role of Sharia in addressing the ethical issues raised by genomics, the published proceedings of the abovementioned conferences and symposia and the works written by individual religious scholars have almost exclusively focused on the discipline of Islamic jurisprudence (*fiqh*). The emphasis was on highlighting specific Sharia-based determinants (*ḍawābiṭ Sharʿiyya*) or juristic rulings (*aḥkām fiqhiyya*) which are supposed to govern specific genomics-related techniques and applications. These works depended heavily on the discipline of *fiqh* and employed the system of "five rulings" (*al-aḥkām al-khamsa*)[18] because like most pioneer studies, they were concerned with providing practical and direct answers for urgent questions. However, maintaining the role of Sharia amid the complex and multidimensional nature of the ethical questions raised by the field of genomics requires a broader and more comprehensive ethical discourse than the widely used one today, which has almost exclusively been employing the tools of *fiqh*. Below, the final section of this chapter will review the strengths of the *fiqh*-centered bioethical discourse and how its weaknesses can be improved in the future.

Concluding and Critical Remarks

This chapter surveys the deliberations, which started as early as the beginning of the twentieth-century, regarding the role of Sharia in addressing the ethical issues raised by modern biomedical sciences, including genetics and genomics. These deliberations, characterized by their almost exclusive dependence on the discipline of *fiqh*, have their own strengths and weaknesses.

With respect to the strengths, the *ijtihād* practiced by individual Muslims religious scholars, and later collectively in collaboration with biomedical scientists, in the area of biomedical ethics, significantly contributed to preserving a role for Sharia in this domain. This was different from the parallel developments in the Western bioethical discourse where the role of religion was in-

18 For more information about this system used for the categorization of human actions, see Ghaly 2016, 39-40.

creasingly marginalized. The most important writings and published works, in terms of quantity, quality,[19] and impact in the public space[20] in the Muslim world and for Muslims in general, are still based on Sharia as a source of reference. Also, contemporary Sharia scholars are still the most active contributors to these discussions within the Muslim world context. Their views and fatwas continue to be considered important references for many of the institutions operating in the field of healthcare in the Muslim world. The limited space here is not sufficient to provide all the instances which illustrate this aspect, but some illustrative examples can be mentioned. At the level of individual Muslims, the fatwas issued by some contemporary Sharia scholars testify to the existence of interaction between them and the general public, including medical doctors. For instance, ʿAbd Allāh Ibn al-Ṣiddīq al-Ghumārī (d. 1992), the Moroccan jurist and Ḥadīth scholar, responded to the questions raised by students in the Alexandria Faculty of Medicine. His answers were later published more than once (Ibn al-Ṣiddīq n.d.; idem n.d.). Also, Shaykh Jād al-Ḥaqq ʿAlī Jād al-Ḥaqq (d. 1996), the former Shaykh of al-Azhar, answered a number of questions raised by female students in the Faculty of Medicine at al-Azhar University, along with some medical explanations prepared by some Faculty of Medicine professors (Jādd al-Ḥaqq 2005). Likewise, a number of fatwas on medical matters were issued by Shaykh ʿAbd al-ʿAzīz Ibn Bāz (d. 1999) during his meetings with the staff at al-Nūr Hospital in Mecca (Ibn Bāz 1999). The final example here is Yūsuf al-Qaraḍāwī's published fatwas, where a separate section was dedicated to bioethical issues in the second and third volumes (Qaraḍāwī 1994, vol. 2, 525-619; Qaraḍāwī 2003, vol. 3, 513-534).

At the institutional level, various questions have been addressed to Sharia scholars by Ministries of Health and other governmental and non-governmental entities, some of which are from outside the Muslim world.[21] In response

19 The databases which catalogue research studies on biomedical ethics in the Muslim world are a witness to this fact. An example of these is the database "Islamic Medical and Scientific Ethics," which is affiliated with one of the most important scientific research institutions in the field of biomedical ethics in the world, namely The Kennedy Institute of Ethics at Georgetown University.

20 This this applies to what happens within the Muslim world, such as the codification of laws that govern techniques like organ transplantation and the practices of *in vitro* fertilization clinics. Additionally, this holds also true for the ongoing deliberations at international forums, such as the United Nations where, for example, the fatwa issued by Aḥmad al-Ṭayyib, the mufti of Egypt at the time, was cited in the UN discussion on cloning. (Eich 2006, 300-301, 305).

21 In a personal interview held in Europe several years ago, Aḥmad Rajāʾī al-Jundī (Assistant Secretary-General Assistant of the Islamic Organization for Medical Sciences) told me

to these questions, Muslim religious scholars issued fatwas, some of which are now published and widely circulated. Just as examples, we refer to the fatwa of Shaykh Ḥasan Ma'mūn (d. 1973), in response to a question from the Egyptian Al-Nūr and Amal Association, on the permissibility of donating the eye of a deceased person, and the fatwa of Shaykh Muḥammad Khāṭir (d. 2004), in response to a question from the Office of the Legal Advisor to the Egyptian Minister of Health, on the use of skin from a deceased person to treat the burned skin of the living (al-Bār 1992, 327-331). The same Ministry sent another question to the Egyptian *Dār al-Iftāʾ* about the possibility of establishing a human milk bank (Jundī 1983, 458). Also, the set of fatwas issued by Shaykh al-Qaraḍāwī's on organ transplantation were originally responses to a list of questions from the Organization of Islamic Medicine in South Africa and the Department of Islamic Medicine at King Abdul Aziz University in Jeddah, which were compiled together and published by the Kuwait Transplant Society (Qaraḍāwī 2010 and 1996). There are also two examples from the Kingdom of Saudi Arabia, the first being the fatwa issued by the Saudi Council of Senior Religious Scholars (*Hay'at Kibār al-ʿUlamāʾ*) in response to a question from the Head of the Pediatrics Department, Faculty of Medicine in Abha on the "procedures of cardiopulmonary resuscitation in some futile instances." The text of this fatwa was published on the website of the General Presidency of Scholarly Research and Ifta (www.alifta.net). The second example is the fatwa issued by the Permanent Committee for Scholarly Research and Issuing Fatwas (*Al-Lajna al-Dā'ima li al-Buḥūth al-ʿIlmiyya wa al-Iftāʾ*) in response to a question posed by the vice-chairman of the North West Armed Forces Hospital on the use of the defibrillator. The text of this fatwa was published on the abovementioned website as well. The hospital drafted its policies in alignment with the purport of the fatwa.[22] This fatwa also became well-known among researchers who wrote on this subject in journals published in English, referring to it as Fatwa No. 12086 (Ayed and Rahmo 2014; Chamsi-Pasha and Albar 2017). Another relevant example comes from the United Arab Emirates, where a fatwa was issued in response to a question which had been occupying the minds of those working in the healthcare sector about the (in)compatibility of the "Good Samaritan"

---

that the former US President Bill Clinton had sent a letter to the Organization asking for the opinion of Islam on the issue of cloning in the wake of the cloning of Dolly. He also informed me that he kept the original copy of Clinton's letter at the Organization's headquarters in Kuwait. There is also the fatwa issued by the Muslim Law (Shariah) Council in the United Kingdom in 1995 on organ donation, in response to a question addressed raised by the British Department of Health (Ghaly 2012a).

22 See http://afhsr.med.sa/cqi_web/docs/Standards/LD/Samples/40.09%20LD%20%20DNRC.pdf (Retrieved August 15, 2017).

principle with Islamic Sharia. This is about individuals who volunteer to save people whose lives are in danger but do not want to be held legally accountable in case the rescue endeavor fails. The questioner also wondered whether differences in gender and religious affiliation would also matter in this case.[23]

The deliberations on genomics do not represent an exception to the above-sketched landscape. The aforementioned symposium organized by the Faculty of Science at the University of Qatar in 1993 adopted a position which was also held later by many who participated in the individual and collective discussions on genomics. The advocates of this position strongly called upon Muslims to contribute to this emerging scientific field. This contribution was presented as a collective duty (*farḍ kifāya*), which is incumbent on Muslims in the present era. Undoubtedly, such a positive position should have had an impact on some Muslim countries which launched national genome projects, such as Qatar and its Qatar Genome Program (QGP) and the Kingdom of Saudi Arabia which established its own Human Genome Project. Both projects were launched at the end of 2013. Qatar Biobank, which has been playing a key role in establishing and managing the QGP, paid attention from the beginning to the significance of addressing relevant ethical questions from the Sharia perspective. To do so, the Qatar Biobank convened an international symposium, which was attended by a number of Sharia scholars. In its pamphlet designed to familiarize the public with the activities of the biobank, a separate section was dedicated to the interplay of Sharia and the activities of the biobank. In this pamphlet, the biobank stressed its adherence to the "Islamic Code of Medical Ethics," also known as the "Kuwait Document," which was originally issued by the World Organization for Islamic Medicine that later became known as the Islamic Organization for Medical Sciences (IOMS). In the same pamphlet, the Qatar Biobank spoke about collaboration with the Research Center for Islamic Legislation and Ethics at Qatar Foundation (Qatar Biobank 2014, 12-13). Additionally, the biobank approached some professors in the College of Sharia and Islamic Studies at Qatar University to seek their opinion about the permissibility of using blood, urine or saliva samples for scientific research purposes, also after the death of the sample donor.

The abovementioned examples demonstrate that the efforts of Muslim religious scholars, often in collaboration with biomedical scientists, could demon-

23 See https://www.clydeco.com/insight/article/good-samaritan-principles-in-the-uae-legal-liabilities-when-administering-f (Retrieved August 16, 2017). I would like to extend here my thanks to my colleague Jothi Ravindran (Legal Adviser to the Sidra Medical and Research Center in Qatar), who drew my attention to this fatwa when she herself was seeking a fatwa from Sharia scholars in Qatar on the same subject.

strate the relevance of Sharia to the bioethical deliberations and the ability to provide answers for the questions raised by individuals and institutions. However, keeping Sharia at the heart of bioethical discourse in the era of genomics will face challenges ahead. Below, two main important points will be highlighted to explore how certain weaknesses in the contemporary Islamic bioethical discourse can be improved. The first point is the need to revise the concept of Sharia itself and its scope. The second point concerns the pool of participants in the deliberations which theorize the presumed role of Sharia in the field of bioethics.

With respect to the challenge at the conceptual level, the term "Sharia" is one of the most frequently used words when discussing Islam, either negatively or positively, in the modern era. This holds especially true for fields like biomedical and financial ethics. Despite the frequent use of this term, rarely do we find researchers or scholars who discuss how the term Sharia should be defined or how its scope should be determined.[24] However, examining available literature on biomedical ethics from an Islamic perspective, whether within the context of individual or collective *ijtihād*, shows that Sharia is almost exclusively seen through the lens of Islamic jurisprudence (*fiqh*). In other words, Sharia is perceived as a set of practical provisions to be extracted directly from the Qurʾān and Sunna or premised on the works of previous jurists throughout the history of Islam. There is no question that the juristic dimension plays a pivotal role in delineating the role of Sharia and it should not be marginalized or dispensed with. However, many of the ethical issues which arise in the genomic era do not fall within the traditional scope of the discipline of *fiqh*. The question of the genome is much more profound and complex than being merely a matter of issuing partial ethical judgements about the use of a specific form of technology in a particular context. The results of genomic research and its current and future applications raise major questions about how to (re)consider some central concepts which shaped the Islamic ethical discourse throughout history. Such as, legal capacity (*taklīf*), human agency, acquisition of acts (*kasb*), determinism and free will in the light of what genomics revealed about the role of genes in determining some aspects of our structure, tendencies, and be-

24 Some of the important exceptions in this regard include the statement made by al-Khādimī in one of his research studies about Sharia guidelines for genomic research. He said that the concept of *al-Sharʿ al-Islāmī* consists of two major parts: the first one manifests itself in the religious Scriptures and their detailed evidences, while the second part consists of the general rules, overall objectives, and governing principles (Khādimī 2007, 4-5). What al-Khādimā has explicitly said here indeed reflects the concept of Sharia as understood by other contemporary scholars who practice *ijtihād* in the field of biomedical ethics, though they might not have stated it so explicitly.

havior, which eventually led to terms like "genetic determinism." These issues, among several others, cannot be dealt with by looking at Sharia through the lens of *fiqh* alone. Integrating insights coming from disciplines like philosophy, theology, Sufism, Qurʾān exegesis, and Ḥadīth commentaries are indispensable to be certain that Sharia can still continue to guide the ethical discussions of the genomic era. Therefore, the *ijtihād* process exploring the role of Sharia in fields like genomics should involve specialists in these disciplines and not only those specialized in *fiqh*. This leads to the second challenge which concerns the identity of the participants in the process of this *ijtihad*.

Throughout Islamic history, jurists have been entrusted with the task of interpreting the foundational texts of Islam (viz. Quran and Sunna) to extract practical rulings which guide the behavior of Muslims in various aspects of life including those related to states of health and sickness. This process was known as *ijtihād*, which literally means exerting one's utmost effort. With time, *ijtihād*, in its technical sense, became the monopoly of those who excel in the discipline of *fiqh*. The prominent religious scholar Muḥammad al-Shawkānī (d. 1839), representing a mainstream position among Muslim jurists including the contemporary ones, argued that even if *ijtihād* was undertaken by specialists in other disciplines like theology, its outcome will not be recognized (Shawkānī 1999, 2, 206; Qaraḍāwī 1996, 12-13). However, the ethical questions raised by modern biomedical sciences, as explained in this chapter, revealed the inability of Muslim jurists to exercise *ijtihād* by themselves. This is due to various reasons, especially the educational background of these jurists which usually focused on "religious" sciences and the Arabic language. This necessitated their collaboration with biomedical scientists so they can have a proper understanding of the biomedical information related to the bioethical issues under discussion. The history of collective *ijtihād* in the field of bioethics, which spans about four decades, shows the involvement of biomedical scientists in their capacity as equal "partners" in the process of *ijtihād* with jurists, simply as "informants" whose task is limited to explaining or simplifying specific biomedical information. This position was explicitly expressed by Shaykh ʿAbd al-ʿAzīz Ibn Bāz in his response to a question on whether patients can accept a fatwa given by a physician or if they should still consult a Sharia scholar. Ibn Bāz, said: "The patient should seek the scholars' feedback about what physicians say regarding religious rulings, because physicians are knowledgeable about their own field, and religious scholarship has its own specialists… The physician's duty is to ask but not to issue fatwas without proper knowledge, because he is not a Sharia scholar." (Ibn Bāz 1999, 24-25). The same tendency is found in the practices of *fiqh* academies. In personal communication with the Saudi physician Muḥammad ʿAlī al-Bār, he told me that the collective deliberations go as following: the physicians explain the scientific aspects to the jurists and together they discuss related issues. But the session dedicated to discussing and voting on the final recommendations and resolutions is attended only by the religious scholars. However, this is not a uniform practice for all institutions which employ the mechanism of collective *ijtihād* in the field of bioethics. The Kuwait-based Islamic Organization for Medical Sciences (IOMS) broke away from this tradition by engaging biomedical scientists in all stages, including drafting and adopting the final recommendations. Some of the religious scholars who participated in the symposia organized by the IOMS made observations about this practice. The critique was directed to the physicians who cross over the borders of their specialization and argue about religious rulings that should otherwise be left to the jurists. ʿAbd al-Qādir al-ʿAmmārī, one of the jurists who participated in the 1983 symposium organized by the IOMS on the beginning and end of human life, said: "I call upon everyone to stick to their specialization. The physician must not deal with anything but with what he sees in front of him. Delving into the interpretation of Prophetic traditions and the discussions of the jurists should be deferred to the jurists and specialists." The late ʿIṣām al-Shirbīnī (d. 2010), one of the participating physicians in this symposium, commented on al-ʿAmmārī's position, explaining how that the process of *ijtihād* should be a shared task between jurists and physicians, and that neither party can accomplish the task alone (Al-Madhkūr a.o. 1985, 221-264). Such disagreements among the participants in these collective deliberations encompass the difficulty of setting clear borders between the task of the physicians and that of religious scholars. As biomedical issues grow in complexity and ramifications, drawing a border between the task of explaining the scientific aspect of biomedical technologies, usually assigned to biomedical scientists, and addressing the ethical issues raised by these technologies, traditionally entrusted to religious scholars, is getting increasingly difficult, if not impossible in many cases. Against this background, the process of *ijtihād* in the age of genomics requires better management and coordination between the two groups. For the group of biomedical scientists, the discussions about genetics and genomics revealed that some jurists and biomedical scientists began complaining that some of those who participated in the meetings organized by the *fiqh* academies are not coming from these scientific fields. They explain that the pool of participating biomedical scientists in the collective discussions hardly witnessed any modifications or updates since the 1980s. However, fields like genetics and genomics are relatively new and many of these scientists did not study these specializations enough. The critics argue that most of them depend only on what they read in some discrete articles but with no concrete scientific contribution to genetics or genomics. The point these jurists want to make is that religious scholars who participate in the *collective ijtihād* are usually required to produce original knowledge and not just transfer the opinions of early jurists. So, the critics argue that such requirement should also apply to the participating biomedical scientists if they want to act as partners in the process of *ijtihād*, i.e., they should also produce knowledge in fields like genetics and genomics and not just translate published material into Arabic because this would not qualify to the level of *ijtihād*, in its technical sense.

References

Abū al-Baṣal, ʿAbd al-Nāṣir. 2004. "Al-Inʿikāsāt al-akhlāqiyya li-al-baḥth fī majāl al-khalāyā al-jidhʿiyya: Ruʾya sharʿiyya. *Hady al-Islām*, 48 (4): 11-18.

Āl Shāfiʿ, Marīʿ ibn ʿAbd Allāh. 2007. *Risālat kharīṭat al-jīnūm al-basharī wa-al-ithbāt al-jināʾī. Dirāsa taʾṣīliyya taṭbīqiyya*. M.A. thesis Riyadh: Naif Arab University for Security Sciences.

Al-Munaẓẓama al-Islāmiyya li al-Tarbiya wa al-ʿUlūm wa al-Thaqāfa (Īsiskū). 1993. *Al-Inʿikāsāt al-akhlāqiyya li al-abḥāth al-mutaqddima fī ʿilm al-wirātha*. Rabat, Morocco: Islamic Educational, Scientific and Cultural Organization & Tripoli, Libya: World Islamic Call Society.

ʿAshmāwī, Muḥammad Saʿīd al-. 2004. *Maʿālim al-Islām*. Beirut: Muʾassasat al-Intishār al-ʿArabī.

ʿAwaḍī, ʿAbd al-Raḥmān al-, and Aḥmad Rajāʾī al-Jundī (eds.). 2000. *Al-Wirātha wa-al-handasa al-wirāthiyya wa-al-jīnūm al-basharī wa-al-ʿilāj al-jīnī: Ruʾya Islāmiyya*. Kuwait: Islamic Organization for Medical Sciences.

ʿAwaḍī, ʿAbd al-Raḥmān al-, and Aḥmad Rajāʾī al-Jundī (eds.). 2008. *Al-Wirātha wa-al-takāthur al-basharī wa inʿikāsātuhā: Ruʾyat al-adyān al-samāwiyya wa wijhat naẓar al-ʿalmāniyya*. Kuwait: Islamic Organization for Medical Sciences.

Ayed, Tareq Al- and Nabil Rahmo. 2014. "Do Not Resuscitate Orders in a Saudi Paediatric Intensive Care Unit". *Saudi Medical Journal*. 35 (6): 561-565.

Bār, Muḥammad ʿAlī al. 1992. *Al-Fashal al-kulawī wa-zarʿ al-aʿḍāʾ: Al-Asbāb wa-al-aʿrāḍ wa-ṭuruq al-tashkhīṣ wa-al-ʿilāj*. Damascus: Dār al-Qalam. Beirut: al-Dār al-Shāmiyya.

Beauchamp, Tom, and James Childress. 2013. *Principles of Biomedical Ethics*. Oxford: Oxford University Press, 6th ed.

Chamsi-Pasha, Hassan and Muhammad Albar. 2017. "Do Not Resuscitate, Brain Death, and Organ Transplantation: Islamic Perspective". *Avicenna Journal of Medicine* 7: 35-45.

Collins, Francis. 2006. *The Language of God. A Scientist Presents Evidence for Belief.* New York: Free Press.

DePamphilis, Melvin, and Stephen Bell. 2011. *Genome Duplication. Concepts, Mechanisms, Evolution, and Disease.* New York: Garland Science.

Eich, Thomas. 2006. "The Debate on Human Cloning among Muslim Religious Scholars since 1997". In *Cross-Cultural Issues in Bioethics: The Example of Cloning*, ed. Heiner Roetz, Amsterdam: Rodopi.

ELSI Research Planning and Evaluation Group (2000). *A Review and Analysis of the Ethical, Legal, and Social Implications (ELSI) Research Programs at the National Institutes of Health and the Department of Energy.* Available at http://www.genome.gov/Pages/Research/DER/ELSI/erpeg_report.pdf (Retrieved August 8, 2017).

Ghaly, Mohammed. 2012. "Milk Banks through the Lens of Muslim Scholars: One Text in Two Contexts". *Bioethics* 26 (2): 111–127.

Ghaly, Mohammed. 2012a. "Religio-ethical Discussions on Organ Donation Among Muslims in Europe: An Example of Transnational Islamic Bioethics". *Medicine, Health Care and Philosophy* 15: 207-220.

Ghaly, Mohammed. 2016. *Genomics in the Gulf Region and Islamic Ethics: The Ethical Management of Incidental Findings.* Doha, Qatar. based World Innovative Summit for Health (WISH). Available at http://www.wish.org.qa/wp-content/uploads/2018/01/Islamic-Ethics-Report-EnglishFINAL.pdf, retrieved 28 May 2018.

Green, Eric, James Watson and Francis Collins. 2015. "Twenty-Five Years of Big Biology". *Nature* 526: 29-31.

Have, Henk ten. 2012. "Potter's Notion of Bioethics". *Kennedy Institute of Ethics Journal* 22 (1): 59-82.

Ḥusayn, Muḥammad al-Khaḍir. 1999. *Al-Sharīʿa al-Islāmiyya ṣāliḥa li kull zamān wa-makān.* Cairo: Nahḍat Miṣr li-al-Nashr wa-al-Tawzīʿ.

Ibn ʿAbd al-ʿAzīz, Ṣāliḥ. 2000. "Al-Jīnūm al-basharī… kitāb al-ḥayāt". *Al-Iʿjāz al-ʿIlmī* 7: 38-42.

Ibn Bāz, ʿAbd al-ʿAzīz. 1999. *Fatāwā ʿājila li-mansūbī al-ṣiḥḥa.* Al-Mamlaka al-ʿArabiyya al-Saʿūdiyya: Wizārat al-Shuʾūn al-Islāmiyya wa-al-Awqāf wa-al-Daʿwa wa-al-Irshād.

Ibrāhīm, ʿAbd Allāh ʿAlī. 2004. *Al-Sharīʿa wa-al-ḥadātha: Jadal al-aṣl wa-al-ʿaṣr.* Cairo: Dār al-Amīn.

Idrīs, ʿAbd al-Fattāḥ. 2003. "Al-Amn al-maṭlūb li-ʿilm-al-kharīṭa al-jīniyya". *Majallat al-Waʿy al-Islāmī* 40 (450): 22-25.

International Islamic Fiqh Academy (IIFA). 1998. "Qarār taṭbīq aḥkām al-Sharīʿa al-Islāmiyya". *Majallat Majmaʿ al-Fiqh al-Islāmī al-Dawlī*, 1988, 4th ed.

Islamic Fiqh Academy (IFA). 2002. *Aʿmāl wa buḥūth al-dawra al-sādisa ʿashar li-al-Majmaʿ al-Fiqhī al-Islāmī fī Makka al-mukarrama.* Makka: Majmaʿ al-Fiqh al-Islāmī.

Ismāʿīl, ʿAbd al ʿAzīz. 1959. *Al-Islām wa-al-ṭibb al-ḥadīth.* Cairo: al-Sharika al-ʿArabiyya li-al-Ṭibāʿa wa-al-Nashr.

Jābirī, Muḥammad ʿĀbid al-. (1996). *Al-Dīn wa al-dawla wa taṭbīq al-Sharīʿa*. Beirut: Markaz Dirāsāt al-Waḥda al-ʿArabiyya.

Jād al-Ḥaqq, ʿAlī Jād al-Ḥaqq. 2005. *Aḥkām al-Sharīʿa al-Islāmiyya fī masāʾil ṭibbiyya.* Cairo: Jāmiʿat al-Azhar, 3rd edition.

Jonsen, Albert. 2006. "A History of Religion and Bioethics". In Guinn, David (ed.). *Handbook of Bioethics and Religion,* edited by David Guinn. Oxford: Oxford University Press.

Kanʿān, Aḥmad Muḥammad. 2003. "Al-Jīnūm al-basharī wa taqaniyyāt al-handasa al-wirāthiyya. Muqārabāt fiqhiyya". *Majallat al-Buḥūth al-Fiqhiyya al-Muʿāṣira* 15(60): 68-101.

Khādimī, Nūr al-Dīn al-. 2003. "Al-Jīnūm al-basharī". *Majallat al-Buḥūth al-Fiqhiyya al-Muʿāṣira,* 15: 47-48.

Khādimī, Nūr al-Dīn al-. 2004. "Al-Khāriṭa al-jīniyya al-bashariyya (al-jīnūm al-basharī): Al-Aḥkām al-Sharʿiyya wa-al-ḍawābiṭ al-akhlāqiyya", *Majallat al-Mishkāt,* 4(2): 59-76.

Khādimī, Nūr al-Dīn al. 2007. *Al-Ḍawābiṭ al-Sharʿiyya li-buḥūth al-jīnūm al-basharī.* Baḥth muqaddam ḍimna aʿmāl al-Muʾtamar al-ʿArabī al-Thānī li-ʿUlūm al-Wirātha al-Bashariyya, Dubai.

Lewis, Ricki. 2014. *Human Genetics. Concepts and Applications.* New York: McGraw-Hill Education, 11th edition.

Madhkūr, Khālid al-, and others (eds). 1985. *Al-Ḥayāh al-insāniyya: Bidāyatuhā wa-nihāyatuhā min manẓūr Islāmī.* Kuwait: Islamic Organization for Medical Sciences.

Qaraḍāwī, Yūsuf al-. 1993. *Sharīʿat al-Islām ṣāliḥa li al-taṭbīq fī kull zamān wa makān.* Cairo: Dār al-Ṣaḥwa li-al-Nashr wa-al-Tawzīʿ.

Qaraḍāwī, Yūsuf al-. 1994. *Fatāwā muʿāṣira.* 2nd volume. Al-Kuwait: Dār al-Qalam. 3rd edition.

Qaraḍāwī, Yūsuf al-. 1996. *Al-Ijtihād fī al-Sharīʿa al-Islāmiyya maʿa naẓarāt taḥlīliyya fī al-ijtihād al-muʿāṣir.* Kuwait: Dār al-Qalam.

Qaraḍāwī, Yūsuf al-. 2003. *Fatāwā muʿāṣira.* 3rd volume. Kuwait: Dār al-Qalam. 3rd edition.

Qaraḍāwī, Yūsuf al-. 2010. *Zirāʿat al-aʿḍāʾ fī ḍawʾ al-Sharīʿa al-Islāmiyya.* Cairo: Dār al-Shurūq.

Qatar Biobank. 2014. *A Healthier Future Starts with You.* Doha: Qatar Biobank.

Quradāghī, ʿAlī, al- and ʿAlī Yūsuf al-Muḥammadī. (. 2006). *Fiqh al-qaḍāyā al-ṭibbiyya al-muʿāṣira. Dirāsa ṭibbiyya fiqhiyya muqārana muzawwada bi qarārāt al-majāmiʿ al-fiqhiyya wa al-nadawāt al-ʿilmiyya.* Beirut: Dār al-Bashāʾir al-Islāmiyya. 2nd edition.

Raysūnī, Aḥmad al-. 2013. *Abḥāth fī al-maydān.* Manṣūra: Dār al-Kalima li-al-Nashr wa-al-Tawzīʿ.

Rajāʾī al-Jundī, Aḥmad, ed. 1983. *Al-Injāb fī ḍawʾ al-Islām.* Kuwait: al-Munaẓẓama al-Islāmiyya li-al-ʿUlūm al-Ṭibbiyya.

Riḍā, Muḥammad Rashīd. 1910. "Muddat ḥaml al-nisāʾ sharʿan wa-ṭibban". *Majallat al-Manār*, vol. 12: 900-904.

Saʿdī, ʿAbd al-Raḥmān. 2011. *Majmūʿ muʾallafāt al-shaykh al-ʿallāma ʿAbd al-Raḥmān ibn Nāṣir al- Saʿdī: Majmūʿ al-fawāʾid wa iqtināṣ al-awābid.* Qatar: Wizārat al-Awqāf wa-al-Shuʾūn al-Islāmiyya.

Shaham, Ron. 2010. *The Expert Witness in Islamic Courts: Medicine and Crafts in the Service of Law*. Chicago: University of Chicago Press.

Sharafī, ʿAbd al-Majīd al-Sūsuwwah, al-. 2013. "Ahammiyyat al-fatwā al-jamāʿiyya wa-ḥujjiyyatuhā". *Majallat al-Ḥaqq* 17: 22-27.

Shawkānī, Muḥammad ibn ʿAlī al-. 1999. *Irshād al-fuḥūl ilā taḥqīq al-ḥaqq min ʿilm al-uṣūl.* Beirut: Dār al-Kitāb al-ʿArabī.

Ṣiddīq, Abū al-Faḍl ʿAbd Allāh al-Ḥasanī al-Ghumārī, al-. N.D. *Ajwiba hāmma fī al-ṭibb.* Cairo: ʿAlī Raḥmī.

Ṣiddīq, Abū al-Faḍl ʿAbd Allāh al-Ḥasanī al-Ghumārī. N.D. *Taʿrīf ahl al-Islām bi-anna naql al-ʿuḍw ḥarām*. Cairo: Dār Miṣr li-al-Ṭibāʿa.

Tawṣiyyāt.1998. "Tawṣiyyāt al-dawra al-ḥādiya ʿashara li-Majlis Majmaʿ al-Fiqh al-Islāmī: Ḥawla nadwat al-Kuwayt bi-shaʾn al-handasa al-wirāthiyya wa al-ʿilāj bi-al-jīn wa al-baṣma al-wirāthiyya". *Majallat Majmaʿ al-Fiqh al-Islāmī al-Dawlī.* No. 11

Tirmanīnī, ʿAbd al-Salām al-. 1977. "Wujūb taṭbīq al-Sharīʿa al-Islāmiyya fī kull zamān wa makān". *Majallat al-Ḥuqūq wa al-Sharīʿa* 1(2): 181-197.

ʿUbaydī, Zaynab ʿAbd al-Qādir, al-. 2017. *Faḥṣ al-jīnūm al-basharī, Dirāsa fiqhiyya taṭbīqiyya*. M.A.-thesis. Doha: Qatar University.

Wiryāshī, ʿAbd al-Kāfī. 2016. Irhāṣāt al-khibra al-ṭibb-sharʿiyya fī al-Sharīʿa al-Islāmiyya. *Majallat al-Ḥuqūq* 30: 11-33.

Yashū, Ḥasan. 2015-2016. "Al-Jīnūm al-basharī wa aḥkāmuh fī al-fiqh al-Islāmī. Ruʾya maqāṣidiyya". *Majallat Kulliyyat al-Sharīʿa wa-al-Dirāsāt al-Islāmiyya*, 33 (1): 17-80.

Zakariyyā, Fuʾād. 1986. *Al-Ḥaqīqa wa-al-wahm fī al-ḥaraka al-Islāmiyya al-muʿāṣira.* Cairo: Dār al-Dirāsāt wa-al-Nashr wa-al-Tawzīʿ.

CHAPTER 2

Islamic Ethics and Genomics: Mapping the Collective Deliberations of Muslim Religious Scholars and Biomedical Scientists

Mohammed Ghaly[1]

When the Human Genome Project (HGP) took off in 1990, experts in the field were aware of the fact that this scientific megaproject would generate ethical questions and conundrums that should be taken seriously.[2] So, an ethical arm for the HGP was established, namely the Ethical, Legal and Social Implications (ELSI) program. Five percent of the total HGP budget was dedicated to the ELSI program, making the project one of the largest-ever investments in bioethics research. Unlike most of the previous bioethics research, the ELSI program worked in conjunction with the scientific research activities. Rather than waiting for the results of the scientific research and their possible ethical implications, the HGP leadership decided to anticipate, identify, analyze and address the ethical concerns early on. The HGP example of conflating genomics with ethics concurrently became a to-be-followed model, sometimes with critical remarks, for subsequent genomics projects conducted elsewhere. Major research funding organizations, such as the Wellcome Trust and the UK Economic and Social Research Council, have also set a financial plan for research on genomics-related ethical issues (Rabinow and Bennett 2009, 106;

1 Professor of Islam and Biomedical Ethics, Research Center for Islamic Legislation & Ethics (CILE), College of Islamic Studies, Hamad Bin Khalifa University, Doha, Qatar, mghaly@hbku.edu.qa

2 This research was made possible by the NPRP grant "Indigenizing Genomics in the Gulf Region (IGGR): The Missing Islamic Bioethical Discourse", no. NPRP8-1620-6-057 from the Qatar National Research Fund, a member of The Qatar Foundation. The statements made herein are solely the responsibility of the author. I hereby submit my due thanks to the Research Assistant of the project, Mrs. Shaimaa Moustafa, and Dr. P.S. Van Koningsveld (Leiden University) who helped editing earlier versions of this chapter.

Jasanoff 2011, 7; Boddington 2012, 24-25; Kaye 2012, 673–674; Green et al 2015, 31; Morrison, Dickenson and Lee 2016, 1-6).

Despite its considerable richness and potential usefulness for addressing many issues, the ELSI program and its resulting literature were less beneficial to the religious, not to mention particularly Islamic, perspectives. The governing moral landscape of the ELSI programs was dominantly, and sometimes even exclusively, secular in nature. The ELSI of the HGP and its subsequent versions in Western countries did not even include the word "religion" in the title. This made the ELSI literature considerably poor when it comes to incorporating the perspective of religious ethics; a shortcoming that was highlighted by those who critically reviewed the ELSI work (ELSI Research Planning and Evaluation Group 2000, ii, iii, 3, 19, C2, C3).

A Genomics in the Age of Collective Reasoning (*al-ijtihād al-jamāʿī*)

Like the ELSI programs, the Islamic discussions on the ethical implications of the HGP and genomics in general were initiated before the completion of the scientific research. In fact, these discussions started even before the establishment of the national genome projects led by Qatar and Saudi Arabia. Furthermore, the ELSI literature was used in these discussions, especially by Muslim biomedical scientists[3], as background information featuring the key ethical dilemmas and the main benefits and risks involved. However, unlike the ELSI work, Islamic ethical deliberations on genomics had their own distinct language, style, modes of reasoning and prioritization of the key ethical concerns, which are all steeped in the religious tradition of Islam and are couched under the key term of independent and critical reasoning (*ijtihād*).[4] The crux of *ijtihād* within the context of genomics is that Muslim religious scholars[5] approach the foundational texts of Islam (viz. Quran and Sunna) and

3 "Muslim biomedical scientists" and, less frequently, "physicians" are used as generic terms referring to the participants in the collective deliberations on bioethical issues, with background in biomedical sciences.

4 See for instance the resolution adopted by the Islamic Fiqh Academy (IFA), affiliated with the Muslim World League on *ijtihād*, issued in January 1985. The resolution stressed the necessity of practicing *ijtihād*, especially in its collective form, in order to address the modern complex issues from an Islamic perspective. See Baʿdānī 2016, 92-94.

5 "Muslim religious scholars" and, less frequently, "jurists" are used in this chapter as generic terms comprising the broad spectrum of those with expertise in Islamic sciences. Sometimes, the term "jurists" is used to make reference to the experts in the discipline of Islamic jurisprudence in particular. Whenever the latter is the case, I indicate this clearly in the text.

their hermeneutics in order to show what these texts would imply with regards to such novel bioethical questions. As for genomics in particular, some Muslim religious scholars gave their own individual insights, through the mechanism of individual reasoning (*al-ijtihād al-fardī*). Nonetheless, the main and rigorous discussions took the form of interdisciplinary discourse between Muslim religious scholars and biomedical scientists, through the mechanism of collective reasoning (*al-ijtihād al-jamāʿī*).

In its individual form, the whole process of *ijtihād*, starting from developing the right perception (*taṣawwur ṣaḥīḥ*) of the issue at hand, which was termed elsewhere the "informative" component, and ending by determining the right action to be taken, which was named elsewhere the "normative" component, is traditionally managed by an individual religious scholar, more particularly the jurist (*al-faqīh*) (Ghaly 2015, 287-288). Within this type of *ijtihād*, which has dominated throughout the history of the Islamic tradition, the jurist can, in principle consult with specialists in fields like medicine or engineering in order to improve the informative component of their *ijtihād*. However, the whole process remains individual in character in the sense that it is one jurist who is responsible for managing this process and, more importantly, seen as the individual issuer of the fatwa (*muftī*). On the other hand, the collective reasoning (*al-ijtihād al-jamāʿī*), as its very name suggests, is collaborative in nature and thus is not based on one single jurist but a group of individuals who collaboratively manage the whole process. This collaboration can take the form of consulting non-*fiqh* specialists, like physicians or scientists, in order to improve the abovementioned informative component or consulting other religious scholars to make sure that the normative component and the resulting fatwa are not flawed. Conventionally speaking, biomedical scientists would be responsible for developing the right perception (*taṣawwur ṣaḥīḥ*), or the informative element of the *ijtihād* process, by explaining, say, what genomics exactly is about to the religious scholars. On their turn, religious scholars will make use of this scientific explanation of genomics in order to construe the religious perspectives, or the normative element of the *ijtihād* process, in conformity with this right perception. However, we shall see below that the process of *al-ijtihād al-jamāʿī*, especially as far as genomics is concerned, is highly dialectical and the arguments and counterarguments go frequently back and forth among these diverse groups of participants. For instance, we will see how biomedical scientists contribute to the discussions on the normative part of the *ijtihād* process.

Historically, *al-ijtihād al-jamāʿī* has its roots back in the early history of the Islamic tradition, where some would date it back to the lifetime of the Prophet of Islam and the subsequent period of the Rightly-Guided Caliphs, but it always remained less widespread than the individual *ijtihād* and was only sporadically practiced (Raysūnī 2010, 59-64). In the twentieth century, the need for reviving the mechanism of *al-ijtihād al-jamāʿī* in general, and particularly when it relates to novel issues, as it is the case in the field of biomedical ethics, was repetitively voiced by both religious scholars and biomedical scientists.

Employing *al-ijtihād al-jamāʿī*, its advocates argued, was indispensable to properly address the complex ethical questions raised by astounding technological advancements, which transformed the nature of many aspects of people's lives. By the beginning of the 1980s, the mechanism of *al-ijtihād al-jamāʿī* started to take an institutionalized form. The Islamic Organization for Medical Sciences (IOMS), based in Kuwait, which was established in 1981 and assumed its current name in 1984, has been the most active and all their symposia exclusively focused on bioethical issues. Shortly before getting its current name, particularly in 1983, the IOMS initiated the series of *al-Islām wa al-mushkilāt al-ṭibbiyya al-muʿāṣira* (Islam and Contemporary Medical Issues), which incorporated a long list of publications on various topics, including genomics. The IOMS coordinates with two other institutions whose interest in bioethics is rather occasional, as part of their broad interest in the role of Sharia in the modern world. One of these two institutions is the Islamic Fiqh Academy (IFA), established in 1977, which is affiliated with the Muslim World League and based in Mecca, Saudi Arabia. The other institution is the International Islamic Fiqh Academy (IIFA), established in 1981, based in Jeddah, Saudi Arabia, and affiliated with the Organization of Islamic Cooperation (Ghaly 2015, 292-294).

It is to be noted that the gravity shift to the collective *ijtihād* did not terminate its individual form. The conclusions resulting from *ijtihad*, whether collective or individual, are not religiously binding and, in principle, they can be challenged by another collective *ijtihād* or even by individual scholars. Individual *ijtihād* creates an opportunity for the religious scholar (*mujtahid*) to contemplate and reflect upon the issue at hand and the related textual references and contextual aspects, making it more prone to error. Alternatively, collective *ijtihād* is more restrictive because each participant in this process has to take into consideration the other participants' thoughts; this makes it less susceptible that flawed conclusions are collectively adopted. In spite of this, collective *ijtihād* can only materialize and flourish when the participating individuals develop their own individual *ijtihād* and then constructively share the resulting conclusions with their peers. Therefore, these two forms of *ijtihād* are not necessarily mutually exclusive to each other (Raysūnī 2010, 59). This holds true for the case of genomics as well. Although collective *ijtihād* dominated the discussions, some Muslim religious scholars made their own individual contributions to this topic (e.g. Khādimī 2003, 7-48; Kanʿān 2003, 68-101; Idrīs 2003, 22-25; Khādimī 2004, 59-76; ʿUthmān 2009; Ghaly 2016, 34).

During the 1990s, the decade of the HGP, the mechanism of *al-ijtihād al-jamāʿī* was already institutionalized more than a decade ago. Using this mechanism for addressing bioethical issues started to be the norm, and a certain legacy started to take form. Before its symposium was held in October 1998, which addressed the ethical issues of genomics, the abovementioned series "Islam and Contemporary Medical Issues" of the IOMS had already organized more than ten interdisciplinary symposia in which Muslim religious scholars and biomedical scientists collaboratively deliberated on a wide range of topics like abortion, beginning and end of human life, organ donation, AIDS, cloning, ...etc. This is also the case for the other two institutions, namely IFA and IIFA, where the mechanism of collective *ijtihād* was adopted to discuss many bioethical discussions, including assisted reproductive technologies, blood transfusion, human milk banks, organ transplantation, and sex reassignment surgery (Baʿdānī 2016, 199-210, 222, 223, 476-485, 491).

The almost two-decade experience of adopting the mechanism of collective *ijtihād* in the field of bioethics, with considerable success, made approaching genomics through this mechanism, an indisputable choice. Collective *ijtihād* was accepted as a recognized and credible mechanism for tackling modern bioethical questions, where they were seen as too complex to be addressed by those specialized in either religious sciences or biomedical sciences alone. If this is valid to issues like human milk banking and assisted reproductive technologies, then it applies in a much stronger sense to the case of genomics. This explains the frequency of collective discussions on genomics and Islamic ethics. Besides the aforementioned IOMS, IFA and IIFA, other institutions also used the mechanism of *al-ijtihād al-jamāʿī* to address the ethical aspects of genomics, as to be outlined below.

Interdisciplinary Deliberations

To my knowledge, the seminar "Ethical Implications of Modern Researches in Genetics" (*Al-Inʿikāsāt al-akhlāqiyya li al-abḥāth al-mutaqddima fī ʿilm al-wirātha*), organized by the Faculty of Science at the University of Qatar during the period 13-15 February 1993, was the first to examine the Human Genome Project and the prospective field of genomics from an Islamic ethical perspective. The proceedings of the seminar were published in both Arabic and English (ISESCO 1993; Īsiskū 1993).[6] During the period 13-15 October 1998, the

6 I hereby submit my due thanks to Dr. Khalid Al-Ali, the former director of the Foundation Program at Qatar University and Chairperson of the UNESCO World Commission on the Ethics of Scientific Knowledge and Technology (COMEST). He thankfully made me aware of this

IOMS organized the symposium "Genetics, Genetic Engineering, Human Genome and Gene Therapy: An Islamic Perspective" (*Al-Wirātha wa al-handasa al-wirāthiyya wa al-jīnūm al-basharī wa al-ʿilāj al-jīnī: Ruʾya Islāmiyya*), henceforth the 1998 symposium. The proceedings of this symposium and its final recommendations remain the most influential document, and subsequent collective deliberations highly depend on them. In its eleventh session held during the period 14-19 November 1998, the IIFA discussed the recommendations of the 1998 symposium, but deferred the resolution to another future meeting because the participants felt the need for conducting further study and research. During the period 5-10 January 2002, the IFA held its sixteenth session, which discussed among other issues, the possible fields in which the DNA fingerprinting can be employed. The seventh resolution of this session made a cursory reference to the human genome, stressing that it cannot be dealt with as a commodity in whatever way. The Faculty of Sharia and Law at the United Arab Emirates University organized the conference "Genetic Engineering between Sharia and Law" (*Al-Handasa al-wirāthiyya bayna al-Sharīʿa wa al-qānūn*) during the period 5-7 May 2002, whose proceedings were published in four dense volumes (Kulliyyat 2002). During the period 6-9 February 2006, the IOMS organized an international Seminar on "Human Genetic and Reproductive Technologies: Comparing Religious and Secular Perspectives". The recommendations of this seminar included a section entitled "Declaration of Principles", which paraphrased specific segments of the recommendations adopted during the 1998 symposium. The attempt here was seemingly to augment the support for these principles by engaging secular and religious voices from outside the Islamic tradition (Awadi and Gendy 2008, 1173-75). The second edition of the conference series, "Pan Arab Human Genetics", organized by the Dubai-based Centre for Arab Genomic Studies (CAGS), included a Public Forum on "The Ethical Perspectives of Human Genetic Applications in the Arab World", which was held on 20 November 2007. Besides the submitted papers, the forum issued the "Dubai Declaration", adopting some standpoints related to genomics. A few years later, and during its twentieth session held during the period 13-18 September 2012, the IIFA rekindled the discussions on the recommendations of the 1998 symposium, but, yet again, the resolution was deferred to a future meeting. During the period 23-25 February 2013, a specialized seminar took place in Jeddah that was jointly organized by the IIFA and IOMS. After an extensive journey of almost 15 years, the recommendations of the 1998

seminar and provided me with its publications. In the following sections, I will make use of the Arabic and English editions of this seminar depending on the original text of submitted articles.

symposium were endorsed, with few modifications and additional points, by the IIFA during its twenty-first session, held on 18-22 November 2013.[7]

In addition to hosting the conference held in 1993, Qatar also hosted some of the recent expert meetings during which both biomedical scientists and religious scholars deliberated on genomics. In collaboration with other Qatar-based institutions, the Research Center for Islamic Legislation & Ethics (CILE) convened two activities. On 2 October 2014, a public seminar entitled "Islamic Ethics in the Era of Genomics" was organized in collaboration with the then Qatar Supreme Council of Health (SCH), now Ministry of Public Health. As part of its 2016 edition, the Doha-based World Innovative Summit for Health (WISH) collaborated with CILE in organizing a Research Forum on "Genomics in the Gulf Region and Islamic Ethics", which focused on the ethical management of incidental findings. The study produced by this Research Forum was published in both Arabic and English (Ghaly 2016; Ghaly 2016a). Finally, CILE organized the international seminar "Islamic Ethics and the Genome Question" during the period 3-5 April 2017, the proceedings of which are published in this volume.

The analyses provided in this chapter are based on a careful review of the abovementioned deliberations, including some of the unpublished papers which were presented during these interdisciplinary meetings. However, the proceedings of the abovementioned 1998 symposium will serve as the main reference in this chapter. This choice has to do with the seminal role played by the proceedings of the seminar in the overall Islamic ethical discussions on genomics. References to other meetings and publications of individual scholars will be made whenever necessary to show certain similarities or differences between the individual and collective forms of *ijtihād.*

Explanatory Remarks

Before delving into the detailed analysis of the deliberations on genomics and Islamic ethics, three explanatory remarks are due in order to understand the analysis to follow:

The first remark deals with the themes and issues discussed in the abovementioned meetings and conferences. The Human Genome Project (HGP) and the field of genomics in general occupied a central place in the discussions. However, almost all of these meetings also discussed many other issues, some of which are closely related to genomics, whereas others are of less relevance

---

7 I hereby submit my due thanks to Dr. Muḥammad ʿAlī al-Bār who provided me with some of the papers presented to this session. It is to be noted that the proceedings of this session have not been published yet.

or may be completely unrelated. In the following section, I endeavored to keep the focus on the field of genomics and the HGP, but it was almost impossible to avoid references to issues related to other themes, especially genetics, genetic engineering, gene therapy and the like. Because of the nature of the discussions during these meetings and conferences, it was sometimes impossible to make clear distinctions between the points and arguments related to genomics and those related to genetic engineering, genetic counseling, etc.

The second remark refers to the geographical scope of these deliberations. The abovementioned meetings and conferences were almost all transnational in character. In other words, an institution based in Qatar or Kuwait, for example could host the event, but the pool of participants usually represented the diversity of the Muslim world in general, and also sometimes Muslims living as religious minorities worldwide. However, one notices that almost all events took place in the Gulf region, which witnessed the key genomics projects in the Muslim world. Furthermore, the countries that hosted many of these events, especially Qatar and Saudi Arabia, also established national genome projects, both in December 2013.[8] This indicates that the conflation of genomics and ethics, which we have seen in the HGP, continued in the initiatives taking place in the Muslim world.

The third remark relates to one of the typical difficulties of practicing *ijtihād* by religious scholars in the field of bioethics, namely the difficulty of grasping the technicalities of scientific information, especially in complex disciplines like genetics, genetic engineering and genomics. Among other reasons, Mainstream Muslim religious scholars hardly have any background information about this type of knowledge or even access to relevant first-hand or primary sources (Ghaly 2015, 288-289). The deliberations on genetic engineering, genomics and the HGP demonstrated how difficult the interdisciplinary communication was between biomedical scientists and religious scholars. In the 1998 symposium, the IOMS president, ʿAbd al-Raḥmān al-ʿAwaḍī, recognized that the participating religious scholars had difficulties understanding the lecture given by the Syrian biomedical scientist Hānī Rizq[9] and asked that scientific information should be presented in a simpler and clearer way[10] (Jundī 1998,

8 For more information about genomics projects and initiatives in the Gulf region, see Ghaly 2016, 7-15.

9 After the symposium, Rizq wrote two key Arabic books that were meant to introduce scientific information, especially in the field of genetics, to the general educated public. Two of his books, the latest of which was on human genome and ethics, (Rizq 2003; Rizq 2007) received awards from the Kuwait Foundation for the Advancement of Sciences (KFAS).

10 The published proceedings of this symposium do not include a paper written by Dr. Rizq. However, Dr. Muḥammad ʿAlī al-Bār, in his published paper in this symposium, spoke

195-196). After the first session of the symposium, the IOMS secretary general assistant, Aḥmad al-Jundī, mentioned that he had received many proposals suggesting that another biomedical scientist, namely Ḥassān Ḥatḥūt, should present his paper earlier than it was planned because of his ability to simplify scientific information. Eventually this happened to clearly save the situation and improve the level of communication with religious scholars (Jundī 1998, 266). Ḥatḥūt himself recognized this problem and criticized the participating physicians for being sometimes inclined to "stretch their muscles" by presenting complex information inaccessible to the religious scholars. Conversely, Ḥatḥūt explained, physicians have to be aware that their exclusive mission is to communicate specialist information to the jurists. Whatever they do, which does not contribute to fulfilling this mission, is nothing but useless effort (Jundī 1998, 321). Ḥatḥūt's presentation was well received and some jurists commended him for his ability to communicate complex information in an easy and accessible way (Jundī 1998, 297, 301, 302, 303). However, some religious scholars, like the Saudi Nāṣir al-Maymān and the Syrian Muḥammad Rawwās Qalʿajī, continued complaining about this problem and demanded that biomedical scientists should use an easier and more accessible language (Jundī 1998, 315, 353). During subsequent discussions within the International Islamic Fiqh Academy (IIFA), al-Bār spoke about the same problem when he commented on what happened during the 1998 symposium. He said that after about a one-hour lecture given by Rizq, the participating jurists said, "Translate to us what he said. We did not understand anything" (IIFA 1998, 11/1112).

Additionally, both religious scholars and biomedical scientists objected the imprecise or vague character of some information presented by biomedical scientists. For instance, there were wide discrepancies in the papers submitted by the scientists about the number of genes in the human body, the accuracy of statistics mentioned by some papers, the right terminology to be used, the very definition of genetic engineering and whether cloning can be part of it, etc. (Jundī 1998, 299, 306, 308-310, 311, 317, 318-319, 684). This can be due to the fact that some of these scientists were not geneticists by specialization,

---

about a certain paper that Dr. Rizq submitted to the symposium and that it included some inaccurate information (Bār 1998, 622). So, it is possible that al-ʿAwaḍī is referring here to a paper which did not find its way to publication, maybe because of its inaccessibility to the religious scholars or because of including mistaken information. However, Dr. Rizq presented two papers written by other scientists, respectively the Saudi Ṣāliḥ al-Kurayyim and the Moroccan Muḥammad al-Yashawī, because they were not available for presenting their papers on the first day of the seminar. So, it is possible that al-ʿAwaḍī here is referring to the presentation that Riqz gave on behalf of these two authors. It is to be noted that al-Yashawī could join the discussions later (Jundī 1998, 1012).

like the internist Muḥammad al-Bār, the gynecologist Ḥassān Ḥatḥūt and the pharmacist Aḥmad al-Jundī. Another possible reason for the confusion around specific scientific information can also be traced to the fact that most of the papers went back to the 1990s when the HGP was still in progress and many issues were unsettled among scientists worldwide. The papers written in the 1990s, by both biomedical scientists and religious scholars, continued to be the main reference for all subsequent discussions with hardly any new updates that could have had tangible impact on the interdisciplinary discussions.[11] A third possible reason is that some religious scholars were simply looking for the impossible, namely having clear-cut (*qaṭʿī*) information all the time. The very nature of an emerging and rapidly developing field like genomics makes it sometimes difficult to have stable and certain information which cannot be challenged by further research. One of the clear examples in this regard is calculating the exact number of genes in the human genome. This issue has always remained controversial, and one of the latest publications shows that scientists still cannot agree on how many genes are in the human genome and sometimes even on how to define a gene (Willyard 2018).

Below, we will notice that this problem made some religious scholars feel that they missed the right perception (*taṣawwur ṣaḥīḥ*) of some issues related to genomics. Additionally, the absence of specific scientific information or the feeling that such information is not clear-cut or conclusive enough makes it difficult for the religious scholar to make a rigorous weighing between possible benefits (*maṣāliḥ*) and expected harms (*mafāsid*), or, in bioethical terms, the so-called benefit-risk assessment. In his comment on the draft of the final recommendations of the 1998 symposium, the UK-based physician ʿAbd al-Majīd Qaṭma said that the possible harms of a technology, like gene therapy cannot be known for sure, and this uncertainty will continue for a long time. That is why, he argued, it is better to wait until we are 100% sure (Jundī 1998, 1005). Another related problem was determining the person(s) who has/have the authority to decide what is beneficial and what is harmful. The Syrian religious

11 As an example, one can check one of the newest papers in this regard, namely the paper of Aḥmad al-Jundī, which he presented during the eleventh session held by the IIFA in September 2012 and then again, but in a much more concise form, in the specialized seminar jointly organized by the IIFA and IOMS in February 2013 (Jundī 2013). Despite some updates included in the paper, we hardly see any influence resulting in modifying or updating the final recommendations adopted by the 1998 symposium. Strikingly enough, some of the ethical issues which emerged after the completion of the Human Genome Project and later dominated the ethical deliberations on genomics worldwide, like the management of incidental findings, hardly received any attention in the parallel Islamic bioethical discussions. See Ghaly 2016, 30.

scholar, Aḥmad al-Kurdī, argued that the referential authority *in toto* should be given to the academics and scholars, each in liaison with one's specialization (Kurdī 1998, 241). On the other hand, the Tunisian religious scholar, Mukhtār al-Sallāmī, argued that such an opinion is factually isolating the jurists from the reasoning process. However, al-Sallāmī added, deliberations on these issues should remain interdisciplinary by facilitating communication between specialists in human genome and genetic engineering on one hand and jurists on the other hand. In his comment on al-Sallāmī's critique, al-Kurdī said that this type of interaction between the two groups is actually what he meant (Jundī 1998, 249, 264). All the preceding difficulties did problematize the process of *ijtihād* to the extent that some religious scholars became even reluctant to express an Islamic ethical position in general.

B Framing Genomics: Two Main Approaches

The participants in the abovementioned deliberations agreed that the world is currently witnessing one of its biggest scientific revolutions ever, especially in fields like genetics and genetic engineering. To them, the Human Genome Project (HGP) is at the very heart of this revolution. In the symposium held in Qatar in 1993, the Pakistani molecular biologist, Anwar Nasim, said that genetic engineering and its related disciplines are advancing at an unparalleled tempo, which was never seen throughout the history of biology. As for the HGP in particular, he said "the current effort to map and sequence the entire human genome is, without doubt, the most significant and ambitious undertaking of biological research in modem times" (Nasim 1993, 63, 70). In his opening speech of the 1998 symposium, the IOMS secretary general assistant, Aḥmad al-Jundī, said that what has been achieved during the last fifty years is equivalent to multiple folds of what humanity could achieve since the beginning of creation. He enumerated giant steps made by the relatively new field of genetics, which are increasingly narrowing the gap between imagination and reality.[12] Similar statements were also expressed by Muḥammad al-Mursī Zahra (the then dean of the Faculty of Sharia at the United Arab Emirates University) in his intro-

12 Al-Jundī dedicated a number of pages to outline the key achievements made by genetics and genetic engineering. He gave examples like using electronic microscopes and computers to fathom out the cell and unearth its secrets, producing human insulin (laboratory-grown synthetic insulin, which mimics insulin in humans) to replace the animal/porcine insulin, trying to overcome the scarcity of human organs for transplantation by producing genetically-engineered porcine hearts so that they will not be rejected when transplanted in human bodies, and DNA fingerprinting (Jundī 1998a, 24-26).

duction to the voluminous publication on genetic engineering between Sharia and law (Kulliyyat 2002, 5-6). Speaking about the Human Genome Project (HGP), al-Jundī said that this is the most serious issue in the field of genetic engineering. Despite possible risks related to autonomy and privacy that must be taken into consideration, al-Jundī argued that the potential of the HGP will go beyond mapping the genes and discovering mutations to eventually open the door for studying the reasons behind these mutations, and how to fix them through gene therapy (Jundī 1998a, 23-28). Recognizing the new scientific revolution and the crucial role played by the HGP therein was a recurrent theme during the 1998 symposium and was repeatedly voiced by many others outside the symposium (ISESCO 1993, 263; Nasim 1993, 63, 70; Anees, 1993, 78; Ḥaffār 1993, 123-137; Jundī 1998, 68, 70, 211, 274, 736, 797, 1024; Ḥatḥūt 1998, 274; Maymān 1998, 797-798). Later on, we will notice that this agreed-upon fact among the participants would have an impact on framing genomics. It would lead to the expression of some theologically tinted explanations of how it was possible that scientific communities based in the West could achieve such a revolution despite their carelessness and negligence of religious guidance and its associated values, whereas Muslim countries have hardly made any substantial contribution in this regard.

Beyond this point of agreement, one can notice two different approaches towards genomics and related issues. Each approach is comprised of two main aspects; one is theological and theoretical in nature, while the other focuses on juristic and practical elements. In other words, each approach is premised on certain theological assumptions, which are further fleshed out and phrased in a juristic and a practical position towards the field of genomics, as epitomized by the Human Genome Project (HGP). Both religious scholars and biomedical scientists contributed to each approach. As explained in the first section of this chapter, the conventional boundaries between the tasks assigned to biomedical scientists and those entrusted to religious scholars were blurred in these discussions. Below, we will see that the biomedical scientists did not restrict their contribution to providing scientific information only, i.e., the informative component of the *ijtihād* process. On various occasions, they additionally contributed to the normative component by giving their insights on theological and juristic aspects of genomics. It is to be noted, however, that the contribution of the religious scholars to the informative component remained quite minimal.

1 *Precaution-Inclined Approach*

In a bid to explain the abovementioned point with regards to the scientific revolution, its exclusive Western leadership and the absence of Muslim countries' contribution, this approach made use of a quasi-determinist, and typically Ashʿarī[13], theological framing. This theological framing was mainly expounded by the Kuwaiti religious scholar, ʿAjīl al-Nashmī, in his paper submitted to the abovementioned 1998 symposium. He started his paper with lamenting the deplorable state of scientific research in the Muslim world and expressing his frustration that today's Muslims abandoned the leading role that their ancestors ever played in advancing sciences. Furthermore, al-Nashmī asked Muslims to stick to the firm belief that all modern scientific advancements in cutting-edge fields like genetic engineering, the HGP and gene therapy, could only materialize because this was God's will. He explained that it was only God who provided these Western scientists with the necessary power and capabilities to accomplish these achievements, and had He willed otherwise, they would never have been able to achieve anything. However, the results of these scientific ventures belong to these scientists' own acts for which they remain responsible, and God gave them the ability to do these acts by way of testing (*ibtilāʾ*) His creatures to see how they will behave (Nashmī 1998, 545-547). This is a typically Ashʿarī position which explains the seemingly problematic phenomena in life by trying to strike a balance between two points. On one hand, there is stress on God's omnipotence and that nothing can take place in the universe against His will. On the other hand, there is recognition of a certain degree of individuals' freedom to act so that humans remain responsible for their acts by way of acquisition (*kasb*). Within such a position, there is little space left for detailed rational argumentations about the theodicy or the possible wisdom behind such problematic phenomena (Ghaly 2010, 24-26).

Against the background of this quasi-determinist theological framing, al-Nashmī moved to the juristic practical aspects of this approach where he gives an overall preference for a casuistic approach. He argued that Muslims should deal with the applications of these cutting-edge scientific ventures and try to evaluate the benefits and harms of each application through of the lens of Sharia. This means that each application should have its separate religious ruling (*ḥukm sharʿī*). As for the Human Genome Project (HGP) in particular, al-Nashmī argued that it is in principle a noble project, or –again reflecting his inclination towards precaution– this is how it should be. Al-Nashmī held the notion that the overall juristic framework, which governs the HGP and its possible applications, is the framework of the five higher objectives of Sharia (*maqāṣid al-Sharīʿā*), namely safeguarding religion, one's life, intellect, prog-

13 The analysis provided by al-Nashmī here is clearly inspired and influenced by the Ashʿarī theory of acquisition (*kasb*). For more information about this theory, see Abrahamov 1989.

eny and property. According to al-Nashmī, the HGP relates more to the third objective, namely the preservation of progeny (*nasl*) whose integrity should always be safeguarded against any possible manipulation or misuse. If this is the case, al-Nashmī elaborated, then the default rule concerning the HGP and its applications should be that all related actions are in principle prohibited unless there are strong arguments to justify an exception to this default rule (Nashmī 1998, 548-550).

Strikingly enough, the position premised on prohibition as the governing rule is not the mainstream position in Islamic jurisprudence (*fiqh*), where humans are generally permitted to make use of what God created unless there is a scriptural reference or compelling reason to move it from the realm of original permissibility (*al-barāʾa al-aṣliyya*) to prohibition. This mainstream position, supported by the majority of Muslim jurists, is based on Quranic verses like "He is the One Who created everything in the earth for you" (Q. 02:29) and "And He has subjected to you whatever is in the heavens and whatever is on the earth" (Q. 45:13). Jurists couched this position in the well-known legal maxim, "permissibility is the original state of things (*al-aṣl fī al-ashyāʾ al-ibāḥa*)" (Wizārat 1983-2006, 1/130, 18/74-75, 103). What made a contemporary jurist like al-Nashmī transfer genomics from this mainstream original permissibility to the realm of original prohibition? Besides the technical juristic reason mentioned by al-Nashmī himself, viz. the relevance of genomics to the objective of safeguarding progeny which dictates more cautiousness, there are other possible reasons related to the scientific and socio-political context of the Muslim world in which the field of genomics was born.

Integral to this approach is the idea that Muslims should be aware of the possible religious perils of this scientific revolution, despite its possible beneficial advancements in fields like genetics and genomics. In various places in al-Nashmī's paper, one easily observes his deep distrust of the Western scientific institutions, which dominate the field of genomics, when it comes to the commitment to religious values. Unsurprisingly, al-Nashmī explained, many of the results of modern scientific research are not in conformity with the Islamic Sharia because the leaders in these fields are not guided by religious values and are mainly motivated by material interest and personal desire (*hawā*). According to him, the absence of divine guidance, as communicated through revealed scriptures, will inevitably lead to misguidedness and deviation from the straight path. He added that this misguidedness got even normalized to the extent that many Muslims believe that this [absence of religious guidance in scientific research activities] is the norm to be followed (Nashmī 1998, 46). Similar concerns were shared by other participants like Aḥmad al-Jundī, the IOMS secretary general assistant and the late Egyptian physician and former

Minister of Health, Ibrāhīm Badrān (d. 2015) (Jundī 1998, 100, 102). In his opening speech for the 1998 symposium, al-Jundī said that one should be alert that most of the scientific researchers have no religion to abide by except their own scientific imagination (Jundī 1998a, 29). Al-ʿAwaḍī, al-Jundī and the Mauritanian religious scholar, ʿAbd Allāh Bin Bayya, referred to the example of scientific research on nuclear energy, which eventually led to catastrophic repercussions by manufacturing the atomic bomb and using it twice. Bin Bayya expressed his fear that biology could move into the same direction that nuclear energy had walked through before, and thus may eventually lead to self-destruction of humanity.[14] In his paper submitted to the 1998 symposium, Aḥmad al-Jundī quoted Oppenheimer (d. 1967), the scientific director of the Manhattan Project, to say, "Now and now only, science has fallen into sin" (*al-ān wa al-ān faqaṭ waqaʿa al-ʿilm fī al-khaṭīʾa*)[15] (Jundī 1998, 30). In his paper submitted to the specialized seminar jointly organized by the IIFA and IOMS in 2013, al-Jundī suggested that a link between this notorious example and the Human Genome Project (HGP) is not too far-fetched. He recalled the history of the US Department of Energy, the main catalyst of the HGP, which goes back to the Manhattan Project and its role in developing the atomic bomb during World War II (Jundī 1998a, 30; Jundī 1998, 194, 197; Jundī 2013, 13).

It seems that the perceived tension between current scientific research activities and religious values also influenced some Muslim religious scholars while weighing possible harms against expected benefits in order to judge the Human Genome Project (HGP) and the field of genomics in general. Ben Bayya spoke about estimations stating that 30% of beneficial resources on earth was exhausted in the twentieth century (Jundī 1998, 197). As for the HGP in particular, al-Nashmī dedicated less than one page to outline its possible benefits, mainly preventing and treating genetic diseases (Nashmī 1998, 551-

14 The case of the nuclear bomb was, for these participants, the most glaring example to show how destructive scientific research can be. The Egyptian physician added other examples which show the severity of possible harms that can result from originally good scientific research and technologies. He referred to the advanced means of transportation that cause the death of 5 million people per year; the industrial revolution which left millions of qualified workers without jobs; and the laser that can be transformed into lethal weapons, making people blind before their death. See Jundī 1998, 101.

15 Oppenheimer's statement has to do with his experience after watching the first atomic bomb test, called Trinity, and naturally with the later atomic bombings of Hiroshima and Nagasaki. It is to be noted that the original statement, "physicists have known sin", is much more nuanced and cautiously formulated than what the Arabic translation given by al-Jundī suggests. According to some, this degree of ambiguity in the phrase was intended by Oppenheimer himself, see (Thorpe 2006, 12, 190).

552). However, possible harms and corruptions of the HGP and gene therapy were discussed in about ten pages and divided into three distinct sections, namely technical, ethical and psychosocial harms. As for the technical harms, al-Nashmī spoke about the risk of cancer, or even death, for the individuals who undergo gene therapy. He also added that genetically modified animals might end up developing abnormal genes, which can put human life and the whole environment at risk. As for the psychosocial risks, al-Nashmī held that sequencing genome could lead to genetic discrimination with negative impact on one's profession and family. For instance, when information about the sequenced genome reveals one's susceptibility to serious diseases, he/she can be discriminated against by having difficulties to find a job or even a future marriage partner. Concerning the possible ethical harms, al-Nashmī said that subjecting body- and germ-cells to laboratory tests can, unnecessarily, undermine human dignity in many cases. Usually, the main aim in such cases is gaining money and celebrity rather than conducting proper scientific research. What is even more concerning, al-Nashmī added, is the risk of compromising people's privacy by exposing sensitive information included in their genomes to unauthorized agencies and institutions like insurance companies (Nashmī 1998, 555-565).

It is clear that a jurist like al-Nashmī feels that he is facing a quite suspicious technological advancement, whose possible harms outweigh its expected benefits, while he himself has no power to control or guide its future course. In such a situation, it is not surprising to resort to the position that everything related to this new advancement is prohibited until it is proven otherwise. According to al-Nashmī, the only exception to be permitted in this regard is gene therapy at the level of body cells rather than germ cells. Al-Nashmī, in line with many other religious scholars, argued that gene therapy falls within the scope of medical treatment (*tadāwī*), whose benefits of treating diseases are to be recognized from an Islamic perspective. Additionally, within the system of *maqāṣid al-Sharīʿa*, gene therapy is more relevant to the objective of protecting one's life (*ḥifẓ al-nafs*) that generally entails permissiveness rather than safeguarding progeny (*ḥifẓ al-nasl*), which usually dictates more cautiousness (Nashmī 1998, 552-554).

Finally, as part of their inclination to cautiousness, the contributors to this approach were reluctant about whether Muslim countries should play a role in this phase of the ongoing scientific revolution or abstain from contributing. Al-Nashmī explicitly stated that Muslims are helpless in this regard. According to him, scientific research will move forward today or tomorrow, and the stakeholders of this research will completely disregard Muslims who will only have to deal with the new reality imposed upon them. He added that Muslims' voices are discounted in this regard, and poor Muslim countries are sometimes even misused as field experiments. Without regaining the scientific and civilizational leadership, al-Nashmī argued, both Muslim jurists and political leaders in Muslim countries will not be in a position to do anything except preparing themselves for the worst possible scenarios by drafting protective religious rulings and ethical safeguards (Nashmī 1998, 545-548, 560). Al-Nashmī's concerns about the missing role of Muslim countries in this regard were commonly shared by others like Ḥamdī al-Sayyid, the then head of the Egyptian Medical Syndicate and the Egyptian physician Ibrāhīm Badrān (Jundī 1998, 191, 201). Some of the participants took this position a step further and asked for extreme cautiousness. The Egyptian religious scholar, Muḥammad Ra'fat 'Uthmān, argued that experiments in the field of genetic engineering seemed too risky and unsafe. Therefore, moratorium would be a good option. According to him, Muslim scientists would better refrain from participating in this field and let scientists in the West continue the work they started until it becomes certain that the final products are free from ethical concerns and physical harms (Jundī 1998, 247).[16] Besides postponing the scientific activity, some religious scholars, like Bin Bayya, also asked for parallel cautiousness when it comes to developing an Islamic ethical position. This certainly relates to the third remark, explained in the first section of this chapter, which elaborated the difficulties of the religious scholars to grasp the scientific technicalities of fields like genetics and genetic engineering. Bin Bayya said that religious scholars are required to issue a fatwa in which the scholar is supposed to deduce what God wants people to do. Bearing in mind this very nature of the fatwa, religious scholars are in need of certainty (*yaqīn*) or preponderant probability (*ẓann ghālib*), about available scientific information before stating anything. That is why it will be too early to issue a general fatwa about these advancements in the light of the current state of uncertainty about specific information (Jundī 1998, 256-257). On the other hand, the UK-based physician, 'Abd al-Majīd Qaṭma[17], argued that Muslim jurists are not yet ready to give fatwas on these complex issues because they are still not sufficiently aware of the relevant scientific discussions and conferences taking place in the UK and Eu-

16 It seems that the further discussions during the 1998 symposium made 'Uthmān change his mind later and express opinions which are closer to the second approach, outlined below. We see this change clearly in his post-symposium published book on genome ('Uthmān 2009). More details about this change in position will be mentioned below. In any case, this is one of the examples which show how collective reasoning and interdisciplinary deliberations (*al-ijtihād al-jamāʿī*) can influence the individual form of *ijtihād*.

17 His family name was sometimes written as "Qaṭāyā", but it seems to be just typo (Jundī 1998, 837).

rope. As for long-term solutions, Qaṭma proposed that Muslim religious scholars would study medicine, as do some lawyers and physicians in the UK who combine between studying medicine and law (Jundī 1998, 839). The Kuwaiti religious scholar, ʿAbd Allāh Muḥammad ʿAbd Allāh, found that the proposed cautiousness of the jurists when they deal with a brand-new issue, such as genetic engineering, is in line with the ideal practice of early religious scholars. The process can be time consuming, sometimes lasting a year, e.g. by studying and verifying the economic reality, including visiting the actual markets to see how people conclude transactions on the ground, before giving their religious advice (fatwa). That is why, ʿAbd Allāh suggested, it might be better if today's jurists would first visit the laboratories and observe in reality what happens there, so that their fatwas would be as precise as possible (Jundī 1998, 259-260).

2 *Embracement-Inclined Approach*

Contrary to the precaution-inclined approach, this approach responds to the success of the scientific revolution led by Western institutions and the failure of Muslim countries in this regard by giving a different theological framing. The main focus of this theological framing is God's justice and wisdom rather than His omnipotence. It is also more open to rational argumentation about the theodicy where human agency occupies a central place. In certain elements, this approach seemed to bear the spirit of Muʿtazilī theology (Ghaly 2010, 26-29).

One of the main advocates of this approach who contributed to its theological framing was the late Egyptian US-based physician, Ḥassān Ḥatḥūt (d. 2009), a prominent and influential figure in *al-ijtihād al-jamāʿī* deliberations on bioethical issues. In his paper submitted to the 1998 symposium[18], the first subtitle reads "paradox". In this section, Ḥatḥūt spoke about God's wisdom that dictated that humans are uniquely gifted with intellect; their main tool to acquire knowledge. Throughout history, Ḥatḥūt explained, humans could employ their intellectual capacities to read the universe and unearth its various secrets, one after the other, to the extent that they could achieve breakthroughs and revolutions. However, some people lagged behind in this human search for, and march to, knowledge because they did not use their intellect as they should have. If they continue to do so, Ḥatḥūt added, their deserved fate will be marginalization and exploitation through the other advanced nations (Ḥatḥūt 1998, 274). The following section of Ḥatḥūt's paper was entitled "Man Explores Man (*Taʿarruf al-insān ʿalā al-insān*)", where he introduced the HGP

18 Selected parts of this paper was published later in an interview form in 2003, see Amīn 2003.

as one of the key stations in man's long journey to know oneself. This journey, Ḥatḥūt explained, started with very basic knowledge about man's external appearance and physical makeup as male or female. As time went by, this knowledge continued to improve and assume complex forms where credit goes to sciences like comparative anatomy, studies exploring the genetic structure of the nucleuses of the cells in human bodies, and later the DNA discovery in 1953 by the two Noble laureates, James Watson and Francis Crick. Ḥatḥūt argued that the HGP is reading and exploring the human being at the molecular level. Besides their benefit for improving self-knowledge or enhancing the "know thyself" value, Ḥatḥūt explained that genomics and the HGP also contribute to having a better knowledge of life and the universe in general. According to him, the four nucleotides found in DNA, namely Adenine, Thymine, Cytosine and Guanine known with the abbreviation ATCG, are the four letters that compose the language of life (*lughat al-ḥayāh*). Here, He continued, that the ATCG plays the same role that dots and dashes do in the telegraph and the figures one and zero do in the computer world (Ḥatḥūt 1998, 275-277).

During the further discussions among the biomedical scientists and religious scholars, Ḥatḥūt elaborated on this point by quoting the Quranic verse "Say, 'Travel throughout the earth and see how He has originated the creation'. Then God will bring the next life into being. Surely, God has power over everything" (Q. 29:20). Ḥatḥūt commented by saying that this is a Quranic command, which applies to the question of genome, genetic engineering and the like (Jundī 1998, 320-321). More Quranic verses in the same spirit were added by the Syrian religious scholar, 'Abd al-Sattār Abū Ghudda, in his paper presented to the same symposium, including "And in your own selves; do you then not behold?" (Q. 51:21) and "Our Lord is He Who gave to each thing its due shape and nature, then guided it aright" (Q. 20:50)[19]. Such scriptural references, Abū Ghudda stated, mean that the whole creation, including the universe and man therein, is governed by consistent and coherent laws (*sunan*) that can be discovered by human intellect (Abū Ghudda 1998, 573). Against this theological background, the results of the scientific revolution are compatible with God's justice and wisdom in the sense that those who used what God gifted them with, viz. human intellect, and worked relentlessly (Western countries) ended up harvesting good results, while those who fell short of the ideal behaviour in this regard (Muslim countries) lagged behind. Consequently, the existing

19 An extensive list of the Quranic verses, which outline the relationship between man and the universe and urge man to look into the wonders of this universe and discover its secrets, was given by the Moroccan religious scholar, Muḥammad al-Rūkī, and the Syrian Aḥmad al-Kurdī. See Rūkī 1998, 218-219; Kurdī 1998, 233-236.

gap in scientific research between the West and the Muslim world is presented through the lens of human agency not in a deterministic framework where the focus is on accepting the status quo as part of God's will.

As we shall see below, this different theological framing will result in different juristic practical perspectives on various issues. It is true that religious scholars and biomedical scientists, who contributed to this approach, accepted certain points advocated by the precaution-inclined approach like the overall preference to a casuistic or case-by-case approach where each application of these cutting-edge technologies is evaluated on the basis of its overall benefits and harms through of the lens of Sharia. Within the paradigm of the five higher objectives of Sharia (*maqāṣid al-Sharīʿa*), they also agreed that these technologies are more relevant to the objective of safeguarding progeny (*nasl*) and, to a certain extent, also to protecting one's life (*nafs*), especially when it has to do with therapeutic applications like gene therapy (Abū Ghudda 1998, 577, 579). Beyond this, the contributors to the two approaches expressed different viewpoints on many issues, as outlined below.

As for the overall governing rule, which applies to the Human Genome Project (HGP) and generally to fields like genetics and genetic engineering, the contributors to this approach opted for the "original permissibility" (*al-barāʾa al-aṣliyya*). The Moroccan religious scholar, Muḥammad al-Rūkī, extensively spoke about this rule and its application to genetic engineering in plants, animals and also in humans but with a higher degree of cautiousness (Rūkī 1998, 216-225). Abū Ghudda argued that this position, especially when it comes to the HGP, should not be a disputable issue (Abū Ghudda 1998, 578). Abū Ghudda defended the relevance of the position of "original permissibility" to these new technologies by referring to the so-called principle of "scripturally unattested or unregulated benefit" (*al-maṣlaḥa al-mursala*) which has its roots in the Islamic legal theory.[20] Abū Ghudda recalled this principle to argue that unprecedented situations, like the issues relating to genetics and the HGP, which entail recognized benefits but are not declared permissible or otherwise by a direct scriptural evidence, should be judged as permissible. According to him, the principle of *al-maṣlaḥa al-mursala* is crucial evidence recognized by Sharia (*dalīl Sharʿī*) when one addresses novel issues (*mustajaddāt*) (Abū Ghudda 1998, 577).

As for the risk-benefit assessment or weighing expected benefits versus possible harms, the mode of reasoning and the resulting conclusions were both different from the precaution-inclined approach. Reference was made to a point that early Muslim religious scholars reiterated, namely the very nature

20 For more information about this principle, see Hallaq 1997, 112-113; Opwis 2010, 165-173.

of this life hardly allows for the existence of things that are purely and exclusively beneficial (*maṣlaḥa maḥḍa*). The normal course of this life is that everything has two inseparable sides, one beneficial, the other harmful. What people should usually do is to weigh between these two sides and see which side is stronger than the other.[21] As for the Human Genome Project (HGP) and technologies related to fields like genetics and genetic engineering, Abū Ghudda stressed the strength of the expected benefits within the scale of Sharia. According to him, such benefits are not luxuries but would rather fall within the highest degree of benefits, namely the necessities (*al-ḍarūriyyāt*). Instead, he recognized that possible harms should be taken seriously because they can eventually disturb one of the higher objectives of Sharia, namely safeguarding progeny (Abū Ghudda 1998, 577-579). Some religious scholars who participated in these discussions, like the Syrian Muḥammad Rawwās Qalʿajī, clearly stated that the argumentation of Abū Ghudda proved to be more convincing than that of al-Nashmī. The Syrian religious scholar, Aḥmad al-Ḥājjī al-Kurdī, added that al-Nashmī was quite uncharitable when he spoke of the possible harms of the HGP, many of which are not necessarily inevitable (Jundī 1998, 601-603). On the other hand, Abū Ghudda explained that such harms can be controlled, regulated or at least minimized through the mechanism of Sharia-based determinants (*ḍawābiṭ Sharʿiyya*) so that one can make sure that the harms will not eventually override the benefits (Abū Ghudda 1998, 577-579). As the interdisciplinary deliberations during the 1998 symposium proceeded, the Egyptian religious scholar, Muḥammad Ra'fat ʿUthmān, despite his conservative opinions expressed by the beginning of the seminar, was convinced that the benefits of the HGP strongly override the possible harms. He tentatively expressed this opinion during the seminar (Jundī 1998, 300-301, 834) but his outspoken opinion was expressed in his book on the genome and DNA, which was published in 2009. For instance, the view that the Human Genome Project (HGP) can eventually lead to genetic discrimination that, for him, is nothing but unjustified fear (ʿUthmān 2009, 79-80). On his turn, the Saudi gynecologist, ʿAbd Allāh Bāsalāma, made use of the very theological framing presented by al-Nashmī to dispel such fears (Jundī 1998, 254-255). As long as one believes that nothing happens in this universe without God's will, Bāsalāma argued, one should not worry about the fate of humans or even their possible ruin. At the end, humans are God's creatures and He is the One who can protect them. It is God who, one

21 In order to give credibility for this premise and its rootedness in the Islamic tradition, Abū Ghudda quoted the prominent religious scholar, al-ʿIzz Ibn ʿAbd al-Salām (d. 1263), who wrote one of the most authoritative and influential works related to the concept of *maṣlaḥa*. See Ibn ʿAbd al-Salām 1991.

day, can burn the factories, stop the flow of knowledge and put the whole life to an end (Jundī 1998, 254-255).

The contributors to the embracement-inclined approach shared the concerns raised by their peers in the precaution-inclined approach about the possible risks or harms that can result from the separation between scientific research conducted by Western institutions and the religious values (Abū Ghudda 1998, 577). As the deliberations of the 1998 symposium advanced, however, some of the participants insisted that Western institutions are still committed to strict standards and regulations despite the absence of outspoken commitment to certain religious values. ʿAbd al-Majīd Qaṭma, who already contributed to the first approach, conceded that the situation in countries like the United Kingdom might be much better than that in the Muslim world, thanks to thousands of civil society associations active in raising public awareness (Jundī 1998, 838-839). In his paper submitted to the specialized seminar jointly organized by the IIFA and IOMS in 2013, Aḥmad al-Jundī gave a somehow different picture about the relationship between genomics and religion in the Western context. He addressed the case of Francis Collins, the director of the HGP, who was said to be an atheist but then turned to be a believer in God because of his research in this field. After two years of contemplation, al-Jundī added, Collins eventually couched his journey of searching for the truth by saying "I found God in the human genome" (Jundī 2013, 24). Al-Jundī's account of Collins's combination of scientific excellence and belief in God missed a few but important nuances. The overall idea that Collins is a prominent scientist and also a committed believer in God is already attested by his own book, *The Language of God,* published in 2006. However, the book shows that Collins was already a committed believer before leading the HGP, as he spoke about spending a long afternoon praying in a little chapel, seeking guidance from God whether he should accept the offer of being the HGP director (Collins 2006, 119). Thus, Collins' religious commitment did not arise because of his involvement in the field of genomics in particular, although it is clear that his unique experience with the HGP had a positive impact on his belief in God. Although it is an individual case, such an account of a scientist of the caliber of Collins shows that the situation in the West is not as gloomy as al-Nashmī and his likeminded peers may think and that scientists with commitment to religious values can still play leading roles in a scientific mega-venture like the HGP. However, one should not overstate the impact of this supposed science and faith harmony on the bioethical reasoning even for Collins himself. His aforementioned book was appended with a section on "The Moral Practice of Science and Medicine: Bioethics". Collins argued that religious values could play a role, although limited, in the current bioethical deliberations despite possible objections from professional bioethicists (Collins 2006, 235-270). But "I hesitate, however, to advocate very strongly for faith-based bioethics", Collins concluded (Collins 2006, 217).

The above-sketched theological framing, the way the contributors to this approach viewed the risk-benefit assessment and the other related juristic and practical aspects, all paved the way to reach the following positive conclusion; Muslims' contribution to fields like genetics, genetic engineering, gene therapy and the like is not only permissible, but is a collective duty (*farḍ kifāya*). Various arguments were advanced to support this conclusion. Contributing to these sciences and related technologies was seen as a positive response to the call of Islam to search for knowledge; whatever knowledge as long as it is beneficial for mankind. Throughout the history of Islamic civilization, Muslims provided significant contributions to science and it is now the turn of today's Muslims to do the same through these emerging fields (Zuḥaylī 1998, 776; Kurdī 1998, 240; Khādimī 2004, 61). Furthermore, the applications of these emerging fields are meant to help humans improve their health through preventive or therapeutic techniques, and all of these fall within the scope of medical treatment (*tadāwī*), whose knowledge is also a collective duty from an Islamic perspective (Zuḥaylī 1998, 777). The third key argument dealt with socio-political dimensions. As explained above, the contributors to Islamic bioethical discourse on genomics agreed that this field makes part of an impressive scientific revolution whose resulting technologies will determine the future, and even the fate, of countries worldwide (Jundī 1998, 13, 71, 251; Ḥatḥūt 1998, 274; Nashmī 1998, 545). The Tunisian religious scholar, Nūr al-Dīn al-Khādimī, spoke about an ongoing civilizational race towards achieving scientific supremacy. Currently, he explained, modern scientific discoveries are under global non-Islamic, sometimes even inhumane, hegemony, which monopolizes the resulting technologies and often deprives Muslim countries from having access to these technologies and their benefits. This context of civilizational competition, al-Khādimī argued, dictates that the whole *umma* (Muslim nation) is under collective obligation (*farḍ kifāya*) to participate in promising fields like genetics and genomics. Political leaders and scientists, who have the ability to participate in exploring the genome, are even under individual obligation (*farḍ ʿayn*) to do so (Khādimī 2004, 63; Khādimī 2007). In this vein, the idea of calling for a moratorium on scientific research in the Muslim world related to promising fields like genetics and genomics was vehemently opposed and seen as considerably harmful for the future of Muslim countries (Jundī 1998, 248, 251, 255, 258,).

C Further Developments

The two main approaches examined in this chapter, with the associated arguments and counter-arguments and internal agreements and disagreements, both contributed to shaping the Islamic discourse on the Human Genome Project (HGP) and genomics in general, in addition to guiding subsequent on-the-ground developments in some Muslim countries.

As far as the overall framing of the HGP and genomics is concerned, the embracement-inclined approach proved to be more appealing and convincing than the precaution-inclined approach. This is clearly reflected in the final recommendations adopted by the conferences and expert meetings outlined in the first section of this chapter. In its Final Report and Recommendations, the 1993 seminar organized by the Faculty of Science at the University of Qatar spoke highly of the HGP and considered it "the most ambitious scientific project in the history of mankind", stressing that "Muslims should not be idle by-standers in this endeavor but should contribute their share to the study of the human biological heritage and to the study of man's future". Consequently, it called upon rich Islamic countries "to generously fund this research, at a level corresponding to the importance and size of the task, so that Muslims may be present in one of humanity's most delicate enterprises and so that we may benefit from its far-reaching results" (ISESCO 1993, 263; Īsiskū 1993, 360-361). The same tone is reiterated in the final recommendations adopted by the 1998 symposium that was organized by the IOMS. The recommendations made no mention of the quasi-determinist theological framing introduced by the precaution-inclined approach, but adopted the other theological framing proposed by the embracement-inclined approach. The HGP was framed as part of man's quest to know oneself and to explore the laws governing God's creation as implied in Quranic verses, such as: "We will show them Our signs in the universe and within themselves until it becomes clear to them that it is the truth" (Q. 14:53). At the practical juristic level, the position of the embracement-inclined approach was also adopted in these recommendations. The HGP was introduced as an added value to the health and medical sciences in their mission to prevent and treat diseases. Thus, the recommendations concluded, reading the human genome falls within the scope of collective duties in society (Jundī 1998, 1048). The same recommendations also included a call for Muslim countries to join the field of genetic engineering by establishing research centers whose activities should be in compliance with the Islamic Sharia (Jundī 1998, 1047).[22] By the end of the conference organized by the Dubai-based Centre for

22 The exact points outlined in these recommendations were quoted verbatim in the recom-

Arab Genomic Studies (CAGS) in 2007, the participants issued the so-called Dubai Declaration (*Bayān Dubayy*). The way the declaration was formulated shows that participating in the field of genomics was no longer a question anymore but a taken-for-granted fact, "Since the Arab World is capable of participating in genome research, there is an urgent need for the formation of national committees where the mission is to define an ethics code for scientific research in each of the Arab countries and, subsequently, to coordinate between them and committees in other States". So, the question to be addressed here is no longer whether these countries should contribute to genomics or not, but rather how their contribution should be regulated from an Islamic ethical and legal perspective (http://www.cags.org.ae/eodubaideclaration.pdf).

The subsequent on-the-ground developments in scientific research, at least in the countries that hosted some of the collective deliberations outlined in this chapter, also proved that the embracement-inclined approach had the upper hand. In December 2013, Qatar and Saudi Arabia declared launching their large-scale national human genome projects, each with a huge budget and strong political support at the governmental level.[23] Available literature indicates that Islamic ethical deliberations, including those examined in this chapter, helped these projects and the associated biobanks in developing their guidelines. This has to do with the fact that "Islam is the dominant religion in these countries, and it affects people's behavior and influences their positions" (Ghiath et al 2015, 53). As for the Saudi Biobank, the two researchers, Ghiath Alahmad and Kris Dierickx, stated that it was "designed in a manner to respect not only international guidelines and Saudi law but also Islamic values, as outlined by the Saudi Biobank governance document" (Alahmad and Dierickx 2014, 682). The Qatar Biobank does not differ much from the Saudi biobank in this regard. In 2014, the biobank released a booklet entitled *A Healthier Future Starts with You*, which addressed questions related to the relation between scientific biomedical research and Islamic values. The booklet also indicated that the Qatar Biobank is keen to make all its current and future activities compliant with Islam, in collaboration with the Research Center for Islamic Legislation & Ethics (CILE) (Qatar Biobank 2014, 12-13). The last conference of the Qatar Genome Program on "Ethics, Regulations, and Best Practices in Genomic Medicine", held on 29-30 April 2018, was jointly organized with CILE,

---

mendations adopted by the participants in the twenty-first session of the IIFA, which was held in November 2013 (http://www.iifa-aifi.org/2416.html).

23 For an overview of these two projects and parallel developments in other countries, see Ghaly 2016, 7-19.

which supervised two distinct sessions of the conference dedicated to discussing relevant ethical issues from an Islamic perspective.[24] The influence of the precaution-inclined approach was most visible in highlighting the urgency of possible risks and harms associated with conducting research within fields like genetics and genomics. The final recommendations adopted by the 1998 symposium and their updated version in 2013 adopted by the IIFA, like many other documents, strongly reflected the fears that this type of biomedical research can violate some Islamic values. The recommendations were included in a relatively short text of about 1230 words. In such a concise text, about ten times the reference was made to the necessity of making sure that all research activities are in compliance with the Islamic Sharia and its core values. In support of this argument, we quote phrases like "No research, therapy or diagnosis related to someone's gene or genome can be undertaken unless a rigorous assessment is conducted beforehand in order to measure the potential risks and benefits associated with these activities, in compliance with the provisions of Sharia" and "It is not permissible to use the genome in a harmful way or in any way that violates the Islamic Sharia" (Jundī 1998, 1046; http://www.iifa-aifi.org/2416.html). Despite such frequent references to the significance of abiding by the so-called Sharia-based determinants (*ḍawābiṭ Sharʿiyya*), unfortunately the final recommendations were usually ambiguous about what exactly these determinants are. However, individual scholars who contributed to these Islamic bioethical deliberations have been trying to clarify some of these *ḍawābiṭ* on specific topics like gene therapy, genetic testing, genetic counseling, DNA fingerprinting or profiling and genetically modified organisms (GMOs) (Jundī 1998, 6-10).

Concluding Remarks

In the bestseller *The Language of God,* the acclaimed scientist and director of the Human Genome Project, Francis Collins, held that ethical dilemmas associated with advances in genomics and related fields should not be left to the scientists alone to speculate. Although they have a critical role to play in the deliberations on such dilemmas, scientists' perspectives should be espoused with a wide variety of other perspectives at the table (Collins 2006, 270-271).

24 To attract high-quality research, CILE published a call-for-papers in both English and Arabic (https://www.cilecenter.org/en/news/call-for-papers-policies-regulations-and-bioethics-of-genomic-research/) and a Background Paper was drafted in order to streamline the discussions in the conference (https://www.cilecenter.org/en/wp-content/uploads/2018/01/Background-Paper-QGP-CILE-Conference-April-2018.pdf).

The review of the discussions on genomics and Islamic ethics presented in this chapter illustrates that this was the case when the ethical aspects of the Human Genome Project (HGP) and genomics were discussed in the Muslim world. Scientists collaborated with Muslim religious scholars through a certain mechanism of collective and interdisciplinary reasoning rooted in the Islamic tradition, known as *al-ijtihād al-jamāʿī*.

By addressing the ethical questions of genomics and related fields through the mechanism of *al-ijtihād al-jamāʿī*, both biomedical scientists and religious scholars could achieve together what each group could not have done alone. Within this interdisciplinary setting, Muslim religious scholars could develop a kind of scientific literacy about genomics and gain scientific information that they otherwise would not have access to. However, these interdisciplinary discussions were not without difficulties. Scientific information provided by biomedical scientists was not always clear enough or delivering the level of certainty that Muslim religious scholars were seeking. What must be done with such incomplete or indecisive information, especially when a rigorous benefit-risk assessment should be performed on cutting-edge technologies like those in the field of genetics and genomics? Where are the borderlines that should distinguish between the role to be played by the biomedical scientists and the one assigned to religious scholars? Whose opinion should weigh heavier when the two groups disagree with each other? These were some of the controversial questions that the contributors to these interdisciplinary deliberations had to grapple with. This chapter reviewed the various ways used by these participants to address such questions and highlighted the key agreements and disagreements. This study has also shown that the Islamic ethical deliberations had their own concerns, which we do not see, or at least do not occupy a central position, in parallel discussions in the West, e.g. the perceived separation between scientific research on one hand and religious guidance and associated values on the other. This made some religious scholars quite suspicious about the intentions, aims and long-term plans of scientific institutions based in the West and concurrently almost obsessed with the fear that the same separation can occur to scientific institutions (to be) based in the Muslim world. This point raises questions about the hypothetical universality of secular ethics and the conviction that non-religious ethics can speak for everybody, hence making it an integral part of the so-called public morality. These discussions showed that putting religious values aside when discussing the (un)ethical character of scientific research can be quite problematic for certain groups of people.

As for the overall position towards the Human Genome Project (HGP) and the field of genomics in general, the chapter analysed two main approaches.

The precaution-inclined approach is leaned towards taking the "safe side" option by requesting to wait and taking time before rushing into joining the on-going scientific research ventures related to emerging fields like genetics and genomics. To be on the safe side, overall preference is given to considering all related activities as prohibited until this is proven otherwise. The embracement-inclined approach is more pre-emptive in nature, where the key governing idea is that Muslims should not remain idle anymore, and immediate pro-active steps must be taken to ensure that Muslim countries will make significant contributions to the on-going scientific revolution in these fields. A great deal of the chapter is dedicated to the detailed arguments and counter-arguments of each approach. The study argues that the embracement-inclined approach proved to be more influential, both at the theoretical level of the ethical discourse and at the practical level of actual genomics initiatives, which took place in some Muslim countries.

Despite the various breakthroughs achieved by the interdisciplinary discussions reviewed in this chapter, these discussions have shown that there are serious challenges ahead. Generally speaking, there is a serious problem of pursuing the recent scientific updates in a rapidly growing field like genomics. The material presented in the conferences and expert meetings held in the 1990s remained to be the only reference in all-subsequent discussions with hardly any significant updates, even after the completion of the HGP. With regard to the informative component of these discussions, which is usually assigned to the biomedical scientists, it is clear that more specialists in genetics and genomics should be involved. One would also add that papers submitted to these meetings and conferences should be solicited from geneticists with a good publication record in the field, not just those who can read works published by others and then translate them into Arabic. Additionally, the overall scientific literacy of religious scholars should improve, and they should not remain exclusively dependent on the papers submitted to each conference. As for the normative component which is generally entrusted to the religious scholars, much more rigorous tools should be developed to manage the benefit-risk assessment, even if no conclusive information is not available yet. The discussions reviewed in this chapter, and also elsewhere (Ghaly 2012, 190-191), demonstrate that religious scholars usually expect biomedical scientists to only come up with information that has been verified and consequently get recognized as certain and conclusive, otherwise this will not be part of proper science. This perception of science can be quite problematic, especially in fields like genomics and genetics. I would suggest addressing this problem by improving the literacy of religious scholars in philosophy of science in general and philosophy of medicine in particular. When it comes to clinical research, clinical medicine and therapeutic interventions in particular, philosophers of medicine indicate that uncertainty is unavoidable, with just a few exceptions like vaccination and antibiotics. The same holds true for preventive medicine where uncertainty proliferates and thus claims of certainty are often baseless. Bearing this mind, decisions and judgements in clinical medicine are usually based on plausibility more than on certainty. William Osler, the renowned Canadian physician known as the "Father of Modern Medicine", recognized this fact when he called medicine the art of probability and the science of uncertainty (Thompson and Upshur 2018, 3, 77, 122, 127, 138, 141, 144).

As for the contributors to these interdisciplinary discussions, only two groups still dominate the discussions, namely biomedical scientists and religious scholars. However, the complexity of the ethical dilemmas raised by fields like genomics and genetic engineering necessitate having various groups with much more diversified backgrounds. The group of religious scholars usually consists, dominantly or exclusively, of specialists in Islamic jurisprudence (*fiqh*); the so-called jurists (*fuqahāʾ*). However, the vast Islamic tradition cannot be reduced to the discipline of *fiqh*, despite its recognized significance in the Islamic bioethical discourse. The absence of specialists in Islamic theology and philosophy in the discussions reviewed in this chapter was reflected in the somehow poor and superficial discourse on the intersection between genomics and Islamic theology and philosophy. Serious ethical dilemmas with crucial theological and philosophical underpinnings were completely missing, including the very concept of soul and its possible relation with the genome.[25] Surely, these interdisciplinary discussions would be much more enriched once the pool of participants get progressively diversified by adding specialists in other related fields depending on the topics to be addressed, e.g. social sciences, medical anthropology, secular bioethics, Jewish and Christian bioethics, medical law, etc. We hope that the material included in this volume will set the suitable base for filling some of the abovementioned gaps.

---

25 The background paper of the CILE seminar, organized on 3-5 April 2017, whose proceedings are published in this volume, outlined some of these issues and questions like: what makes us distinctively human? How to determine the boundaries between what is normal/natural and abnormal/unnatural? How should the controversy on determinism and free will be revisited in the age of genomics? (https://www.cilecenter.org/en/wp-content/uploads/2016/06/Genomics-Background-Paper-English.pdf).

References

Abrahamov, Binyamin. 1989. "A Re-Examination of al-Ashʿarī's Theory of *Kasb* According to *Kitāb al-Lumaʿ*". *The Journal of the Royal Asiatic Society of Great Britain and Ireland* 2: 210-221.

Abū Ghudda, ʿAbd al-Sattār. 1998. "Al-Muwākaba al-Sharʿiyya li muʿṭayāt al-handasa al-wirāthiyya". In *Al-Wirātha wa al-handasa al-wirāthiyya wa al-jīnūm al-basharī wa al-ʿilāj al-jīnī: Ruʾya Islāmiya*, edited by Jundī, Aḥmad Rajāʾī, 573-594. Kuwait: Islamic Organization for Medical Sciences.

Alahmad, Ghiath and Kris Dierickx. 2014. "Confidentiality, Informed Consent and Children's Participation in the Saudi Biobank Governance: A Comparative Study". *Eastern Mediterranean Health Journal* 20 (11): 681-689.

Alahmad, Ghiath, Mohammed Al Jumah and Kris Dierickx. 2015. Confidentiality, Informed Consent, and Children's Participation in Research Involving Stored Tissue Samples: Interviews with Medical Professionals from the Middle East". *Narrative Inquiry in Bioethics* 5(1): 53-66.

Al-Munaẓẓama al-Islāmiyya li al-Tarbya wa al-ʿUlūm wa al-Thaqāfa (Īsiskū). 1993. *Al-Inʿikāsāt al-akhlāqiyya li al-abḥāth al-mutaqddima fī ʿilm al-wirātha*. Rabat, Morocco: Islamic Educational, Scientific and Cultural Organization & Tripoli, Libya: World Islamic Call Society.

Amīn, Ṭāhā. 2003. "Qirāʾa īmāniyya li al-jīnūm al-Basharī". *Al-Waʿy al-Islāmī* 39 (447): 37-41.

Anees, Munawar. 1993. "Biological Sciences: Moral Mediators in the Making? In Islamic Educational, Scientific and Cultural Organization (ISESCO)". *Ethical Implications of Modern Researches in Genetics*, 77-88. Rabat, Morocco: Islamic Educational, Scientific and Cultural Organization & Tripoli, Libya: World Islamic Call Society.

Baʿdānī, Muḥammad Nuʿmān al-. 2016. *Tabwīb qarārāt majmaʿayy al-fiqh* (*al-dawlī wa al-rābiṭa*) *ilā al-dawra al-thānya wa al-ʿishrīn*. Available at http://www.saaid.net/book/open.php?cat=4&book=11261 retrieved on 26 May 2018.

Bār, Muḥammad ʿAlī al-. 1998. "Naẓra fāḥiṣa lī al-fuḥūṣāṭ al-ṭibbiya al-jīniyya (al-faḥṣ qabl al-zawāj wa al-istishāra al-wirāthiyya)". In *Al-Wirātha wa al-handasa al-wirāthiyya wa al-jīnūm al-basharī wa al-ʿilāj al-jīnī: ruʾya Islāmiya.*, edited by Jundī, Aḥmad Rajāʾī, 621-662. Kuwait: Islamic Organization for Medical Sciences.

Boddington, Paula. 2012. *Ethical Challenges in Genomics Research: A Guide to Understanding Ethics in Context*. New York: Springer.

Collins, Francis. 2006. *The Language of God: A Scientist Presents Evidence for Belief*. New York: Free Press.

ELSI Research Planning and Evaluation Group. 2000. *A Review and Analysis of the Ethical, Legal, and Social Implications (ELSI) Research Programs at the National Institutes of Health and the Department of Energy*. Available at http://www.genome.gov/

Pages/Research/DER/ELSI/erpeg_report.pdf (retrieved on 25 May 2018)
Ghaly, Mohammed. 2010. *Islam and Disability: Perspectives in Theology and Jurisprudence*. London: Routledge.
Ghaly, Mohammed. 2012. "The Beginning of Human Life: Islamic Bioethical Perspectives". *Zygon: Journal of Religion and Science* 47 (1): 175-213.
Ghaly, Mohammed. 2015. "Biomedical Scientists as Co-Muftis: Their Contribution to Contemporary Islamic Bioethics". *Die Welt des Islams* 55: 286-311.
Ghaly, Mohammed. 2016. *Genomics in the Gulf Region and Islamic Ethics: The Ethical Management of Incidental Findings*. Doha, Qatar. based World Innovative Summit for Health (WISH). Available at http://www.wish.org.qa/wp-content/uploads/2018/01/Islamic-Ethics-Report-EnglishFINAL.pdf, retrieved 28 May 2018.
Ghaly, Mohammed. 2016a. *ʿIlm al-jīnūm fī minṭaqat al-Khalīj: idārat al-natāʾij al-ʿaraḍiyya min manẓūr Islāmī.* Doha, Qatar: World Innovative Summit for Health (WISH). Available at http://www.wish.org.qa/wp-content/uploads/2018/01/Islamic-Ethics-Report-Arabic.pdf, retrieved 28 May 2018.
Green, Eric, James Watson and Francis Collins. 2015. "Twenty-Five Years of Big Biology". *Nature* 526: 29-31.
Ḥaffār, Saʿīd al-. 1993. "Al-Inʿikāsāt al-qiyamiyya wa al-khlāqiyya wa al-qānūniyya wa al-insāniyya li abraz munjazāt al-thawra al-iḥyāʾiyya". In Al-Munaẓẓama al-Islāmiyya li al-Tarbya wa al-ʿUlūm wa al-Thaqāfa (Īsiskū) (1993). *Al-Inʿikāsāt al-akhlāqiyya li al-abḥāth al-mutaqddima fī ʿilm al-wirātha*, 123-139. Rabat, Morocco: Islamic Educational, Scientific and Cultural Organization & Tripoli, Libya: World Islamic Call Society.
Hallaq, Wael. 1997. *A History of Islamic Legal Theories: An Introduction to Sunni Uṣūl Al-Fiqh*. Cambridge: Cambridge University Press.
Ḥatḥūt, Ḥassān. 1998. "Qirāʾat al-jīnūm al-basharī". In *Al-Wirātha wa al-handasa al-wirāthiyya wa al-jīnūm al-basharī wa al-ʿilāj al-jīnī: ruʾya Islāmiya.*, edited by Jundī, Aḥmad Rajāʾī 273-286. Kuwait: Islamic Organization for Medical Sciences.
Ibn ʿAbd al-Salām, al-ʿIzz. 1991. *Qawāʿid al-aḥkām fī maṣāliḥ al-anām*. Edited by Ṭāhā ʿAbd al-Raʾūf Saʿd. Cairo: Maktabat al-Kulliyyāt al-Azhariyya.
Idrīs, ʿAbd al-Fattāḥ. 2003. "Al-Amn al-maṭlūb li al-kharīṭa al-jīniyya". *Majallat al-Waʿy al-Islāmī* 40 (450): 22-25.
International Islamic Fiqh Academy (IIFA). 1998. *Majallat Majmaʿ al-Fiqh al-Islāmī*, (11). Jeddah: International Islamic Fiqh Academy.
Islamic Educational, Scientific and Cultural Organization (ISESCO). 1993. *Ethical Implications of Modern Researches in Genetics*. Rabat, Morocco: Islamic Educational, Scientific and Cultural Organization & Tripoli, Libya: World Islamic Call Society.
Jasanoff, Sheila. 2011. *Reframing Rights: Bioconstitutionalism in the Genetic Age*. Cambridge: Massachusetts Institute of Technology.
Jundī, Aḥmad al-.1998a. "Lamḥa ḥawla al-handasa al-wirāthiyya wa al-jīnūm al-basharī

wa al-ʿilāj al-jīnī: Ruʾya Islāmiyya". In *Al-Wirātha wa al-handasa al-wirāthiyya wa al-jīnūm al-basharī wa al-ʿilāj al-jīnī: Ruʾya Islāmiya*, edited by Jundī, Aḥmad Rajāʾī, 23-30. Kuwait: Islamic Organization for Medical Sciences. .

Jundī, Aḥmad Rajāʾī. 1998. *Al-Wirātha wa al-handasa al-wirāthiyya wa al-jīnūm al-basharī wa al-ʿilāj al-jīnī: ruʾya Islāmiya*. Kuwait: Islamic Organization for Medical Sciences.

Jundī, Aḥmad. 2013. "Al-Jīnūm al-Basharī min al-naẓariyya ilā al-taṭbīq". In *Buḥūth wa tawṣiyyāt al-nadwa al-ʿilmiyya ḥawla al-wirātha wa al-handasa al-wirāthiyya wa al-jīnūm al-basharī min manẓūr Islāmī*, edited by Abū ʿAlyū, 14- 24. Jeddah: International Islamic Fiqh Academy. An online version can be accessed via https://www.imamu.edu.sa/elibrary/Documents/Genetic_Engineering.pdf (retrieved 21 June 2018)

Kanʿān, Aḥmad Muḥammad. 2003. "Al-Jīnūm al-basharī wa taqniyyāt al-handasa al-wirāthiyya: muqārabāt fiqhiyya". *Majallat al-Buḥūth al-fiqhiyya al-Muʿāṣira* 15 (60): 68-101.

Kaye, Jane et al. 2012. "ELSI 2.0 for Genomics and Society". *Science* 336: 673–674.

Khādimī, Nūr al-Dīn al-. 2003. "Al-Jīnūm al-basharī". *Majallat al-Buḥūth al-fiqhiyya al-Muʿāṣira* 15 (58): 7-48.

Khādimī, Nūr al-Dīn al-. 2004. "Al-Kharīṭa al-jīniyya al-bashariyya (al-jīnūm al-basharī): al-aḥkām al-Sharʿiyya wa al-ḍawābiṭ al-akhlāqiyya". *Majallat al-Mishkāh* 2: 59-76.

Khādimī, Nūr al-Dīn al-. 2007. "Al-Ḍawābiṭ al-Sharʿiyya li buḥūth al-jīnūm al-basharī". A paper presented to the second conference on "Pan Arab Human Genetics" held in November 2007, available at http://www.cags.org.ae/e1khadami.pdf (Retrieved 20 June 2018).

Kulliyyat al-Sharīʿa wa al-Qānūn. 2002. *Al-Handasa al-wirāthiyya bayna al-Sharīʿa wa al-qānūn*. Al-Ain, UAE: United Arab Emirates University.

Kurdī, Aḥmad al-Ḥājjī al-. 1998. "Al-Handasa al-wirāthiyya fī al-nabāt wa al-ḥayawān wa ḥukm al-Sharīʿa al-Islāmiyya fīhā". In *Al-Wirātha wa al-handasa al-wirāthiyya wa al-jīnūm al-basharī wa al-ʿilāj al-jīnī: ruʾya Islāmiya.*, edited by Jundī, Aḥmad Rajāʾī, 229-242. Kuwait: Islamic Organization for Medical Sciences.

Maymān, Nāṣir al-. 1998. "Al-Irshād al-jīnī: ahammiyyatuh – āthāruh -maḥādhīruh". In *Al-Wirātha wa al-handasa al-wirāthiyya wa al-jīnūm al-basharī wa al-ʿilāj al-jīnī: Ruʾya Islāmiya*, edited by Jundī, Aḥmad Rajāʾī, 797-824. Kuwait: Islamic Organization for Medical Sciences.

Morrison, Michael, Donna Dickenson and Sandra Soo-Jin Lee. 2016. Introduction to The Article Collection "Translation in Healthcare: Ethical, Legal, and Social Implications". *BMC Medical Ethics* 17 (74): 1-6.

Nashmī, ʿAjīl al-. 1998. *Al-Waṣf al-Sharʿī li al-jīnūm al-basharī wa al-ʿilāj al-jīnī*. In Jundī, Aḥmad Rajāʾī (ed.) *Al-Wirātha wa al-handasa al-wirāthiyya wa al-jīnūm al-basharī wa al-ʿilāj al-jīnī: ruʾya Islāmiya*, 545-570. Kuwait: Islamic Organization for Medical

Sciences.

Nasim, Anwar. 1993. "Genetic Manipulations and Ethical Issues: Challenges for the Muslim World. In Islamic Educational, Scientific and Cultural Organization (ISESCO)" *Ethical Implications of Modern Researches in Genetics*, 59-74. Rabat, Morocco: Islamic Educational, Scientific and Cultural Organization & Tripoli, Libya: World Islamic Call Society.

Opwis, Felicitas. 2010. Maṣlaḥa *and the Purpose of the Law: Islamic Discourse on Legal Change from the 4th/10th to 8th/14th Century*. Leiden: Brill.

Qatar Biobank. 2014. *A Healthier Future Starts with You*. Doha: Qatar Biobank.

Rabinow, Paul, and Gaymon Bennett. 2009. "Synthetic Biology: Ethical Ramifications 2009". *Systems and Synthetic Biology* 3 (1-4): 99-108.

Raysūnī, Aḥmad al-. 2010. *Abḥāth fī al-maydān*. Cairo: Dār al-Kalima li al-Nashr wa al-Tawzīʿ.

Rizq, Hānī. 2003. *Mūjaz tārīkh al-kawn: min al-infijār al-ʿaẓīm ilā al-istinsākh al-basharī*. Damascus: Dār al-Fikr.

Rizq, Hānī. 2007. *Al-Jīnūm al-basharī wa akhlāqiyyātuh*. Damascus: Dār al-Fikr.

Rūkī, Muḥammad al-. 1998. "Al-Istifāda min al-handasa al-wirāthiyya fī al-ḥayawān wa al-nabāt wa ḍawābiṭuhā al-Sharʿiyya". In *Al-Wirātha wa al-handasa al-wirāthiyya wa al-jīnūm al-basharī wa al-ʿilāj al-jīnī: ruʾya Islāmiya*, edited by Jundī, Aḥmad Rajāʾī, 211-225. Kuwait: Islamic Organization for Medical Sciences.

Thompson, R. Paul and Ross E. Upshur. 2018. *Philosophy of Medicine: An Introduction*. New York: Routledge.

Thorpe, Charles. 2006. *Oppenheimer: The Tragic Intellect*. Chicago: The University of Chicago Press.

ʿUthmān, Muḥammad Raʾfat. 2009. *Al-Mādda al-wirāthiyya: al-jīnūm*. Cairo: Maktabat Wahba.

Willyard, Cassandra. 2018. Expanded Human Gene Tally Reignites Debate. *Nature* vol. 558: 354-355.

Wizārat al-Awqāf wa al-Shuʾūn al-Islāmiyya bi al-Kuwayt. 1983-2006. *Al-Mawsūʿa al-fiqhiyya*. Kuwait: Ministry of Endowments and Islamic Affairs.

Zuḥaylī, Muḥammad al-. 1998. Al-Irshād al-jīnī. In *Al-Wirātha wa al-handasa al-wirāthiyya wa al-jīnūm al-basharī wa al-ʿilāj al-jīnī: ruʾya Islāmiya,* edited by Jundī, Aḥmad Rajāʾī, 773-793. Kuwait: Islamic Organization for Medical Sciences.

CHAPTER 3

Transformation of the Concept of the Family in the Wake of Genomic Sequencing: An Islamic Perspective

Ayman Shabana[1]

The twentieth century witnessed many life-changing scientific and technological achievements that touch almost all aspects of human life both at the individual and collective levels.[2] One of the most fascinating and impactful discoveries has been the identification of the human genetic structure in the form of deoxyribonucleic acid (DNA). Subsequent efforts aimed to decipher the entire human genetic makeup, which was successfully achieved with the completion of the Human Genome Project. The human DNA has become the main marker of personal identity, if not even destiny, with its ability to reveal important information about one's current as well as future health conditions. Consequently, it has opened up a new chapter in the history of medicine with the introduction of personalized medicine, which aims to evaluate individuals' healthcare needs on the basis of their genetic structures. It has also acquired a metaphysical status with its comparison with the soul and its identification as the locus of human personhood, although unlike a soul it has a physical existence (Chadwick 2006, 256).

On the other hand, the availability of this genetic information has raised serious ethical, legal and social questions that concern not only the individuals whose DNA is being examined but also their families. Increasingly physicians and life scientists are trying to come to terms with the fact that having a genetic condition (disease or mutation) is a family experience, rather than an individual one. To what extent then does this new medical and scientific state

1 Associate Research Professor at Georgetown University's School of Foreign Service in Qatar (SFS-Q), as2432@georgetown.edu

2 This publication was made possible by NPRP grant # NPRP8-1478-6-053 from the Qatar National Research Fund (a member of Qatar Foundation). The statements made herein are solely the responsibility of the author.

of affairs change/challenge our perception of family relationships and the very concept of the family itself? What does it mean to be related to someone? Does this rest solely on biological or genetic factors? And finally, to what extent does this biological revolution impact relationship with future family members?

Modern genomic technology has inspired many technical applications that have forced reconsideration of many aspects of the family. The most striking feature of these applications is not only their ability to impact existing family relationships but also, more poignantly, to influence important traits and characteristics in prospective offspring. This chapter aims to highlight some of the numerous vexing questions that these applications raise for both existing as well as prospective family relationships and to explore the range of Islamic responses to these questions. More particularly, it discusses the extent to which these technologies challenge an ideal Islamic model of the family as well as the distinctive characteristics of such a model. At the core of these discussions lies a central question on the permissibility/desirability of utilizing these new technologies. In other words, should they be celebrated as a gift of the God-given human intellect or avoided due to their involvement of uncalculated risks that threaten to disrupt the original order of divine creation? The chapter examines the extent to which various applications of genetic technology are transforming some of the most important aspects governing the structure of the family in terms of its formation through marriage and also individuals' ability to control the reproductive process by influencing basic genetic characteristics of their prospective children. In exploring Islamic discourses on these issues, particular focus is placed on the normative pronouncements as well as related discussions of several transnational institutions such as the Islamic Organization for Medical Sciences, the Islamic Fiqh Council of the Muslim World League, and the International Islamic Fiqh Academy of the Organization of Islamic Cooperation. Several recent studies aimed to provide Islamic perspectives and guidelines on cutting edge genetic and genomic research but they have not paid close attention to this normative body of literature that was generated by these institutions (Al Aqeel 2007; El Shanti et al 2015). This chapter, therefore, aims to contribute to existing scholarship by highlighting the potential role that this literature can play in this regard.

Genetic Revolution and Genetic Testing

The history of modern genetic testing goes back to 1953, when Francis Crick and James Watson identified the basic genetic structure (the deoxyribonucleic acid or DNA) in the form of a double helix, which comprises the chemical compounds responsible for the production and maintenance of all living organisms. This major scientific discovery was the prelude to the successful completion of the Human Genome Project in 2003. This major scientific achievement is said to have ushered a transition from an industrial age to a biotech age (Rifkin 2006, 46). Enthusiastic depictions characterize the human genome as the book of life, the code of codes, or the human blueprint (Rose 2006, 252). The entire human genetic repository (genome) consists of 20.000-25.000 genes, which exist in the form of extended segments (of varying length) on the base pairs that make up the spiral staircase or double helix. A human genome consists of a total of 3 billion base pairs within the nucleus of each cell, which are bundled into 46 chromosomes. They are arranged into 23 pairs, out of which 22 are the same for both males and females and only one (the sex chromosome) varying between a male and a female. Sequencing of the entire human genome has inspired scientists to develop various types of tests to screen genetic disorders and devise means to fix them or preempt their occurrence. Genetic disorders occur as a result of mutations or alteration in one's genetic structure, which can then be passed down to subsequent generations. Genetic alterations are responsible for as many as 4000 hereditary diseases and genetic tests are now available for over 1000 diseases, which are expected to increase in the future (Vaughn 2010, 460-1). Some of the most common types of genetic tests include: newborn screening for detection and early treatment of certain diseases; carrier testing to determine whether a person is a carrier of a particular disease; predictive testing, especially in case of family history; diagnostic testing for confirmation purposes; prenatal testing to screen fetuses for certain disorders such as the Down Syndrome; and pre-implantation genetic diagnosis (PGD) to screen IVF embryos prior to implantation into the mother's uterus (Vaughn 2010, 462). Despite their immense potentials, these tests are quite complex to the extent that some researchers question their utility. Part of this complexity is due to the nature of their results, which are usually probabilistic rather than conclusive.[3] Moreover, tests can hardly confirm whether a particular genetic disorder is linked to a single gene mutation, multiple mutations, or yet as a combination of gene mutations and other environmental factors. Scientists cannot identify all possible mutations responsible for a particular disease (apart from available tests) or even potential mutations that may occur in the future. Also, severity of symptoms may vary from one case to another depending on interaction with other factors. Finally, the most challenging aspect about genetic testing is availability of effective treatment. Testing can

3 On the probability rather than certainty of genetic test results, see Emslie and Hunt 2006, 104.

only confirm a particular diagnosis but this does not mean that there is a cure for every (known) disorder.[4]

Impact of Genetic Technology on Existing and Prospective Family Relationships

Many researchers note that genomic technology is changing medicine in significant ways. For the purpose of this chapter, one remarkable feature of these genomics-driven changes is the growing realization of one's biological ties and connections with family members. With reference to the different types of genetic testing mentioned above, it is often noted that revelation of test results is not always a blessing because the process usually comes with a psychological toll regardless of the outcome of the testing process. Most importantly, these results often do not pertain to the individual being tested but they may be relevant to family members as well, thereby raising the question whether it would be necessary to share this information with related family members who are likely to be affected. For individual patients, the situation may vary depending on availability of a cure. In other words, one's decision to share testing results with family members who are expected to develop similar symptoms may depend on whether a medical treatment already exists or not. Some may find that revelation of distressing information in the form of susceptibility to develop an untreatable condition such as Alzheimer's disease would be of little use if not outright harmful. On the other hand, treating physicians may find themselves torn between a patient's right to autonomy (in case they do not wish to reveal test results to family members who are likely to be affected) and the duty to prevent harm to others (by sharing such information) (Vaughn 2010, 464-5). While some may argue that the revelation of testing results should be the norm, others argue that revelation of test results is not always useful especially when susceptibility to genetic disorder may lead to genetic discrimination in the form of bias by an employer or an insurance company.[5] From another perspective incidental or inadvertent findings during the testing process could have serious social implications as is the case with

4 For example a 2002 study showed that 81 percent of respondents wanted to undertake genetic testing when a cure is available. See Vaughn 2010, 463.

5 In the United States, the Genetic Information Nondiscrimination Act (2008) prohibits discrimination by an employer or an insurance agency on the basis of genetic information. Undocumented cases of discrimination, however, are difficult to account for. See Vaughn 2010, 466

misattributed paternity (Reilly 2006, 67). The use of genetic testing for paternity verification is particularly interesting in light of the fact that it has been adopted in many jurisdictions as the main method for the legal ascertainment of paternity. Within the Muslim context it has stirred heated debates due to significant ramifications on Islamic family regulations.

Modern advances in genetic technology have not only impacted existing family relationships but they have also allowed the possibility of predetermining the nature and shape of these relationships. For example, recent genomic advances have given rise to a wide array of procedures that aim at screening and even manipulating human genetic structure for therapeutic or non-therapeutic enhancement purposes through various types of genetic testing and genetic engineering. The outcome of these procedures could have lasting consequences for prospective family members. Some of the most important examples include carrier or predictive genetic testing, which can be undertaken to ensure proper matching for marital purposes. Prospective couples may undergo these tests to circumvent certain genetic disorders in future offspring, especially in societies where consanguineous marriages are common. In these cases, testing aims to screen couples to determine whether one or both individuals are carriers of a genetic disorder.[6] Carriers possess one copy of a gene mutation and this does not mean that they do or will have the disease but when two carriers get married their children will inherit two copies of the mutated gene responsible for a particular disease (e.g. cystic fibrosis or Tay-Sachs), which will significantly increase their chance of having the disease. Apart from legal enforceability of these tests, which would depend on particular jurisdictions, these tests pose a series of moral concerns for these prospective couples and also for their families such as necessity to submit to these tests, sharing test results, possibility of concluding marriage despite positive test results, and finally impact on and responsibility towards future offspring (Vaughn 2010, 462).

Another example of tests that affect prospective family members is prenatal testing, which is undertaken during pregnancy to screen fetuses for particular diseases such as the Down Syndrome, which is found to be common when pregnancy occurs after the age of 35. While these tests can be useful in assuring parents about the health status of a fetus they raise the problem of moral de-

6 Premarital genetic screening has been used in places like Cyprus, where Thalassemia is a major public health issue, to reduce birth rate of affected babies. Some studies show that a large percentage of prospective couples who learn that they are both carriers before marriage continue with marriage. The majority of those couples use prenatal diagnosis in every pregnancy and resort to abortion when fetuses are affected. See Modell 2006, 119.

cision-making in case an abnormality is detected, for which the main solution is often abortion. Legality of abortion would depend on several factors such as the age of the fetus, the moral-religious perspective, and also the legal jurisdiction in question. This particular case of selective abortion, however, raises several additional ethical questions pertaining to perception of and attitude towards disability and disabled persons. Another problem has to do with the certainty of diagnosis or even percentage of accuracy. Ultimately this testing raises a question about the extent to which it can be used to screen for common disorders that can be treated with drugs? (Vaughn 2010, 466-7) Similarly, preimplantation genetic diagnosis (PGD) shares a great deal with prenatal testing because it also aims to screen embryos or pre-embryos for genetic disorders prior to implantation into the uterus. While this type of testing also highlights the two issues of sanctity of human life and moral status of embryos, it also poses a set of additional moral concerns associated with the nature and objective of this testing. Equally problematic is preimplantation genetic screening (PGS), when it is undertaken for fetal sex selection to ensure pregnancy with a fetus of a desired sex, as it raises questions of gender discrimination and natural gender balance.

In the same vein, different applications of genetic engineering raise similar questions pertaining to prospective offspring. Ability to decipher human genes and identify their functions inspired efforts to repair mutated or faulty genes through gene therapy or genetic engineering. This could take the form of replacing, fixing, or activating particular genes. Gene therapy may target regular body (somatic) cells or germline (ovum or sperm) cells. While somatic gene therapy aims to fix a disorder within a person's body, germline gene therapy impacts one's offspring. Although this latter type of gene therapy is not yet available, it raises the question of manipulating the genetic structure of prospective children, which is sometimes referred to as "designer babies." While somatic gene therapy undertaken for therapeutic purposes is usually praised as a commendable undertaking, germline gene therapy, similar to PGD and PGS, raises a question about the merit of enhancement as well as the boundaries of legitimate and illegitimate intervention. Most importantly, it also raises a question about making important and lasting decisions on behalf of future generations and whether this is warranted or even desirable (Vaughn 2010, 468; Barry 2012, 254). Finally, stem cell research also raises questions about enhancement and boundaries of proper and improper use of stem cells. Stem cells are particularly important due to their high therapeutic potential and also their ability to develop into any type of body cells. The main ethical problem associated with stem cell research has been the need to destroy embryos in the process of extracting them. Although this problem has been resolved after the development of a technique to reprogram adult cells to function as embryonic stem cells, some argue that it is premature to judge the extent to which this technique can actually match or replace the need for embryonic stem cells. Prior to the emergence of this technique in 2007, the two main sources to extract stem cells were IVF surplus embryos and embryos specifically created for research. Two main methods are used for the creation of embryos for research: parthenogenesis (stimulation of unfertilized eggs from which stem cells can be extracted); and somatic cell nuclear transfer (SCNT), which is also known as cloning. The latter process involves the extraction of the nucleus of an egg and replacing it with the nucleus of a regular somatic cell (Barry 2012, 266). Therapeutic cloning involves the creation of embryos just for the purpose of research. On the other hand, reproductive cloning involves implantation of the created embryo in the uterus, consequently resulting in a copy of the nucleus donor. The birth of Dolly the sheep in 1997 marked the success of the procedure in animals, although to date the procedure has not been tried in humans. This scientific feat, however, stirred reverberating waves of anxiety worldwide, which inspired global consensus on the prevention of human reproductive cloning and the need to develop appropriate research guidelines on these procedures. Nonetheless, the scientific possibility of developing human clones raises important questions not only about human exceptionalism but also about the integration of such clones within families and the social order in general (Barry 2012, 282-5).

Islamic Family Regulations and Genomic Technology

Since the emergence of the various applications of modern genetic technology, ethicists worldwide, both religious and secular, have been grappling with the moral quandaries that they have engendered. Within the Muslim context, responses have come mainly from Muslim jurists, which reflects the continuing influence of Sharia law for the definition of Islamic normativity. The Islamic legal corpus includes detailed regulations on various aspect of family affairs, which continue to inform related legislation in most Muslim majority countries. In light of the remarkable diversity and plurality within the Islamic legal tradition, these juristic regulations were often debated and even contested among jurists of the various legal schools. With the development of biomedical technology, however, Muslim jurists have been forced to revisit certain legal opinions and doctrines that were based mainly on pre-modern medical knowledge and experience. One of the earliest examples is the ruling pertaining to the maximum duration of a viable pregnancy. In the modern period it has been fixed to one single year instead of the extended periods that classical jurists accommodated in order to preserve, to the extent possible, the sanctity of marriage and reputation of married women. It is important to keep in mind that legal integration of technical applications or revision of legal opinions and doctrines in light of advances in scientific knowledge remains subject to a process of negotiation that often requires extensive deliberation and scrutiny by both legal and scientific experts within a particular social and cultural context. Each of the above-mentioned procedures is already being subjected to this process of deliberation and scrutiny. Below I explore Islamic discourses on three main issues: premarital genetic testing, fetal sex selection, and germline genetic modification.

Islamic Law and Medical Suitability for Marriage

Contemporary juristic deliberations on premarital genetic testing are often placed within the context of classical juristic discussions on health-related concerns and their role in either facilitating the conclusion of marriage or warranting its termination.[7] This includes different rules, principles, and general injunctions that aim to prevent diseases or to encourage their treatment. For example, this would cover criteria for the choice of marital partners; pronouncements on guarding against diseases in general and avoiding their causes; and health-related defects sanctioning dissolution of marriage.[8]

Islamic injunctions pertaining to marriage often emphasize the choice of marital partners on the basis of their moral character. Several pronouncements also address physical suitability for marriage. In this context explicit warning against consanguineous marriage is meant to avoid any negative impact on the health of future children, which has been confirmed by modern genetic research. The Islamic tradition includes several references discouraging this practice on the basis of the observation that consanguineous marriages often result in weaker offspring. The most important reference is attributed to the second Caliph ʿUmar who advised the clan of al-Sāʾib to marry outside close family circles (Abū Ghuddah 2000, 1:585; Mahrān 2002, 226-7). Several

7 Both this section on premarital genetic testing and the following one on fetal sex selection draw heavily on an earlier publication, see Shabana 2017, 201-213.

8 Researchers often distinguish between general premarital medical testing, which screens for particular viral or contagious diseases and premarital genetic testing, which screens for genetic conditions that are likely to affect prospective children. While the first type is not contested due to its immediate benefits for the couples themselves, the second is debated due to its hypothetical or inconclusive nature. See ʿAbd Allāh 2007, 9-13; ʿIbādah 2010, 17.

Prophetic narratives are also reported but their authenticity have been questioned. The ban on marriage with the prohibited degrees mentioned in the Qur'ān (4:23) supports this attitude favoring strangers as marital partners over close family members, which is said to be rooted in both physical and psychological considerations (al-Muḥammadī 2005, 321-3).

Apart from these general injunctions on the selection of the marital partner, the Islamic normative tradition includes numerous references emphasizing the importance of taking preventive measures to avoid diseases either by contagion or any other means. The issue of contagion has been particularly problematic due to several competing references that seem to give different connotations on the exact relationship between contagion and actual occurrence of particular diseases. This causal connection between contagion and illness has often triggered larger theological questions pertaining to the efficacy of independent causes and the extent to which a belief in such efficacy would conflict with God's omnipotence.[9]

With few exceptions, contemporary jurists often do not oppose premarital genetic testing but they disagree, however, on the extent to which it can be enforced.[10] In general scholars can be divided into two main groups: those who argue for the enforcement of premarital genetic testing and those who argue that it should remain optional. Jurists who argue for enforceability emphasize physical fitness as an important condition for the achievement of the ideal objectives of marriage, which include sexual gratification and emotional fulfillment. They also emphasize the right of progeny to a healthy life, which involves protecting them against harmful or dysfunctional genes (al-Muḥammadī 2005, 321; al-Zuhaylī 2000; Shabīhunā 2000). Jurists who argue for optionality, on the other hand, link their attitude to the question of medical treatment in general which, according to this line of reasoning, is considered permissible rather than obligatory ('Uthmān 2000).

Supporters of the enforcement of premarital testing point out the importance of exercising discretion when it comes to the choice (*takhayyur*) of the marital partner as indicated in several Prophetic reports. These injunctions on the proper selection of marriage partners can lend support to premarital genetic testing, which would equip prospective couples with valuable information regarding their own health as well as the health of their children (Abū

9 For a discussion on the question of contagion in the Islamic tradition, see Stearns 2011.

10 Some scholars, such as the late Saudī jurist 'Abd al-'Azīz bin Bāz, argued against premarital genetic testing on the grounds that results can be inaccurate. It is also argued that in principle couples are presumed to be free from genetic diseases, which obviates the need for genetic testing. See al-Shuwayrakh 2007, 128-9.

Ghuddah 2000, 1:583).[11] In this regard statements against consanguineous marriages are read in light of modern biomedical findings linking consanguineous marriage with increased likelihood for genetic disorders in the second generation. For example, the Syrian jurist Muḥammad al-Zuḥaylī argues that if tests show more than 50% chance of serious genetic disorders, legal action can be taken to prevent marriage in this case (al-Zuḥaylī 2000, 2:784; Abū Ghuddah 2000, 1:584; ʿAbd Allāh 2000, 2:740-1).[12] With regard to the issue of physical health, supporters of this attitude highlight the importance of guarding against all types of diseases, whether by contagion or any other means.[13] On the issue of contagion, they rarely question its influence and they largely do not think that such influence would imply contradiction with religious faith. In this regard textual references supporting the influence of contagion are emphasized and other competing references are interpreted. For example, the text of the famous Prophetic ḥadīth negating the influence of contagion is said to mean either no contagion can be effective without God's permission or no one should cause contagion to befall others. In this context genetic testing is seen as an important preventive measure to avoid contagion by protecting offspring against potential genetic disorders (Abū Ghuddah 2000, 1:582). With regard to the legal status of medical treatment, although it is recognized that the general ruling is permissibility, this ruling may change to recommendation and even obligation if it affects others as is the case with communicable or genetic diseases (al-Zuḥaylī 2000, 2:779-80).

One of the important contexts within which premarital genetic testing is often placed is the premodern discussions on health defects that justify annulment of marriage (Abū Ghuddah 2000, 1:582-4; Shabīhunā 2000, 2:944).[14] These discussions address the reproductive function of marriage but they also address other aspects such as mental capacity as well as medical conditions affecting one's ability to interact normally with others.[15] Supporters of

---

11 For an overview of this issue in Arabic culture and literature, see Van Gelder 2005.

12 Supporters of enforcing premarital testing argue that it is permissible for the governing authority to enforce these tests in order to protect progeny, which is one of the main objectives of sharīʿah, see also Buḥālah 2010, 301-2.

13 For example, according to the Syrian jurist ʿAbd al-Sattār Abū Ghuddah, premarital genetic testing can be subsumed under a general principle that can be referred to as guarding against all types of diseases. See Abū Ghuddah 2000, 1:582.

14 Subsequent studies often adopt this approach as well. See, for example, Shubayr 2001.

15 This list includes conditions that are common for men and women such as madness (*junūn*) and different types of skin diseases such as leprosy (*baraṣ* and *judhām*); two conditions specific for men, which are castration (*jabb*) and impotence (*ʿunnah*); and three conditions specific for women, which are *fatq*, *qaran*, and *ʿafal*. Some jurists note that

the enforcement of genetic testing do not see the list of defects that pre-modern jurists discuss as exhaustive, which means that any other condition that jeopardizes the continuity of marriage can be added as well (Shabīhunā 2000, 2:952-3).[16] In general, while the discussion on the possibility of enforcing premarital testing presumes that such enforcement is to be done by the government, some also indicate that prospective couples may stipulate such testing if they wish.[17]

While supporters of the enforceability of premarital genetic testing emphasize the reproductive function of marriage, supporters of optionality of genetic testing insist that procreation is not the sole objective of marriage. They also discuss other objectives such as lawful fulfilment of the sexual desire as well as establishment of loving and merciful cohabitation (Mahrān 2002, 227). Similarly, while they do not question the considerable advantages of genetic testing, they also emphasize its limitations. After all, genetic testing does not

these three terms refer to a blockage in the female genital part that obstructs normal sexual relationship. Other jurists make further distinctions between them. See Ibn Qudāmah 1997, 10: 57. Some jurists expand the list to include up to 18 conditions while others argue that any defect that would defeat the original purposes of marriage could be included, see Ibn al-Qayyim, 1998, 5: 166; al-Qaradāghī and al-Muḥammadī 2008, 276-8.

16 By way of analogy to the health conditions that pre-modern jurists list as sanctioning the dissolution of marriage, several jurists argue that a genetic disease can serve as a valid reason for the annulment of marriage provided that such a disease was not known or confirmed before the conclusion of the marital contract, see al-Shuwayrakh 2007, 202-4. Some researchers, therefore, suggest premarital genetic testing as a precautionary preventive measure that should be undertaken by prospective couples. For example, Shubayr notes that undertaking premarital testing does not conflict with sharīʿa or with the objectives of marriage because marriage of healthy couples is likely to last more than that of ill couples. He argues that such tests should be facilitated and administered by the government without charge. He also suggests that such tests should be a standard procedure for all individuals once they reach 15 years of age. This medical statement should then be submitted at the time of concluding the marriage contract, see Shubayr 2001, 336. Some researchers suggest that AIDS can be one of the diseases that may warrant annulment of marriage, see al-Qaradāghī and al-Muḥammadī 2008, 279.

17 The European Council for Fatwa and Research supports this opinion, see al-Qaradāghī and al-Muḥammadī 2008, 297; Mahrān 2002, 212. Classical jurists discuss the possibility of adding an additional stipulation (*sharṭ zāʾid*), which is neither prohibited nor permitted under the rubric of *al-sharṭ al-jaʿlī*. The majority of jurists admit such stipulations as long as they accord with the intent of the contract in question. The Ḥanbalis in particular are known for their acceptance of this type of stipulations as long as they do not violate the intent of the contract. Including premarital genetic testing as a precondition for the marital contract would then depend on whether it is deemed in support of the overall intent of the marital contract, see al-Ludaʿmī 2011, 262-3; Buḥālah 2010, 230-5.

by itself involve any therapeutic value and all it does is reveal existing, latent, or potential risks. Some actually question its utility in case of incurable conditions. They note that negative results in genetic testing do not necessarily mean that the tested individuals are free from genetic diseases in general, but only from the particular genetic diseases for which they are tested. In light of the increasing number of genetic disorders, it is almost impossible to find out one's status regarding all possible genetic diseases. But, apart from these immediate and direct disadvantages that genetic testing may involve, it can also result in several adverse moral, legal, and also economic consequence that may affect not only the person being tested but also other members of the family (al-Bārr 2000, 2:630-1).[18] For example, they argue that enforcing genetic testing would open the door for corruption in case someone wants to obtain a certificate without being tested. Moreover, they denounce general condemnation of consanguineous marriage as the percentage of genetic disorders that can be linked directly and exclusively to this type of marriage can hardly be determined beyond any doubt. Even in the case of genetic disorders, the role of the environment as well as other causal factors cannot be excluded (al-Bārr 2000, 644-9).[19] After all, the Qur'ān includes references to marriage with first cousins (33:50).

Apart from the two main attitudes mentioned above (supporters of enforceability and optionality), some scholars argue that in principle premarital genetic testing should not be made compulsory unless there is a dire necessity for it, in which case the government should take appropriate action (al-Qaradāghī and al-Muḥammadī 2008, 285-8; al-Maymān 2000, 2:821; al-Madḥajī 2011, 2:935). Otherwise, it should be left to individual discretion and more efforts should be made to raise public awareness about its importance.[20] On the

18 See also al-Shuwayrakh 2007, 92-5 (speaking about the false impression that genetic testing may give and also possible adverse consequences). These considerations prompted some religious scholars to question the benefit of genetic testing and argue against it.

19 Moreover, supporters of this attitudes argue that ascertainment of the medical condition of a prospective spouse is not one of the conditions of a valid marriage. They also note that marriage is not necessarily meant for reproduction. See Buḥālah 2010, 308.

20 Islamic organizations and juristic councils issued different statements on this topic. For example, the Kuwait-based Islamic Organization for Medical Sciences supported the attitude to keep premarital genetic testing optional. This is included in the statement that was issued at the conclusion of its seminar on this and related issues in 1998. With regard to genetic testing and counseling, the statement called for: raising public awareness about the importance of genetic counseling, especially for prospective couples; preserving the privacy of individuals and confidentiality of their test results; ensuring that the process remains optional; and raising awareness about genetic risks associated with consanguin-

other hand, non-genetic regular medical tests can be enforced under certain conditions. First, tests should aim to screen for a particular list of dangerous or contagious diseases. Second, failure to submit to testing should not affect the validity of the marital contract. This condition is meant to avoid the possibility of changing the stipulated Sharia-based conditions for the validity of a marital contract either by addition or deletion. Alternatively, non-compliance can be penalized by the payment of a fee or any other similar means but it should not affect the validity of the marital contract (al-Qaradāghī and al-Muḥammadī 2008, 285-8; ʿAbd Allāh 2007, 95, 125). In light of the increasing significance of premarital testing in general and genetic testing in particular they have often been incorporated within family law legislation throughout Muslim-majority countries. However, while some countries make them compulsory, others keep them only optional.[21]

---

eous marriage, see *Ruʾyah Islāmiyyah,* 2:1050-2. The topic was also addressed by both the Islamic Fiqh Council, IFC (affiliated with the Muslim World League in Mecca) and the International Islamic Fiqh Academy, IIFA (affiliated with the Organization of Islamic Cooperation in Jeddah). The IFC decision, which was issued in its 17th session (held in Mecca in 2003), focused on the possibility of stipulating premarital medical testing as a precondition for the conclusion or authentication of the marital contract. It declared that this would be impermissible (*al-ilzām bi al-fuḥūṣ al-ṭibbiyyah wa rabṭ tawthīq al-ʿaqd bihā amr ghayr jāʾiz*). It still called for raising public awareness about premarital tests; encouraging them; and facilitating them for those who wish to undertake them. Finally, it also noted that test results should remain confidential. See *Majallat al-Majmaʿ al-Fiqhī al-Islāmī* 15: 17(2004), 305. On the other hand, the IIFA decision, which was issued in its 21st session (held in Riyadh in November 2013), emphasized the permissibility of premarital genetic testing and also included a statement giving authorities the power to enforce it if this is deemed of a considerable pubic interest (*maṣlaḥah muʿtabarah ʿāmmah*), see http://www.iifa-aifi.org/2416.html (accessed March 2017). A similar indication was also included in IIFA's decision with regard to the rights of the disabled in its 22nd session that was held in Kuwait in March 2015. This decision emphasized the importance of methods that can remove causes of disability such as premarital medical testing and vaccination against polio, see http://www.iifa-aifi.org/3998.html (accessed March 2017).

21 Some Muslim-majority countries already require premarital medical testing but these tests are generally perceived as routine and ineffective. See, for example, Abū Ghuddah 2000, 1:584 (noting that these tests often do not include genetic testing); al-Bārr 2000, 2:631 (noting that in most cases formal medical statement certifying physical fitness for marriage can be obtained easily without actual testing); Mahrān 2002, 213 (noting that in Egypt marriage registrars are required to obtain a written statement from the couples indicating that they do not suffer from concealed diseases, *amrāḍ sirriyyah.* This statement, however, does not have any legal impact and its absence does not have any effect); ʿIbādah 2010, 46 (referring to an Egyptian law that was issued in 2008 requiring premarital medical and genetic tests for the registration of marriage. Many observers, however, com-

Genetic testing is often discussed in relationship to the larger process of genetic counseling, which aims to provide detailed information about the tests being undertaken as well as advice on possible options in light of test results according to best practices. In the Muslim context this also includes elucidating possible implications and consequences of medical decisions in light of Islamic norms and regulations (al-Ḥāzimī 2000, 682). In case of positive results genetic counseling can provide possible options depending on the exact circumstances of the couple and whether both of them carry the same copies of mutated genes. These options range from cautioning them against marriage in severe circumstances to detailed advice on further steps if they choose to proceed with marriage indicating recommended procedures after marriage, whether before pregnancy, during pregnancy, or after birth (al-Ḥāzimī 2000, 680-1). In these cases they would be in a better position when it comes to anticipation of potential disorders as well as necessary arrangements to address them.

Possible options for couples whose offspring are at great risk for genetic disorders prior to pregnancy include preimplantation genetic diagnosis (PGD). This is one of the techniques that assisted reproductive technologies have made possible and by means of which IVF embryos can be tested for potential genetic disorders prior to implantation in the mother's womb. Apart from the high cost of the procedure as well as the limited success rate, the technique raises other ethical questions regarding the moral status of the embryo as well as proper disposal of surplus embryos (al-Bārr 2000, 2:634-5). A range of other options can also be explored during the process of genetic counseling along with related moral as well as religious evaluation of these options (e.g. tempo-

---

plain that it is usually implemented in a formalistic and routine manner. Several media reports indicate that a statement can be issued easily without undergoing any type of testing, see for example http://www.almasryalyoum.com/news/details/42427 (accessed April 2017). Some countries made pre-martial genetic testing compulsory such as Jordan (2004), Algeria (2005), Qatar (2006), and Kuwait (2008), see Buḥālah 2010, 53, 313-32; al-Qaradāghī and al-Muḥammadī 2008, 269. It is also worth mentioning that legislations that make premarital testing mandatory emphasize the confidentiality of the testing results and also freedom of prospective couples to conclude their marriage regardless of the testing results. For example the Algerian law indicates that the certificate issued to the prospective couples only indicates that the prospective couple completed the testing (without including the testing results), see Buḥālah 2010, 75. In Saudi Arabia a ministerial decree was issued in 2004 to enforce premarital testing however, without any limitation on the freedom of prospective couples to conclude the marital contract regardless of the testing results, see al-Madḥajī 2011, 2:936; al-Yābis 2012, 1:220.

rary or permanent prevention of pregnancy or aborting deformed or defective fetuses).

Islamic Reproductive Ethics and Boundaries of Genetic Intervention: Case of Fetal Sex Selection

One of the options that a couple may explore after PGS is fetal sex selection, particularly when a genetic disorder is associated with one sex more than the other. In this case this technique is proposed as a therapeutic procedure rather than an exercise of preference for a particular sex over the other. Apart from the technical possibility of the procedure, fetal sex selection often raises two main theological questions: potential conflict with God's will; and potential risk of unsettling original balance of male and female distribution. The first question emanates from several scriptural sources indicating that knowledge of embryonic life belongs solely to God.[22] These scriptural references imply that this divine knowledge controls one's sex during the early stages of embryonic life.

Juristic discussions on the possibility of fetal sex selection can be traced back to a seminar that the Islamic Organization for Medical Sciences organized under the general theme of "reproduction in light of Islam" held on May 24, 1983 (al-Jindī 1983). Within this seminar the discussion centered around a brief paper that the late Egyptian physician Ḥassān Ḥatḥūt presented in order to explain the medical and scientific nature of the procedure and how it can be implemented. According to Ḥatḥūt, two methods could be used. The first involves extracting a sample of the amniotic fluid surrounding the fetus and analyzing it in order to find out its sex. A decision then can be made either to retain the embryo or to get rid of it depending on the desired sex, which would raise the question of (im)permissibility of abortion as well. The second method, which at the time was undertaken only in animal breeding, depends on sperm sorting by subjecting extracted semen to a technical process by means of which Y-chromosome (male) and X-chromosome (female) are separated and then later injected in order to increase the likelihood of obtaining an embryo with the desired sex (Ḥatḥūt 1983, 37-8).

22 For example, see "God knows what each female carries (whether male or female) and what the wombs decrease or increase (of the pregnancy term) and everything with him is according to a (precise) measure." [Q 13:8] and "with God is knowledge of the Hour (Day of Judgment), He causes rain to descend, He knows what is in the wombs, no single soul knows what it will acquire the following day or where it will perish " [31:34]

In general the opinions expressed during the seminar were indicative of three main orientations, which could be termed as liberal, restrictive, and intermediary. The liberal orientation emphasized the religious merit of discovering the secrets of the universe, which is the explicit goal of science. Ultimately this search for the hidden secrets of the universe cannot escape divine knowledge and this remains a matter of theological belief. For example, in his commentary, the Egyptian jurist Yūsuf al-Qaraḍāwī questioned the possibility that such a procedure would place limitation on God's comprehensive knowledge, which includes every aspect of one's life. He argued that whatever man knows (e.g. in the case of fetal sex determination) is facilitated in the first place by God's knowledge, not despite it. Moreover, human ability to control fetal sex is no different because this also cannot escape God's will or command. As much as human knowledge is facilitated by divine knowledge, so also is human will, which is facilitated by the divine will.[23] As far as the question whether humans should interfere, al-Qaraḍāwī argued that they should not unless in cases of necessity, which should be treated on an individual rather than collective basis. According to this argument, humans are better off maintaining natural gender distribution, which has always been established from the beginning of human existence in the universe (al-Jindī 1983, 94-5). [24] Scientists will continue to explore all types of natural phenomena and Muslims should participate in this effort and explain the boundaries of the permissible and impermissible. In this example, one of the main restrictions would be total ban on the mixing of the gametes of unmarried couples.

Several normative precedents are used to bolster this argument such as the example of Prophet Zakariyya (Zechariah) and his appeal to God to bless him with a baby boy. Another precedent believed to be supportive of this attitude is the Prophet's remark concerning coitus interruptus (*al-ʿazl*). While the Proph-

23 Similarly, the Syrian scholar ʿAbd al-Sattār Abū Ghuddah refers to a distinction that the famous theologian and jurist Ibn Taymiyyah (d. 1328) made between two types of divine will: cosmic and legislative. The former governs what happens in the universe independently of human will or action while the latter leaves room for human voluntary action. Regardless of what man chooses, divine cosmic will ultimately reflects and manifests what God wants. See Abū Ghuddah 1983, 162.

24 This is with reference to this Qurʾānic verse "and you can only will what God has willed" Q 76:30 and 81:29. In support of this argument some participants also elaborated on the notion of ultimate supremacy of the divine will by pointing out the role of the original God-given condition into which man is created (*fiṭrah*). Part of this original condition is individual inclination towards a particular sex, which differs from one person to another and which, even in case of human intervention, would eventually maintain even distribution of the sexes, see Ibid, 96.

et did not forbid his companions, he noted that if a pregnancy is decreed it would occur regardless whether coitus interruptus is practiced or not (al-Jindī 1983, 97). Some reports also urge prospective couples to keep in mind future offspring while choosing marital partners (*takhayyarū li nuṭafikum*) (al-Jindī 1983, 112). One of the main religious concerns that fetal sex selection raises is that it may involve changing the original form of God's creation (*taghyīr khalq Allah*). Proponents of this attitude, however, retort by noting that this process does not entail changing the original nature of either the sperm or the ovum. All what it does is to facilitate the merger of certain gametes, not refashioning them in order to modify their basic characteristics. As long as Sharia rules are observed (e.g. preservation of the marital framework), the process should be subject to ijtihad (al-Jindī 1983, 103). Moreover, in certain individual circumstances the procedure could be quite helpful in satisfying the need for a baby of a particular gender. Such would be the case, for example, of spouses who have four or five girls and they want to have a baby boy. This need should not be dismissed as trivial or insignificant (al-Jindī 1983, 104).

On the other hand, a more restrictive orientation was expressed, mainly by some of the participating physicians in light of practical experience with the early phases of the fetal sex detection technology. They noted that this technology was used almost always to facilitate aborting female fetuses (even in the West), which demonstrates a global trend of anti-female bias.[25] This technology, therefore, raises the risk of facilitating modern forms of female infanticide (*maw'ūdah*), in comparison to the pre-Islamic Arabian practice. Accordingly, Islam's attitude on this issue, should be total rejection of technical intervention in fetal sex selection. Any form of explicit or implicit bias against women would conflict with the original spirit of Islamic legislation and should therefore be condemned, whether before or after birth. Some proponents of technical intervention argue that it could assist in the process of readjusting the balance of gender distribution in certain circumstances such as post-war situations, during which the male-female ratio is usually disturbed in favor of the female side due to the fact that more men are killed during wartime. Opponents, however, refer to some demographic studies indicating that in these circumstances gender balance is usually readjusted through natural means, obviating, therefore, the need for human or technical intervention (al-Jindī 1983, 101). While the proponents of the liberal attitude deemphasized the theological implications of fetal sex selection, proponents of the restrictive attitude insisted that this issue has clear theological implications on the grounds that scriptural

25 For a discussion on how the technology is used to reinforce anti-female attitudes in India, see also Davis 2006, 291-5.

references relegate choice of the gender of babies exclusively to God (e.g. Q 42: 49-50). Moreover, fetal sex selection involves changing God's creation, which covers any type of intervention in the natural process (al-Jindī 1983, 109-11). In particular, this procedure opens the door for more questionable procedures, especially those involving messing with human sperm.

This exchange of views shows that one of the main grounds for disagreement between proponents of the restrictive and liberal orientations is whether fetal sex selection is fundamentally a theological or merely a legal/jurisprudential question. While the former emphasized the theological implications of the issue, the latter insisted that this should be pursued as a regular legal/jurisprudential question. Between these two main orientations, a third attitude also emerged, which urged caution and advised against rushing into premature conclusions. Proponents of this attitude noted that more time was needed in order to be able to judge on the basis of actual results in the real world, in light of the fact that up until then technical means for sex selection had not yet been implemented in human reproduction (al-Jindī 1983, 102). The recommendations issued at the conclusion of the seminar included a statement on the question of fetal sex selection indicating its impermissibility at the collective level. On the other hand while some participants argued for permissibility at the individual level to satisfy the need for gender balancing, others argued for impermissibility lest this should lead to unsettling natural gender balance (al-Jindī 1983, 349).

Following this collective discussion on the issue during the IOMS seminar, normative Islamic discussions usually distinguish two main methods for sex selection: natural and technical.[26] The natural methods include a range of procedures that are meant to increase the likelihood of pregnancy with a fetus of the desired gender such as: timing of the intimate relationship relative to the ovulation process; following a specific diet; or use of special types of herbal or chemical solutions.[27] While these natural methods are generally considered

26 This distinction can be traced back to the initial discussions during the seminar that the IOMS organized in 1983, see al-Jindī 1983, 114. See also al-Rashīdī 2011, 567-624; al-Qaradāghī and al-Muḥammadī 2008, 556-62.

27 For example, some studies indicate that sexual intercourse during the early phase of ovulation increases the chance of obtaining a male fetus in comparison to later phases, which increase the possibility of obtaining a female fetus. Similarly some studies point out the connection between certain types of food and pregnancy with a fetus of a particular gender. Accordingly, a diet rich in potassium and sodium increases the chance of obtaining a male fetus while a diet rich in magnesium and calcium increases the chance of obtaining a female fetus. The sex of a fetus may also depend on the acidic or alkaline environment within the uterus. While an acidic environment is said to be more likely for obtaining a

permissible, some scholars include certain stipulations for their permissibility such as: pursuing such methods should not conflict with belief in divine omnipotence, they should not result in greater harm either for the man or the woman; and they should not involve uncovering of the private parts of the body. The technical methods also include several procedures that are meant to achieve pregnancy with a fetus of the desired gender, which require artificial insemination either internally through direct injection of properly sorted sperm or externally through in vitro fertilization (al-Rashīdī 2011, 590).[28] In general, juristic opinions on the use of technical means to achieve fetal sex selection can be divided into three main attitudes: permissibility as long as there is a justified psychological, social, or medical need; total prohibition; and restricted permissibility in cases of necessity for therapeutic purposes only. The third attitude, representing the majoritarian view, captures the view of the Islamic Fiqh Council of the Muslim World League. Its resolution on this issue during its 19th session, held in Mecca in November 2007, indicates that natural methods for sex assignment are permissible as long as they do not include any questionable procedure from the Sharia perspective. Technical methods, on the other hand, could be used only for medical necessity as is the case with genetic diseases affecting a particular gender. Cases of medical necessity should be evaluated individually on the basis of a consensus evaluation of a specialized committee consisting of a minimum of three physicians.[29]

Fetal sex selection is just one example that illustrates the extent to which the interaction between assisted reproduction and genetic technologies may pose considerable challenges to the traditional structure of the nuclear family. Other examples include use of DNA fingerprinting for paternity verification and also use of gamete donation as well as surrogate motherhood to overcome infertility problems. While DNA paternity testing questions the utility and continuity of the marital presumption, gamete donation and surrogacy arrangements question the traditional definitions of both paternity and maternity.

male fetus, and alkaline environment is more likely for obtaining a female fetus. See al-Rashīdī 2011, 583-6.

28 Some researchers refer to another technical procedure involving genetic intervention to change the sex of the fetus during pregnancy, which has been experimented on animals. In light of the various complications that such a procedure results in, it is deemed impermissible, see Mahrān 2002, 374.

29 See *Majallat al-Majmaʿ al-Fiqhī al-Islāmī* 20:23 (2008): 359–360.

Between Genetic Intervention and Genetic Enhancement: Case of Germline Modification

The unlimited capabilities of the genetic revolution has inspired scientific efforts not only to treat complex and untreatable conditions but also to modify human genetic structure to influence the outcome of the reproductive process. With the development of synthetic biology and its goal to modify and engineer living organisms, scientists are not only trying to understand or explain these living organisms but also to shape them (Lustig 2013, 15). From a religious perspective this explains why genetic manipulation raises the question of "playing God," which has been one of the most commonly used metaphors in bioethical literature with regard to issues ranging from reproductive technologies, genetics, and end of life issues. Within this context, it is used to signify several connotations but at the most basic level it stands for the notion that certain boundaries should not be crossed because they are believed to belong to God's domain (Lustig 2013, 24). From the perspective of Islamic foundational sources, creation is one of the divine acts, which is also captured in the divine names and attributes, one of which is the creator (*al-khāliq*). Islamic scriptural sources often place creation within a restrictive list of actions such as provision (*rizq*), death (*mawt*), and resurrection (*baʿth*) that are usually attributed to God. (e.g. 30:39; 45:26). This exclusive ascription of the act of creation to God within the Islamic moral universe explains the theological as well as ethical-legal problems associated with the notion of changing God's creation. In part, changing God's creation implies a challenge to divine will. Also, the Qurʾān draws a connection between efforts aiming at changing God's creation and the evil plots of the devil (4:119), which is also reinforced in several Prophetic reports characterizing such efforts as deserving divine curse.

In light of its impact on the constitution of future offspring, germline genetic modification often invokes religious reservations associated with the notion of changing God's creation. Moreover, similar to procedures such as reproductive cloning, germline genetic modification raises serious concerns about potential impact on the structure of the family as well as larger social order in the future. Although these procedures have not yet been tested in humans, many researchers argue that it may only be a matter of time.

Muslim discussions on these questions often address potential scenarios, associated ethical-legal implications, and also guidelines that should govern research in this area. Germline genetic modification was already highlighted in some of the collective discussions mentioned above, especially within the various IOMS meetings since the 1980s.[30] In 2006 IOMS convened an international seminar to address the ethical implications of modern genetic and reproductive technologies from an Islamic but also secular as well as interreligious perspectives. The seminar touched on several applications of genetic and reproductive technologies with a particular focus on the family and social institutions.[31] Germline genetic modification was one of the issues that the seminar identified and it featured in different papers as well as group discussions following the main panels. This collective and interdisciplinary discussion revealed the range of theological, legal, or larger social questions that genetic modification raises.

From a theological perspective, germline genetic modification inspires reflections on the concept of creation (*khalq*) and whether such genetic intervention can challenge the exclusive attribution of this act to God. In general, contributions by Muslim scholars during the seminar emphasized the Islamic conception of creation as a divine act. They also emphasized the special place that man occupies within the divinely created universe in his capacity as a vicegerent of the creator, entrusted with the responsibility to develop it following divine instructions and guidelines. Man is created in the best shape and is also endowed with honor and dignity. Other creatures are made subservient to man, who is instructed to interact with them in a careful and conscientious manner. On the basis of this theological background, human scientific efforts should not aim to change this original order of divine creation (al-Saḥmarānī 2008, 1:245-57). Consequently, one of the important guidelines that several participants emphasized with regard to genetic research is the distinction between preventive and therapeutic intervention on the one hand and purely enhancing intervention on the other. Genetic research involving humans should concentrate on preventive and therapeutic purposes only. As far as enhancement efforts are concerned, another distinction is made between enhancement in agricultural and animal research and enhancement involving humans. The first type is considered acceptable in light of its anticipated benefit in improving the quantity and quality of food products. Such research, however, has to comply with standard regulations and research ethics concerning proper treatment of animals and environmental balance. In the case of humans, genetic intervention to manipulate specific characteristics such as

30 See, for example, Sharaf al-Dīn 1983, 136-147; Abū Ghuddah 1983, 148-163 (published also *in Majallat Majmaʿ al-Fiqh al-Islāmī*, 8:3 (1994), 165-178).

31 The seminar had four main themes: humanity and creation; genes, reproductive technologies and the family; social implications of genetic and reproductive technologies; and where boundaries should be drawn. See the summary report in al-Jindī 2008, 115-75.

sex type or skin color would not be acceptable (al-Saḥmarānī 2008, 1:266; Wāṣil 2008, 1:337-55; al-Salāmī 2008, 1:509-30; al-ʿAwaḍī 2008, 605).

Another relevant theological concept that the issue of genetic modification raises is human nature and its uniqueness. Genetic modification raises important questions on possible impact on the original God-created human nature and on the boundaries of acceptable and unacceptable enhancement. For example, the Sudanese scholar Jaʿfar Shaykh Idrīs argues that human nature, as created by God, is both distinct and unmalleable. Genetic modification, if it is implemented in humans for non-therapeutic purposes, would not tamper with this God-created human nature but would end up developing different types of beings (*kāʾināt jadīdah*) (Idrīs 2008, 2:472).[32] The question that he asks is whether it would be of the best interest of humans to allow this to happen. He answers in the negative by exploring all the distinctive features of human nature, which, similar to other creatures in the universe, instinctively recognizes its creator through devotional submission and glorification. Most importantly, the distinctive human nature is marked by its combined (material-spiritual) composition of a body and a spirit, which defines the beginning and end of human life both in this world and the next. This metaphysical dimension of human nature is coded in the inborn disposition (*fitrah*) with which all humans are created and which in essence is inclined towards what is good. In other words, this inborn disposition is not created as neutral but its initial positive inclination is either reinforced or altered through social factors and influences. Although the human body shares many features of the animal nature, it remains also distinct due to its endowment by God with dignity and sanctity, which should be protected and preserved against all types of manipulation. With regard to genetic engineering, a fundamental difference, therefore, should be emphasized between possible therapeutic efforts and non-therapeutic enhancement efforts. While the former should be encouraged, similar to regular medical treatment, the latter, similar to all types of unnecessary tampering with nature, should be banned. Not only these enhancement efforts signify unjustified affront against God but experience shows that such efforts tend to cause more harm than good (Idrīs 2008, 2:479).[33] In addition to common defi-

---

32 On the other hand the Algerian scholar ʿAmmār al-Ṭālibī argues that genetic enhancement could be accommodated as long as it does not result in distorting God's creation. For example, enhancement that aims to improve stamina or immunity should be celebrated, see al-Ṭālibī 2008, 1:423.

33 Some scholars had issues with the notion of challenging divine will and whether man is actually capable of that. For example, the former Muftī of Tunisia Muḥammad Mukhtār al-Salāmī denies this possibility and cites the statement of the second Caliph ʿUmar when he was accused of fleeing from God's decree after his refusal to enter the city afflicted by

nition of *fiṭrah* as a set of inborn characteristics, the Moroccan thinker Ṭāha ʿAbd al-Raḥmān argues that the term can also signify values (*qiyam*) in light of his interpretation of the Qurʾānic verse indicating the original ability that God endowed Adam with (to know all the names, 2:31). ʿAbd al-Raḥmān notes that the etymological root of the term *name* and its meaning in Arabic could refer not only to the ability to distinguish (*tumayyiz*) things but also to evaluate (*tuqayyim*) them. In other words, this original knowledge includes the moral values that are incorporated within the inborn state in which man is created (al-Ṭālibī 2008, 1:423).

From a legal perspective, Muslim scholars often place questions pertaining to reproductive issues within the context of the ultimate objectives of Sharia. Preservation and protection of progeny is considered one of the five essential values that Sharia aims to protect and preserve, along with religion, life, intellect and property (al-Bārr 2008, 1:653-4; Abū Ghuddah 2008, 1:687). On the other hand, with the availability of the various applications of genetic technology, jurists point out a distinction between two types of a Sharia-based ruling (*al-ḥukm al-sharʿī*): *al-ḥukm al-taklīfī* and *al-ḥukm al-waḍʿī* (Abū Ghuddah 2008, 1:688; Abū Ghuddah 1983, 162). While the former delineates assessment of a particular issue along the continuum of the five main categories: obligation, recommendation, neutrality, reprehensibility, and prohibition, the latter specifies relevant causes, conditions, or stipulations. In particular, jurists point out the importance of the second type of rulings regardless of its status on the continuum of the five main categories, which is usually the case with procedures that are deemed prohibited or even reprehensible. They argue that in case a prohibited act is undertaken, this type of ruling will still regulate the implications and consequences of the prohibited act in order to preserve the rights of all parties concerned. This becomes clear in juristic discussions on issues such as reproductive cloning and surrogacy. From an ethical-legal perspective, genetic modification raises the question of guardianship and the boundaries of prospective parents' authority over their prospective children. In other words, would parental guardianship in this case include the ability of parents to manipulate the genetic structure of future children? Would such authority ultimately have implications on the identity of these children and their character? And, to what extent would it change the perception of prospective children as unique and idiosyncratic individuals rather than being another type of consumer goods? In light of these questions, jurists also emphasize the distinction between a preventive or therapeutic intervention, which is seen as warranted and permitted and a non-therapeutic intervention, which is seen as unwar-

---

the plague "we flee from God's decree to God's decree." See al-Jindī 2008, 2:496.

ranted and unpermitted (Abū Ghuddah 2008, 1:698-703).[34] One important point that scholars use to judge the propriety or morality of a particular procedure is the intended objective that it aims to achieve. This criterion is used to distinguish therapeutic or corrective efforts from other non-therapeutic ones. Morality of an objective is also extended to the means through which it will be achieved (Abū Ghuddah 1983, 153). In other words, a moral objective has to be pursued only through moral means. A clear example would be an objective to overcome infertility through artificial insemination. This objective, however, cannot be achieved through a religiously questionable procedure such as gamete donation due to its conflict with the sanctity of the marital relationship.

From a social perspective, genetic modification raises several issues pertaining to social justice and accessibility. If such an option becomes available, critics argue, it would widen the gap between those who can afford it and those who cannot. Moreover, on the long run, enhancement, or at least certain types of it, may even become a standard procedure for the economically privileged, which would further put poorer segments of the society at a disadvantage (Athar 2008, 2:280). The most concrete social implication that many participants raise is the risk of rejuvenating eugenic tendencies and the impact this may have on the definition of what is normal or average, let alone perception of disability and the disabled persons (Badrān and Shāhīn 2008, 2:459; Ebrahim 2008, 2:689). With this new version of eugenics, however, the aspiration to order or design one's babies would not be a far-fetched imagination (Fataḥ Allāh 2008, 2:719). Ultimately, enhancement procedures, similar to other procedures such as reproductive cloning, would seriously impact conventional understanding of family ties and wider social relationships (Ḥatḥūt 2008, 1:198).[35] At the conclusion of the seminar a declaration of principles was issued, which included several points with regard to genetic modification. First, it emphasized the distinction between preventive and therapeutic applications and enhancement applications. Second, another distinction is emphasized between non-human applications and others involving humans. Third, it proscribed any effort aiming at manipulating the human genetic structure, which would adversely affect human personality or distinctive characteristics.[36]

34 Abū Ghuddah uses the term *istibdāl* (replacement) to refer to genetic modification that aims to change basic human qualities or characteristics. See Abū Ghuddah 1983, 154.

35 From another perspective genetic modification also raises all types of speculations about transhumanism and hybrid combinations, see Moazzam 2008, 2:389-408.

36 al-Jindī 2008, 2:747-9. Over the past few decades a number of similar resolutions and statements were also issued to clarify Muslim perspectives on genetic research and various applications of genetic technology. Important examples include the following: recommendations of the 21st IOMS seminar, which was held in 1998 on the theme of ge-

Concluding Remarks: Structure of the Family between Genetic Determinism and Genetic Manipulation

The above analysis shows the power of the genetic technology and its various applications in influencing both existing and prospective family relationships. Such power is expected to only increase in the future, thereby necessitating careful moral evaluation of these various applications. Within the Muslim context, the process of moral assessment is highly dependent on the Islamic normative framework, which is often dominated by the juristic discourses. The examples explored in this chapter, however, reveal the theological vision underlying these juristic discourses. Any systematic effort to provide answers to the questions that the various applications of genetic technology raise would, therefore, need to address the theological assumptions associated with issues such as divine creation, human nature, original disposition, and also scope of human freedom. Moral assessment of the various genetic applications involves meticulous balancing of their anticipated benefits against their potential harms. At this stage of scientific research, genetic testing offers remarkable diagnostic advantages with only limited therapeutic options. The unprecedented diagnostic power of genetic testing is sometimes invoked to support arguments promoting genetic essentialism, determinism, or reductionism, which often ignore environmental as well as other types of factors. On the other hand, genetic applications can enable greater levels of intervention and manipulation of living organisms, as the above discussions concerning impact on prospective family members clearly illustrate. As much as the genetic technology is constantly evolving, resulting in ever increasing number of applications and capabilities, Muslim discourses seek to keep up with these developments by offering at least tentative assessments, most notably as a result of collective deliberations within transnational institutions such as the ones covered in this chapter. Judging from the sizeable volume of publications citing these deliberations (mostly in Arabic), this cumulative body of moral insights would

---

netic engineering and genetic therapy form the Islamic perspective (published in 2000); a decision by the International Islamic Fiqh Academy (of the Organization of Islamic Cooperation, Jeddah) on cloning in its 10th session in 1997, see *Majallat Majmaʿ al-Fiqh al-Islāmī* 10:3 (1997), 417-32; a decision by the Islamic Fiqh Council (of the Muslim World League, Mecca) on genetic engineering in its 15th session in 1998, see *Qarārāt al-Majmaʿ al-Fiqhī al-Islāmī*, 313-5; a decision by the Islamic Fiqh Council on genetic fingerprinting in its 16th session in 2002, see *Qarārāt al-Majmaʿ al-Fiqhī al-Islāmī*, 345-8; a decision by the Islamic Fiqh Council on genetic blood diseases in its 17th session in 2003, see *Qarārāt al-Majmaʿ al-Fiqhī al-Islāmī*, 345-6; a decision by the International Islamic Fiqh Academy on genetics, genetic engineering and human genome in its 21st session in 2013.

be indispensable for exploring Islamic perspectives on these new questions. Already some broad lines can be identified in light of this literature. For example, on the issue of premarital genetic testing, Muslim scholars are unanimous on the importance of educating the public and raising awareness about this procedure, especially in places where consanguineous marriage is a common practice. Jurists, however, are divided on whether it should be enforced but even those who argue for its enforceability insist that this should not impact prospective couples' decision to proceed with marriage even in case of positive results. With regard to fetal sex selection, most jurists approve it for medical purposes. They disagree, however, when the procedure is undertaken for family balancing. Finally, with regard to germline genetic modification, another distinction is made between preventive and therapeutic intervention on the one hand and non-therapeutic enhancement on the other. A near consensus exists on the permissibility of the former case only.

References

ʿAbd Allāh, ʿAbd Allāh Muḥammad. 2000. *"Naẓarāt Fiqhiyyah fī al-Jīnūm al-Basharī, al-Handasah al-Wirāthiyyah, al-ʿIlāj al-Jīnī."* In *Ruʾyah Islāmiyyah li-baʿḍ al-Mushkilāt al-Ṭibbiyyah al-Muʿāṣirah, Thabt Kāmil li-Aʿmāl Nadwat al-Wirāthah wa al-Handasah al-Wirāthiyyah wa al-Jīnūm al-Basharī wa al-ʿIlāj al-Jīnī – Ruʾyah Islāmiyyah*, edited by ʿAbd al-Raḥmān al-ʿAwaḍī and Aḥmad Rajāʾī al-Jundī, 2 vols., 2:727-70. Kuwait: al-Munaẓẓamah al-Islāmiyyah li al-ʿUlūm al-Ṭibbiyyah.

ʿAbd Allāh, Ḥasan Ṣalāḥ al-Ṣaghīr. 2007. *Madá Mashrūʿiyyat al-Ilzām bil-Faḥṣ al-Ṭibbī qabl al-Zawāj, Dirāsah Muqāranah.* Alexandria: Dār al-Jāmiʿah al-Jadīdah.

Abū Ghuddah, ʿAbd al-Sattār. 1983. "Madá Sharʿiyyat al-Taḥakkum fī Muʿṭayāt al-Wirāthah." In *al-Islām wa al-Mushkilāt al-Ṭibbiyyah al-Muʿāṣirah, al-Injāb fī ḍawʾ al-Islām*, edited by Aḥmad Rajāʾī al-Jundī. Kuwait: al-Munaẓẓamah al-Islāmiyyah li al-ʿUlūm al-Ṭibbiyyah.

Abū Ghuddah, ʿAbd al-Sattār. 2000. "al-Muwākabah al-Sharʿiyyah li-Muʿṭayāt al-Handasah al-Wirāthiyyah." In *Ruʾyah Islāmiyyah li-baʿḍ al-Mushkilāt al-Ṭibbiyyah al-Muʿāṣirah, Thabt Kāmil li-Aʿmāl Nadwat al-Wirāthah wa al-Handasah al-Wirāthiyyah wa al-Jīnūm al-Basharī wa al-ʿIlāj al-Jīnī – Ruʾyah Islāmiyyah*, edited by ʿAbd al-Raḥmān al-ʿAwaḍī and Aḥmad Rajāʾī al-Jundī, 2(1):573-94. Kuwait: al-Munaẓẓamah al-Islāmiyyah li al-ʿUlūm al-Ṭibbiyyah.

Abū Ghuddah, ʿAbd al-Sattār. 2008. "al-Wirāthah al-Bashariyyah (wa al-Jīnāt) wa Tuknulūjyā al-Takāthur wa Mawqif al-Sharīʿah al-Islāmiyyah minhā." In *al-Nadawah al-ʿĀlamiyyah ḥawla al-Wirāthah wa al-Takāthur al-Basharī wa Inʿikāsātuhā, Ruʾyat al-Adiyān al-Samāwiyyah wa wijhat naẓar al-ʿAlmāniyyah*, edited by Aḥmad

Rajāʾī al-Jundī, (2), 1:685-714. Kuwait: Islamic Organization for Medical Sciences.

Al Aqeel, Aida. 2007. "Islamic Ethical Framework for Research into and Prevention of Genetic Diseases." *Nature Genetics* 39: 1293-8.

Athar, Shahid. 2008. "Tiqniyāt al-Tanāsul al-Basharī wa al-Handasah al-Wirāthiyyah min al-Manẓūr al-ʿAlmānī." In *al-Nadawah al-ʿĀlamiyyah ḥawla al-Wirāthah wa al-Takāthur al-Basharī wa Inʿikāsātuhā, Ruʾyat al-Adiyān al-Samāwiyyah wa wijhat naẓar al-ʿAlmāniyyah*, edited by Aḥmad Rajāʾī al-Jundī, 2(2): 271-80. Kuwait: Islamic Organization for Medical Sciences.

al-ʿAwaḍī, Ṣiddīqah. 2008. "al-Wirāthah al-Bashariyyah wa Tuknūlūjiyā al-Takāthur wa Inʿikās dhalika ʿalá al-Usrah." In *al-Nadawah al-ʿĀlamiyyah ḥawla al-Wirāthah wa al-Takāthur al-Basharī wa Inʿikāsātuhā, Ruʾyat al-Adiyān al-Samāwiyyah wa wijhat naẓar al-ʿAlmāniyyah*, edited by Aḥmad Rajāʾī al-Jundī, 2(1): 603-10. Kuwait: Islamic Organization for Medical Sciences.

Badrān, Ibrāhīm and Muḥammad ʿAbd al-Ḥamīd Shāhīn. 2008. "Khalq al-Insān wa Insāniyyatuh: al-Khuṭūṭ al-Ḥamrāʾ matá wa ayna Tuḍaʿ." In *al-Nadawah al-ʿĀlamiyyah ḥawla al-Wirāthah wa al-Takāthur al-Basharī wa Inʿikāsātuhā, Ruʾyat al-Adiyān al-Samāwiyyah wa wijhat naẓar al-ʿAlmāniyyah*, edited by Aḥmad Rajāʾī al-Jundī, 2(2): 449-60. Kuwait: Islamic Organization for Medical Sciences.

al-Bārr, Muḥammad ʿAlī. 2000. "Naẓrah Fāḥiṣah lil-Fuḥūṣāt al-Ṭibbiyyah al-Jīniyyah: al-Faḥṣ qabla al-Zawāj wa al-Istishārah al-Wirāthiyyah." In *Ruʾyah Islāmiyyah li-baʿḍ al-Mushkilāt al-Ṭibbiyyah al-Muʿāṣirah, Thabt Kāmil li-Aʿmāl Nadwat al-Wirāthah wa al-Handasah al-Wirāthiyyah wa al-Jīnūm al-Basharī wa al-ʿIlāj al-Jīnī – Ruʾyah Islāmiyyah*, edited by ʿAbd al-Raḥmān al-ʿAwaḍī and Aḥmad Rajāʾī al-Jundī, 2(2): 621-62. Kuwait: al-Munaẓẓamah al-Islāmiyyah li al-ʿUlūm al-Ṭibbiyyah.

al-Bārr, Muḥammad ʿAlī. 2008. "Baḥth ʿan Tiqniyāt al-Wirāthah wa al-Takāthur al-Basharī min al-Manẓūr al-Islāmī." In *al-Nadawah al-ʿĀlamiyyah ḥawla al-Wirāthah wa al-Takāthur al-Basharī wa Inʿikāsātuhā, Ruʾyat al-Adiyān al-Samāwiyyah wa wijhat naẓar al-ʿAlmāniyyah*, edited by Aḥmad Rajāʾī al-Jundī, 2(1): 651-82. Kuwait: Islamic Organization for Medical Sciences.

Barry, Vincent. 2012. *Bioethics in a Cultural Context.* Boston: Wadsworth.

Buḥālah, al-Ṭayyib. 2010. *al-Fuḥūṣāt al-Ṭibbiyyah qabla al-Zawāj.* Manṣurah: Dār al-Fikr wa al-Qānūn.

Chadwick, Ruth. 2006. "Personal Identity: Genetics and Determinism." In *Living with the Genome: Ethical and Social Aspects of Human Genetics*, edited by Angus Clarke and Flo Ticehurst, 255-8. New York: Palgrave.

Davis, Dena. 2006. "Sex Selection." In *Living with the Genome: Ethical and Social Aspects of Human Genetics*, edited by Angus Clarke and Flo Ticehurst, 291-5. New York: Palgrave.

Ebrahim, Abou El-Fadl Muhsin. 2008. "al-Ḥaml al-Badīl min al-Manẓūr al-Akhlāqī al-Qānūnī al-Ijtimāʿi wa al-Islāmī." In *al-Nadawah al-ʿĀlamiyyah ḥawla al-Wirāthah wa al-Takāthur al-Basharī wa Inʿikāsātuhā, Ruʾyat al-Adiyān al-Samāwiyyah wa wijhat naẓar al-ʿAlmāniyyah,* edited by Aḥmad Rajāʾī al-Jundī, 2(2): 669-93. Kuwait: Islamic Organization for Medical Sciences.

El Shanti, Hatem, Lotfi Chouchane, Ramin Badii, Imed Eddine Gallouzi, and Paolo Gasparini. 2015. "Genetic Testing and Genomic Analysis: a Debate on Ethical, Social, and Legal Issues in the Arab World with a Focus on Qatar." *Journal of Translational Medicine* 13: 358-68

Emslie, Carol and Kate Hunt. 2006. "Genetic Susceptibility." In *Living with the Genome: Ethical and Social Aspects of Human Genetics,* edited by Angus Clarke and Flo Ticehurst, 102-7. New York: Palgrave.

Fataḥ Allāh, Maḥmūd. 2008. "Tiqniyyāt al-Injāb wa al-Jīnūm al-Basharī wa al-Usrah." In *al-Nadawah al-ʿĀlamiyyah ḥawla al-Wirāthah wa al-Takāthur al-Basharī wa Inʿikāsātuhā, Ruʾyat al-Adiyān al-Samāwiyyah wa wijhat naẓar al-ʿAlmāniyyah,* edited by Aḥmad Rajāʾī al-Jundī, 2(2): 697-723. Kuwait: Islamic Organization for Medical Sciences.

Ḥatḥūt, Ḥassān. 1983. "al-Taḥakkum fī Jins al-Janīn." In *al-Islām wa al-Mushkilāt al-Ṭibbiyyah al-Muʿāṣirah, al-Injāb fī ḍawʾ al-*Islām, edited by Aḥmad Rajāʾī al-Jundī, 37-8. Kuwait: al-Munaẓẓamah al-Islāmiyyah li al-ʿUlūm al-Ṭibbiyyah.

Ḥatḥūt, Ḥassān. 2008. "Tiqniyyāt ʿIlm al-Wirāthah wa al-Tanāsul ʿinda al-Insān bayna al-Manẓūr al-Dīnī wa al-Manẓūr al-ʿAlmānī." In *al-Nadawah al-ʿĀlamiyyah ḥawla al-Wirāthah wa al-Takāthur al-Basharī wa Inʿikāsātuhā, Ruʾyat al-Adiyān al-Samāwiyyah wa wijhat naẓar al-ʿAlmāniyyah,* edited by Aḥmad Rajāʾī al-Jundī, 2(1): 179-206. Kuwait: Islamic Organization for Medical Sciences.

al-Ḥāzimī, Muḥsin bin ʿAlī Fāris. 2000. "al-Istirshād al-Wirāthī: Ahammiyyat al-Tawʿiyyah al-Wiqāʾiyyah wa Maḥādhīruh al-Ṭibbiyyah wa al-Akhlāqiyyah." In *Ruʾyah Islāmiyyah li-baʿḍ al-Mushkilāt al-Ṭibbiyyah al-Muʿāṣirah, Thabt Kāmil li-Aʿmāl Nadwat al-Wirāthah wa al-Handasah al-Wirāthiyyah wa al-Jīnūm al-Basharī wa al-ʿIlāj al-Jīnī – Ruʾyah Islāmiyyah,* edited by ʿAbd al-Raḥmān al-ʿAwaḍī and Aḥmad Rajāʾī al-Jundī, 2(2): 665-700. Kuwait: al-Munaẓẓamah al-Islāmiyyah li al-ʿUlūm al-Ṭibbiyyah.

ʿIbādah, Ḥātim Amīn Muḥammad. 2010. *al-ʿIlāj al-Jīnī wa al-Fuḥūṣ al-Wirāthiyyah.* Alexandria: Dār al-Fikr al-Jāmiʿī.

Ibn al-Qayyim. 1998. *Zād al-Maʿād fī Hady Khayr al-ʿIbād,* 6 vols., edited by Shuʿayb al-Arnaʾūṭ and ʿAbd al-Qādir al-Arnaʾūṭ. Beirut: Muʾāssasat al-Risālah.

Ibn Qudāmah, ʿAbd Allāh ibn Aḥmad. 1997. *al-Mughnī,* edited by ʿAbd Allāh ibn ʿAbd al-Muḥsin al-Turkī and ʿAbd al-Fattāḥ Muḥammad al-Ḥulw, 15 vols. Riyaḍ: Dār ʿAlam al-Kutub.

Idrīs, Jaʿfar Shaykh. 2008. "Insāniyyatunā Jawhar Thābit wa Wāqiʿ Yataghayyar." In *al-Nadawah al-ʿĀlamiyyah ḥawla al-Wirāthah wa al-Takāthur al-Basharī wa Inʿikāsātuhā, Ruʾyat al-Adiyān al-Samāwiyyah wa wijhat naẓar al-ʿAlmāniyyah,* edited by Aḥmad Rajā'ī al-Jundī, 2(2): 471-80. Kuwait: Islamic Organization for Medical Sciences.

al-Jundī, Aḥmad Rajā'ī, ed. 1983. *al-Islām wa al-Mushkilāt al-Ṭibbiyyah al-Muʿāṣirah, al-Injāb fī ḍaw' al-Islām.* Kuwait: al-Munaẓẓamah al-Islāmiyyah li al-ʿUlūm al-Ṭibbiyyah.

al-Jundī, Aḥmad Rajā'ī, ed. 2008. *al-Nadawah al-ʿĀlamiyyah ḥawla al-Wirāthah wa al-Takāthur al-Basharī wa Inʿikāsātuhā, Ru'yat al-Adiyān al-Samāwiyyah wa wijhat naẓar al-ʿAlmāniyyah*, 2 vols. Kuwait: Islamic Organization for Medical Sciences.

al-Ludaʿmī, Tamām Muḥmmad. 2011. *al-Jīnāt al-Bashariyyah wa Taṭbīqātuhā.* Herndon, VA: The International Institute of Islamic Thought.

Lustig, Andrew. 2013. "Appeals to Nature and the Natural in Debates about Synthetic Biology." In *Synthetic Biology and Morality* edited by Gregory E. Kaebnick and Thomas H. Murray. Cambridge, MA: MIT Press.

al-Madḥajī, Muḥammad ibn Hā'il ibn Ghaylān. 2011. *Aḥkām al-Nawāzil fī al-Injāb.* 3 vols. Riyad: Dār Kunūz Ishbīlyā lil-Nashr wa al-Tawzīʿ.

Mahrān, Al-Sayyid Maḥmūd ʿAbd al-Raḥīm. 2002. *al-Aḥkām al-Sharʿiyyah wa al-Qānūniyyah lil-Tadakhkhul fī ʿAwāmil al-Wirāthah wa al-Takāthur.* Cairo: NP.

al-Maymān, Nāṣir ibn ʿAbd Allāh. 2000. "al-Irshād al-Jīnī: Ahmiyyatuh, Āthāruh, Maḥādhīruh." In *Ru'yah Islāmiyyah li-baʿḍ al-Mushkilāt al-Ṭibbiyyah al-Muʿāṣirah, Thabt Kāmil li-Aʿmāl Nadwat al-Wirāthah wa al-Handasah al-Wirāthiyyah wa al-Jīnūm al-Basharī wa al-ʿIlāj al-Jīnī – Ru'yah Islāmiyyah*, edited by ʿAbd al-Raḥmān al-ʿAwaḍī and Aḥmad Rajā'ī al-Jundī, 2(2): 797-824. Kuwait: al-Munaẓẓamah al-Islāmiyyah li al-ʿUlūm al-Ṭibbiyyah.

Moazzam, Farhat. 2008. "al-Handasah al-Wirāthiyyah wa al-ʿAdālah al-Ijtimāʿiyyah wa Mustaqbal al-Insāniyyah: Iltiqā' al-Makhāwif al-Dīniyyah wa al-ʿAlmāniyyah." In *al-Nadawah al-ʿĀlamiyyah ḥawla al-Wirāthah wa al-Takāthur al-Basharī wa Inʿikāsātuhā, Ru'yat al-Adiyān al-Samāwiyyah wa wijhat naẓar al-ʿAlmāniyyah*, edited by Aḥmad Rajā'ī al-Jundī, 2(2): 389-408. Kuwait: Islamic Organization for Medical Sciences.

Modell, Bernadette. 2006. "Carrier Screening for Inherited Hemoglobin Disorders in Cyprus and the United Kingdom." In *Living with the Genome: Ethical and Social Aspects of Human Genetics*, edited by Angus Clarke and Flo Ticehurst, 114-21. New York: Palgrave.

al-Muḥammadī, ʿAlī Muḥammad Yūsuf. 2005. *Buḥūth Fiqhiyyah fī Masā'il Ṭibbiyyah Muʿāṣirah.* Beirut: Dār al-Bashā'ir al-Islāmiyyah.

al-Qaradāghī, ʿAlī Muḥyī al-Dīn and ʿAlī Yūsuf al-Muḥammadī. 2008. *Fiqh al-Qaḍāyā al-Ṭibbiyyah al-Muʿāṣirah.* Beirut: Dār al-Bashā'ir al-Islāmiyyah.

Qarārāt al-Majmaʿ al-Fiqhī al-Islāmī bi-Makkah al-Mukarramah: al-Dawarāt min al-Ūlá ilá al-Sādisah ʿAshrah, al-Qarārāt min al-Awwal ilá al-Khāmis wa al-Tisʿīn – 1398-1422/1977-2002 (Mecca: Rābiṭat al-ʿĀlam al-Islāmī, n.d).

al-Rashīdī, Fahd Saʿd. 2011. "Ikhtiyār Jins al-Janīn bi al-Wasā'il al-Ṭabīʿiyyah wa al-Mikhbariyyah, Dirāsah Fiqhiyyah Ṭibbiyyah." *Majallat al-Sharīʿah wa al-Dirāsāt al-Islāmiyyah* 26: 567-624.

Reilly, Philip R. 2006. "Informed Consent in Human Genetic Research." In *Living with the Genome: Ethical and Social Aspects of Human Genetics*, edited by Angus Clarke and Flo Ticehurst, 64-9. New York: Palgrave.

Rifkin, Jeremy. 2006. "Patenting of Genes: A Personal View." In *Living with the Genome: Ethical and Social Aspects of Human Genetics*, edited by Angus Clarke and Flo Ticehurst, 46-8. New York: Palgrave.

Rose, Steven P. R. 2006. "Genetic Reductionism and Autopoiesis." In *Living with the Genome: Ethical and Social Aspects of Human Genetics*, edited by Angus Clarke and Flo Ticehurst, 251-4. New York: Palgrave.

al-Saḥmarānī, Asʿad. 2008. "al-Insāniyyah wa al-Khalq, al-ʿĀlam ʿalá al-Fiṭrah." In *al-Nadawah al-ʿĀlamiyyah ḥawla al-Wirāthah wa al-Takāthur al-Basharī wa Inʿikāsātuhā, Ru'yat al-Adiyān al-Samāwiyyah wa wijhat naẓar al-ʿAlmāniyyah*, edited by Aḥmad Rajā'ī al-Jundī, 2(1): 245-77. Kuwait: Islamic Organization for Medical Sciences.

al-Salāmī, Muḥammad al-Mukhtār. 2008. "al-Wirāthah al-Bashariyyah wa al-Takāthur al-Basharī wa Inʿikāsātuhā, Ru'yat al-Adiyān al-Kitābiyyah al-Thalāthah wa al-Taṣawwur al-ʿAlmānī." In *al-Nadawah al-ʿĀlamiyyah ḥawla al-Wirāthah wa al-Takāthur al-Basharī wa Inʿikāsātuhā, Ru'yat al-Adiyān al-Samāwiyyah wa wijhat naẓar al-ʿAlmāniyyah*, edited by Aḥmad Rajā'ī al-Jundī, 2(1) :509-30. Kuwait: Islamic Organization for Medical Sciences.

Shabana, Ayman. 2017. "Empowerment of Women Between Law and Science: Role of Biomedical Technology in Enhancing Equitable Gender Relations in the Muslim World." *Hawwa: Journal of Women of the Middle East and the Islamic World* 15:193-218.

Shabīhunā, Ḥamdātī Mā' al-ʿAynayn. 2000. "al-Amrāḍ al-Latī yajibu an yakuna al-Ikhtibār al-Wirāthī fīhā Ijbāriyan." In *Ru'yah Islāmiyyah li-baʿḍ al-Mushkilāt al-Ṭibbiyyah al-Muʿāṣirah, Thabt Kāmil li-Aʿmāl Nadwat al-Wirāthah wa al-Handasah al-Wirāthiyyah wa al-Jīnūm al-Basharī wa al-ʿIlāj al-Jīnī – Ru'yah Islāmiyyah*, edited by ʿAbd al-Raḥmān al-ʿAwaḍī and Aḥmad Rajā'ī al-Jundī, 2(2): 943-56. Kuwait: al-Munaẓẓamah al-Islāmiyyah li al-ʿUlūm al-Ṭibbiyyah.

Sharaf al-Dīn, Aḥmad. 1983. "Asālīb Diktāturiyyat al-Bayūlūjiā fī al-Mīzān al-Sharʿī." In *al-Islām wa al-Mushkilāt al-Ṭibbiyyah al-Muʿāṣirah, al-Injāb fī ḍaw' al*-Islām, edited by Aḥmad Rajā'ī al-Jundī, 136-47. Kuwait: al-Munaẓẓamah al-Islāmiyyah li al-ʿUlūm al-Ṭibbiyyah.

al-Sharīf, Muḥammad ʿAbd al-Ghaffār. "Ḥukm al-Kashf al-Ijbārī *ʿan al-Amrāḍ al-Wirāthiyyah.*" In *Ru'yah Islāmiyyah li-baʿḍ al-Mushkilāt al-Ṭibbiyyah al-Muʿāṣirah, Thabt Kāmil li-Aʿmāl Nadwat al-Wirāthah wa al-Handasah al-Wirāthiyyah wa al-Jīnūm al-Basharī wa al-ʿIlāj al-Jīnī – Ru'yah Islāmiyyah*, edited by ʿAbd al-Raḥmān al-ʿAwaḍī and Aḥmad Rajā'ī al-Jundī, 2(2): 959-74. Kuwait: al-Munaẓẓamah al-Is-

lāmiyyah li al-ʿUlūm al-Ṭibbiyyah.
Shubayr, Muḥammad ʿUthmān. 2001. "Mawqif al-Islām min al-Amrāḍ al-Wirāthiyyah." In *Dirāsāt Fiqhiyyah fī Qaḍāyā Ṭibbiyyah Muʿāṣirah*, 2(1) :333-47. Ammān: Dār al-Nafāʾis.
al-Shuwayrakh, Saʿd ibn ʿAbd al-ʿAzīz ibn ʿAbd Allāh. 2007. *Aḥkām al-Handasah al-Wirāthiyyah.* al-Riyad: Dār Kunūz Ishbīlyā lil-Nashr wa al-Tawzīʿ.
Stearns, Justin. 2011. *Infectious Ideas: Contagion in Pre-modern Islamic and Christian Thought in the Western Mediterranean.* Baltimore: John Hopkins University Press.
al-Ṭalibī, Ammar. 2008. "al-Insān, al-Fitrah, al-Ṭabīʿah wa la-Tiqniyyah." In *al-Nadawah al-ʿĀlamiyyah ḥawla al-Wirāthah wa al-Takāthur al-Basharī wa Inʿikāsātuhā, Ruʾyat al-Adiyān al-Samāwiyyah wa wijhat naẓar al-ʿAlmāniyyah*, edited by Aḥmad Rajāʾī al-Jundī, 2(1): 371-430. Kuwait: Islamic Organization for Medical Sciences.
ʿUthmān, Muḥammad Rǻfat. 2000. "Naẓrah Fiqhiyyah fī al-Amrāḍ al-Latī Yajibu an yakuna al-Ikhtibār al-Wirāthī fīhā Ijbāriyan kama tara baʿḍ al-Hayʾāt al-Ṭibbiyyah." In *Ruʾyah Islāmiyyah li-baʿḍ al-Mushkilāt al-Ṭibbiyyah al-Muʿāṣirah, Thabt Kāmil li-Aʿmāl Nadwat al-Wirāthah wa al-Handasah al-Wirāthiyyah wa al-Jīnūm al-Basharī wa al-ʿIlāj al-Jīnī – Ruʾyah Islāmiyyah*, edited by ʿAbd al-Raḥmān al-ʿAwaḍī and Aḥmad Rajāʾī al-Jundī, 2(2): 915-39. Kuwait: al-Munaẓẓamah al-Islāmiyyah li al-ʿUlūm al-Ṭibbiyyah.
Van Gelder, Geert Jan. 2005. *Close Relationships: Incest and Inbreeding in Classical Arabic Literature.* London: I.B. Tauris.
Vaughn, Lewis. 2010. *Bioethics: Principles, Issues, and Cases.* Oxford: Oxford University Press.
Wāṣil, Naṣr Farīd. 2008. "Khalq al-Insān bi-Ṣuratihi al-Ṭabīʿiyyah," In *al-Nadawah al-ʿĀlamiyyah ḥawla al-Wirāthah wa al-Takāthur al-Basharī wa Inʿikāsātuhā, Ruʾyat al-Adiyān al-Samāwiyyah wa wijhat naẓar al-ʿAlmāniyyah*, edited by Aḥmad Rajāʾī al-Jundī, 2(1) :337-55. Kuwait: Islamic Organization for Medical Sciences.
al-Yābis, Haylah bint ʿAbd al-Raḥmān ibn Muḥammad. 2012. *al-Amrāḍ al-Wirāthiyyah: Ḥaqīqatuhā wa Aḥkāmuhā fī al-Fiqh al-Islāmī.* 2 vols. Ryadh: Dār Kunūz Ishbīliyyā.
al-Zuhaylī, Muḥammad. 2000. "al-Irshād al-Jīnī." In *Ruʾyah Islāmiyyah li-baʿḍ al-Mushkilāt al-Ṭibbiyyah al-Muʿāṣirah, Thabt Kāmil li-Aʿmāl Nadwat al-Wirāthah wa al-Handasah al-Wirāthiyyah wa al-Jīnūm al-Basharī wa al-ʿIlāj al-Jīnī – Ruʾyah Islāmiyyah*, edited by ʿAbd al-Raḥmān al-ʿAwaḍī and Aḥmad Rajāʾī al-Jundī, 2(2): 773-93. Kuwait: al-Munaẓẓamah al-Islāmiyyah li al-ʿUlūm al-Ṭibbiyyah.

PART 2

Genomics and Rethinking Human Nature

CHAPTER 4

Conceptualizing the Human Being: Insights from the Genethics Discourse and Implications for Islamic Bioethics

Aasim I. Padela[1]

Introduction[2]

Bioethics is a complicated field of inquiry. For one, its subject matter is vast as the "bio-" in bioethics pertains to a broad range of human activities involving the biomedical sciences such as clinical care, research, and policy-making. At the same time, the "-ethics" part of the term brings in multiplicity as it draws in many different moral theories and reasoning modalities utilized by experts to assess the morality of practices and policies relevant to biomedicine. In addition, bioethics as a field of study breeds further variety as it engages multi- and inter-disciplinary perspectives on any particular issue. Consequently, reflecting a field that sits at the interface of many different areas of knowledge, and a discipline that has ambiguous boundaries, bioethics discourse is complex and multi-faceted.

Although there are a great number of issues discussed in the bioethics literature and a great number of disciplinary experts offer perspectives on these issues, contemporary bioethics discourse fundamentally deals with the ethical (and legal) issues that arise from the interaction of the biomedical sciences with society. Thus bioethicists, in their normative mode, make claims about to

1 Aasim I. Padela, MD MSc, The University of Chicago, 5841 S Maryland Ave, MC 5068, Chicago, IL 60637 USA, apadela@uchicago.edu

2 Deep gratitude is due to Dr. Mariel Kalkach Aparacio who assisted in the conduct of this research project. The author also expresses gratitude to Profs. Oliver Leaman Hub Zwart for their insightful review of earlier versions of this manuscript and to participants in the seminar at which it was presented. This paper was presented at the Islamic Ethics and the Genome Question Seminar at the Research Center for Islamic Legislation and Ethics (CILE) in Doha, Qatar in April 2017.

how we ought to behave and structure society. Yet in as much as bioethicists (and bioethics discourse in general) speak to the ethico-legal, they also comment, directly or indirectly, on the ontological. Arguably, the identification of a bioethical problem, the framing of the dimensions of that problem, the ethical concepts deployed to work through the problem, and even the interpretation of the biomedical science or technology that is at the center of the problem are connected to understandings of what sorts of things human beings are. In other words, ontological claims about the nature of things can be carried with bioethical arguments.

By way of example consider the moral assessment of a physician intentionally acting against a patient's informed and voluntary decisions about healthcare. One may formulate ethical problem(s) in many different ways, and use various philosophical concepts to solve the problem. For some the dominant issue is the harm to a patient inflicted by an undesired therapy (a patient focus), others may consider the core issue to be the lack of respect accorded to a patient's choice (a focus on the relationship between patients and doctors), and still others may consider the physician's paternalism to be morally objectionable (a focus on the physician). Clearly all of these different vantage points are related. Yet, the starting point for analyses and the accompanying ethical concepts invoked can belie an ontological claim about the nature of patients. Illustratively, a version of that claim could proceeds as follows: Patients are persons, and persons are members of the human species. Humans, in turn, are living organisms with a certain set of distinguishing capacities (or essential features) that include the potential for ratiocination and will making. These capacities are realized when making healthcare decisions. Physicians likewise are humans and equivalent to patients in essence. Therefore, all else equal, acting against a patient's will is wrong because it disrespects the essence of the patient. Another formulation might invoke the notion of humans being endowed with dignity, and that having dignity differentiates things that cannot be instrumentalized from things that can be. Consequently, should a physician intentionally act against a patient's informed and voluntary decision they are, in effect, instrumentalizing the patient in the pursuit of an interest that the patient has rejected. Consequently, the patient's dignity has been assaulted. The point here is not the logic of these claims, rather it is to highlight that ontologies of the human can undergird bioethical views. Within the discourse these ontologies might be implicit or unspoken but can be nonetheless fundamental.

As a result, religious scholars engaging with the contemporary (and largely secular) bioethics discourse need to not only understand the biomedical science and technology involved with the moral issue at hand and the societal and legal implications of that science and/or technology, but also need to evaluate whether particular ontologies are impressing upon ethical arguments presented in the discourse. If a "secular" ontology prefigures ethical debates, scholars of religion may need to address that ontology alongside the ethics of the matter because religious ontologies can shape distinctive moral worldviews.

With this thesis as an introduction, this paper presents several ontologies of the human that inhabit the "genethics" literature.[3] These ontologies were identified through qualitative content analyses of a systematic literature review of the bioethics literature. Prior to discussing these methods and our findings (the ontologies), a few provisos are in order. First there is considerable debate as to whether ontology precedes ethical deliberation, or whether ethical notions are contained within social structures that define relationships between beings. In other words do (and must) we know what and whom we are before defining ethical obligations related to other entities, or do ethical obligations define us even before we recognize who and what we are. This debate is found within the medical ethics literature where some bioethicists ground theories of medical ethics within ontologies of the living body (Pellegrino and Thomasma 1981), and others hold that a moral philosophy for medicine starts with a "living heteronomy (that) constitutes the basis of patient-physician relation..." (Tiemersma 1987, 133). This paper operates out of the view that ontology and ethics are related, and that for religious traditions (particularly Islam) this linkage is important and fundamental. However the paper does *not* assert that ethics and ontology must *always* be related, or that all of the contemporary bioethics literature proceeds from ontological claims. Thus those seeking to find the bioethical literature replete with discussions on ontology will be disappointed, as will those seeking to find this paper to offer plentiful snippets of text demarcating how ontology informs the ethical in genethics debates. Rather this paper presents conceptualizations of the human being that emerged during our examination (details below) of the bioethics literature; that can be theorized up from the bioethical deliberation contained therein. Our study further proposes that connections between ontology and ethical concepts can

---

3 Genethics initially referred to "the study of the ethical issues that arise out of the science of genetics and the uses of genetic technologies" but presently also encompasses ethical issues relating to genomics. See BM Knoppers. From medical ethics to genethics. Lancet 2000:356, T Lewens. What is genethics? J Med Ethics 2004(30): 326-328, D Heyd. Genethics: Moral Issues in the Creation of People, University of California Press 1992. J Burley and J Harris (eds). A Companion to Genethics. Blackwell Publishing 2004.

reasonably be inferred, even if these links are subliminal to the authors of specific articles. We will demonstrate these particulars below.

Another important point is that the paper asserts that the Islamic tradition has distinctive ways of explaining reality, describing relationships between human beings and other entities, and of moral reasoning. To be sure there is no singular Islam; Islamic theological doctrines and ethico-legal schools are many and a pluralistic orthopraxy constitutes the intellectual tradition. Yet as a tradition that is distinguished from other systems of thought and practice by its scripture and authority structures, one can make claims about there being "Islamic" theological, ethical, and ontological frameworks. While the hallmarks of an "Islamic" system can be debated, and a multiplicity of "Islamics" may be advanced, the paper contends there to exists Islamic worldviews arising out of its scriptural texts. This claim bears underscoring because this paper discusses ontologies of the human and one might argue that the reality of the human being is singular and shared by religious adherents and secularists alike. Our point is not to suggest that there are different types of humans inhabiting the earth some Islamic and some not, rather that the Islamic tradition might offer descriptions of human nature and its essence that are in some way distinctive and different, and that these differences are morally significant for genethics discourses.

Bearing these qualifications in mind, the paper proceeds as follows. The next section details the sources of study and methods by which these ontologies were identified. The subsequent section discusses how the ethical terms and concepts contained within articles match up with a particular ontological stance about the human being. The final section of the paper outlines how these ontologies implicate Islamic theology and ethics, and outlines critical questions Islamic bioethicists must address as they provide Islamic perspectives on genetic and genomic interventions.

Sources and Methods

Several standard social scientific methods common to health research, each with their particular research aim, were utilized in this study. First, a systematic literature review was conducted to identify the major themes and topics of the genethics discourse. Once the ethics topics and concepts were thematically grouped, qualitative content analytic methods were used to identify potential ontologies that could provide explanatory links between ethics concepts and discussion themes (Miles and Huberman 1994, Crabtree and Miller 1999, Corbin, Strauss, and Strauss 2008). Greater details are provided below.

Literature Search Strategy

The genethics discourse is vast, contains multiple different sources for study, and a variety of research approaches can be taken to canvass the discourse. Our primary goal was to identify the major topics of discussion (domains) within the genethics literature, and to catalog the ethical concepts deployed in these debates. Accordingly a systematic literature review was undertaken as a method to capture scholarly discussions among bioethicists. We decided to focus upon journal articles because these are often more timely and contain more concise arguments than books, and are more scholarly than public pieces. Furthermore the MEDLINE database was purposively selected because it contains the greatest number of peer-reviewed journal articles on the life sciences, is globally accessible, and is the primary literature source for clinicians, biomedical researchers, and bioethicists (Falagas et al. 2008)

The OVID interface was used to carry out a systematic literature review of MEDLINE on December 19th 2016. A conjunction of two search terms was used to retrieve pertinent articles, the first term representing genomics/genetics, and the second term restricting articles to those contained within leading bioethics journals. These terms were combined using the Boolean operator "and." Specifically "genomics" as the first term was exploded using MESH headings, keywords, and subheadings to include the following terms "genomics", "genetic therapy", "genetics", "human genome project", "gene therapy" and" genetic research." All of these terms were entered into the search string using the Boolean operator "or". Also using the Boolean operator "or" the second search term limited the literature retrieval to articles contained within the top ten bioethics journals according to the h5-index score. These journals were the *Journal of Medical Ethics*, the *American Journal of Bioethics, Journal of Bioethics,* the *Hastings Center Report, BMC Medical Ethics,* the *Journal of Clinical Ethics, Cambridge Quarterly of Healthcare Ethics,* the *Kennedy Institute of Ethics Journal,* the *Journal of Bioethical Inquiry,* and *Theoretical Medicine and Bioethics*. We restricted the search to these leading bioethics journals in order to capture an "in-group" conversation among bioethicists, although we acknowledge that journals focused on the science and practice of genomics and genetics likely contain ethical debates as well. Finally, for reasons of practicality the search was also restricted to articles published in the past 5 years and in the English language.

Literature Review

After completing the MEDLINE search, two researchers independently reviewed manuscript titles and abstracts for relevance- the primary focus of the paper had to be on ethical issues related to genetics and genomics. Accordingly, book reviews, editorial introductions to special volumes, animal studies, and historical articles were removed from the database, as were duplicate titles. Articles without abstracts were automatically placed into database for full-text review. With this final list of articles for full-text review, the two researchers jointly developed a standard abstraction instrument that classified articles by type (empirical study, case report, bioethical analysis, literature review, commentary, and other).

Moving from conventional systematic literature review methods to a more qualitative grounded-theory based content analysis approach, data abstraction also involved "open coding." (Strauss and Corbin 1998). The two researchers independently described several paper characteristics through open coding including the principal ethical question or topic(s) addressed by the paper and the ethical concepts deployed within the arguments. Each of these labels applied by the researchers functioned as qualitative "codes" for subsequent qualitative analysis.

In order to assure consistency of data abstraction a set of ten abstracts and articles was independently reviewed and "double coded" by the researchers. Discrepancies in data abstraction were resolved by consensus. Subsequently, each researcher independently reviewed approximately half of the remaining abstracts and articles. Combining the two researcher's databases created a final database containing the bibliographic and abstraction data.

Development of Findings

While standard qualitative analysis techniques were used to develop our findings (described below), our approach was also inspired by a critical discourse analysis (CDA) theory. While there are many different ways in which CDA is applied in research and many different techniques, at the core of strategies is an acknowledgement of discourse being both socially constructed and socially conditioned. CDA seeks to make "visible the interconnectedness of things" in order to open up understandings of how social practices inform dominant forms of language use and marginalize others, and emphasizes multidisciplinary approaches (Wodack and Meyer 2009, Fairclough 2009).

Although we do not perform the sorts of in-depth textual and semiotic analyses often used by those employing CDA (we used qualitative analyses methods instead) our project applies a CDA lens to unpacking the often-implicit conceptualizations of the human that undergird the genethics literature. To be sure, we view contemporary bioethics discourse as social constructed and conditioned for it advances ethical pluralism and operates under secular conventions that privilege philosophy over theology, and reason-based arguments over scriptural-based hermeneutics. Our qualitative content analyses sought to make visible the connections between ethical concepts and genethics-related domains of study and fundamental ontologies regarding the human. Finally, we bring multidisciplinary approach to our development of findings for both researchers are situated in multiple bioethics-related practices. Both are practicing clinicians, bioethics researchers, participate in ethics committees, and both have deep religious commitments (MKA is Catholic Christian, and AIP Sunni Muslim).

As noted above the data abstraction instrument allowed the researchers to apply codes representing the principal ethical questions of and ethical concepts mentioned within the papers. For example one article might be entered in the database as focused on analyzing whether parents have a duty to select the best genetic traits for their children (principal question), and used the concepts of beneficence and Parfit's grounds for complaint principle to offer its argument (ethical concepts). Based on these codes the researchers sought to develop higher-order themes to classify the articles by topic of study. Based on consensus the bioethical topics discussed by the papers were grouped into six higher-order domains of study [plus an other category as detailed below], and each article was subsumed under one of these domains via researcher consensus. Once these domains were identified, the next task was to assess relationships between the open-coded ethical concepts, the principal questions of the article, and the domain. In conventional qualitative methods terminology, we were using a grounded theory approach to develop an overall conceptual link between all of the codes within a particular domain. We hypothesized that this link, if present, would represent a specific ontology of the human. In other words, the analyses sought to identify how all of the ethical concepts and questions contained within the articles within that particular domain conveyed a distinctive conceptualization of the human being.

Results

Descriptive Results of the Systematic Literature Review

277 articles were retrieved from MEDLINE, and after discarding duplicates, book reviews, editorial introductions to special volumes, animal studies, historical articles and articles not relevant to genethics, 203 unique articles underwent data abstraction [See Figure 1]. These articles could be grouped into six domains of ethical study: (1) Information Disclosure & Data-Ownership, (2) Human Enhancement & Modification, (3) Ethical Structures & Moral Responsibility in Genomics Research, (4) Human Reproduction Related-Ethics, (5) Special Considerations for Research in Vulnerable Communities and Popula-

tions, (6) Environment & Epigenetics. Articles not fitting within these domains were marked as "other."

Most articles (n=108) related to Information Disclosure and Data-Ownership and covered bioethical issues such as the ethics of conveying incidental findings, and notions of ownership of data in bio banking. The next most represented category was Human Reproduction Related Ethics (n=16) in which articles discussed moral issues related to reproductive cloning, gender and trait-selection in the to-be-born, and other bioethical issues related to pre-implantation genetic diagnoses. Human Enhancement & Modification and Special Considerations for Research in Vulnerable Communities & Populations comprised of a nearly equal number of articles (10 and 11 respectively), and these domains covered articles discussing issues such as the genetic enhancement and the cultural significance of genetic data in minority communities. Eight articles were placed into category 3 and pertained to ethical structures such as informed consent, while 5 articles detailed bioethical issues related to epigenetics and the environment.

As the ethical concepts discussed by, and principal questions of, the articles were linked to specific domains, three ontologies of the human were found to underlie three domains. These ontologies can subsume all of the ethical concepts and questions contained within a particular domain. The articles within the Information Disclosure and Data-Ownership appear to conceptualize the human being as a source of information about the past, present, and future and the ethical arguments and questions pertain to this ontology. The discussions contained within the domain of Human Reproduction-Related Ethics revolve around the ontology of a human as a being whose essence is defined by the capacity to reproduce. Undergirding the ethical concerns contained within the articles about Human Enhancement and Modification was the vision of a human being as an ever-evolving entity.

In what follows we will describe these ontologies with reference to the ethical concepts and questions dealt with by individual articles (in qualitative methods terms we will describe the codes and links). Our intention, however, is not to describe all of the articles along with all of the associated ethical concepts and questions, rather it is to describe the ontologies in sufficient enough terms so as to evidence the validity of our findings.

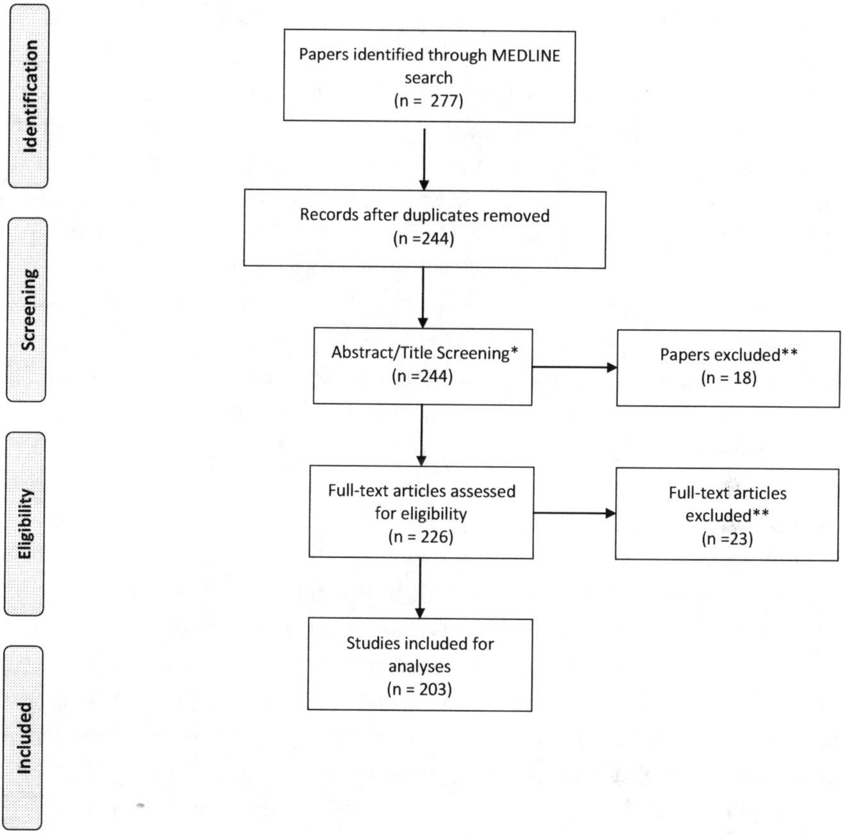

FIGURE 1: *Genethics Literature Review Flow Diagram*
(ADAPTED FROM: MOHER D, LIBERATI A, TETZLAFF J, ALTMAN DG, THE PRISMA GROUP (2009). PREFERRED REPORTING ITEMS FOR SYSTEMATIC REVIEWS AND META-ANALYSES: THE PRISMA STATEMENT. PLOS MED 6(7): E1000097. DOI:10.1371/JOURNAL.PMED1000097)

* 98 articles without abstracts were automatically carried through to full-text review
** Exclusion criteria were same for both stages as outlined in the methods section

The Human Being as a Source of Information about the Past, Present, and Future

Genes are segments of DNA within chromosomes, which assist in the production of proteins that are critical to the physiological functioning of an organism. In a sense, genes are the instruction manuals by which organisms develop, grow, maintain themselves, and reproduce because they contain the blueprint for all proteins needed by a cell. These proteins then play a role in all functions of the body. Genes are also integral to human heredity as they are passed down from parents to children in chromosomes contained within male and female gametes (sex cells). In this way parents and children bear resemblance to one another physically and psychologically, and rare disorders can afflict members of a family because of their similar genetic composition. Given the critical role genes play in the existence, maintenance, and propagation of human life the popular notion that genes make us who we are is understandable.

Given the important role genes play in generating the human being, and their importance in passing on information from one generation to the next, the ontology of a human being as a data store that houses information about the past, present and future is implicit in the genethics literature. This conceptualization prefigures bioethical questions and concepts related to genetic and genomic technologies. The genetic code contained within human cells provides information about the past in that it reveals ancestral linkages, e.g. paternity. It contains data about the present because it reveals data about a person's present physiological and psychological state and disorders that may be present. Genetic data also reveals dispositions and likelihoods related to future disease. In a sense then, genes contain essential knowledge about humans, individually and collectively, relevant to the past, present and future.

When the human being is seen as a repository of such information, it follows that ethical concerns revolve around the moral duties related to protecting, and regulating, the use of these data. To begin with bioethicists debate ownership because the information gleaned from genes is relevant not only to the individual it was procured from, but also to other individuals such as relatives and children because of the shared genetic composition. Such debates were found in many different articles (Dheensa, Fenwick, and Lucassen 2016, Rothstein 2013, Milner, Liu, and Garrison 2013). For example, in their interviews of patients, Dheensa and colleagues found that participants perceived "genetic information as essentially familial;" in the words of one participant shared "this isn't my information, I don't own the gene." (Dheensa, Fenwick, and Lucassen 2016, 2). This view was based on the idea that relatives "have a right to know about their potential risks" for disease, and that patients and clinicians have a moral duty to disclose pertinent information provided sufficient potential for significant harms exist. (Dheensa, Fenwick, and Lucassen) Notably, participants introduced some ambiguity by viewing genetic information related to one's own day-to-day health to be personal and not communal (Dheensa, Fenwick, and Lucassen 2016).

While Dheensa's paper presents qualitative data pertaining to the moral duties of disclosure, other articles present direct normative arguments. The standard view regarding genetic data is that they "belong" to the individual that supplied the physical material from which they were gleaned. The ownership relationship coheres with the ontology of a human as a data-store where genes are its inner core. Accordingly, usage of the data requires authorization from the individual. Bioethicists assessed whether forgoing individual consent in order to disclose genetic information to the other individuals is legitimate. For example the reflection of a moral "duty to warn" relatives that have a genetic marker for disease into legal structures was analyzed by Weaver (2016). She asserts that physicians' "ethic of care" can help inform professional standards and reconcile legal statues so that a particular patient's consent might not be necessary for disclosing risks of disease to that patient's relatives (Weaver 2016). Similarly a series of articles discussed the ethics of re-contacting the family members of a deceased research participant in order to convey genetically determined risks for disease (Milner, Liu, and Garrison 2013, Rothstein 2013, Shah et al. 2013, Taylor and Wilfond 2013). The case study analyzed was complicated by the fact that the research study consent forms inadequately covered such scenarios.

Moral duties and obligations related to genetic data and issues of ownership are "collectivized" in the context of genomic data and biobanking. For example, in the research arena, genetic/genomic data may disclose information about an individual's genetic composition that is unrelated to the particular focus of the research study. The ethics of these "incidental" findings is hotly debated (Viberg et al. 2016, Hofmann 2016, Costain and Bassett 2013, Kleiderman et al. 2014, Gliwa and Berkman 2013, Garrett 2013, Borgelt, Anderson, and Illes 2013, Price 2013, Anastasova et al. 2013, Ross and Reiff 2013, Parens, Appelbaum, and Chung 2013, Greenbaum 2014, Appelbaum et al. 2014). The relevance and limitations of the concepts of "duty to rescue" and "right not to know" to incidental findings is also canvassed by many articles (Berkman, Hull, and Biesecker 2015, Zuradzki 2015, Fenwick et al. 2015, Wachbroit 2015, Meagher 2015, Jecker 2015, Garrett 2015, Parsi 2015, Ulrich 2013). Illustratively, Garrett argues that a rescue paradigm grounded in beneficence insufficiently relates to genomics research because the traditional rescue paradigm was developed for short-term situations where risks were unpredictable and unanticipated. Genomics data have a much longer horizon and allow for the calculation of statistical probabilities for disease-related harms and thus a duty to rescue paradigm does not apply (Garrett 2015, Ulrich 2013). With respect the notion of "right not to know" Hofman investigated the many arguments for, and against, the right of individuals to remain ignorant of data gleaned from their own genome (2016). These arguments include ones that differentiate types of knowledge, ones that analyze the question in terms of ownership of data, and others that focus on the consequences of disclosure. Hofman holds that individuals "should be able to stay ignorant of incidental findings of uncertain significance" if they so choose (Hofmann 2016, 1).

If data about risks to relatives and progeny speak to the conceptualization of a human as containing data about the future, the incidental findings that Hofman discusses speak to both the individual's present and her future. An interesting case discussing the right not to know and one's present stage was covered by Wiesemann (2011). She uses the case of Caster Semenya, a world-class runner who was forced to submit to sex verification via genetic testing in order to compete as a female in competitive sports, to argue that individuals have a right to not know their genetically determined biological sex.

The genome contained within humans also stores information about the past. A particularly intriguing piece information that can be uncovered by genomics is paternity. The bioethical debate here hinges on the merits of truth telling in the context of clinical work balanced against duties of non-malfeasance and the respect for individual autonomy. In their article, Adlan and ten Have analyzed the relevance of these concepts in the context of the Islamic faith and Saudi culture where paternity is part biological and part socially-constructed (Adlan and ten Have 2012). At least one commentator felt that disclosure of biological non-paternity in a Muslim context carries the risk of significant harm to the child and mother, thus he advocates for nondisclosure based a duty not to harm (Zabidi-Hussin 2012). Other bioethicists also reflect upon notions of autonomy, data ownership and the regulation of knowledge in the context of direct-to-consumer genetic testing (Bunnik 2015, Hoffman 2016).

In summary a significant proportion of the genethics literature focuses on the regulation of genomic/genetic data. These data are seen as containing inner knowledge about the human being, and are relevant to the individual's relatives and progeny. Underlying the debates about data ownership and protection, the right to know or not know such data, and the moral duties researchers have to patients, research participants, and the wider society is a conceptualization of the human as a source of information about the past, present, and future.

The Human Being as a Reproductive Organism

One of the characteristics that distinguish living organisms from non-living matter is their capacity to reproduce. There are two forms of reproduction: asexual and sexual, where sexual reproduction requires bringing together two living organisms. Conventional human reproduction is of the sexual type, and allows for the propagation of genes (and the traits mediated through genes) from parents to their children. Genetic and genomic technologies provide humankind with greater knowledge about, and ability to intervene upon, reproduction. Consequently, these newfound capacities generate bioethical debates about whether we ought to manufacture, and otherwise intervene upon, human reproduction events. Obviously, the genethics literature addressing human reproduction operates out of an ontological view of the human as a reproductive organism. The striking feature of the discourse, however, is that it pushes the conventional limits of human reproduction to debate the facilitation of asexual, and artificial sexual, reproduction.

This literature thus views reproduction not as one of the many essential features of the human being, but rather as a defining characteristic that biomedicine might be morally obligated to service. Furthermore, since successful reproduction among the human species generates biological ties between offspring and progenitors, this subgenre also contends with a host of thorny issues about what constitutes, and what moral duties emerge from, the social construct that lies at the intersection of reproduction and genetic resemblance: parenthood.

Stemming from the idea that members of the human species might have a right to reproduce, Fries comments on a request made by a patient's family to harvest oocytes from their brain-dead family member. Such harvest was, at least in part, argued for on the basis of allowing for the individual's genome to be passed on to others post-death (Fries 2016). Several articles also argue over the moral significance and social ramifications of using genetic technology to assist homosexual couples reproduce. For example, Pennings discusses whether ova donation from one partner to another (ROPA) in lesbian couples is analogous to embryo donation, egg donation or gestational surrogacy (2016). He finds each analogy to be imperfect and argues that "ROPA can be seen not only symbolically but physically creating an equal contribution [to parentage]: one partner becomes the birth mother, and the other the genetic mother." (Pennings 2016, 255).

Scientists also have the ability to create synthetic gametes (ova and sperm) from either component. This technology is particularly useful for gay and lesbian couples as it facilitates the creation of a zygote that contains genes from each of the two partners. As genetic material from the sperm can be used to create a synthetic ovum for gay couples and vice versa for lesbian couples, a quasi-natural form of biological reproduction becomes possible. Murphy debunks arguments against synthetic gamete creation for same-sex couples in order to suggest that homosexual individuals have a right to reproduce, and that clinicians are morally bound to assist such reproductive efforts (2014b). The human right to reproduction, notions of parenthood, and their association with genetic resemblance of parent to child is also a central feature in articles discussing the ethics of reproductive cloning and mitochondrial replacement therapies (MRTs) (Harris 2014, Wrigley, Wilkinson, and Appleby 2015, Harris 2015). Reproductive cloning, an asexual form of reproduction not natural to the human species, may be considered to be a useful reproductive technology enabling individuals to reproduce without the need for a partner (genetic or otherwise). On the other hand, MRT is a technique that produces a child with three "parents" since genetic material carried within the mitochondria of a donor ova also becomes part of the zygote. Thus the genetic material in the zygote now comes from the procreative couple as well as the egg donor. As noted above, the idea that parenthood is ascribed solely on genetics is debatable. Furthermore, and with respect to MRT, some bioethicists argue that since the mitochondrial DNA comprises of less than 5% of the total DNA in the zygote, and because the genes contained therein do not contribute to physical resemblance, a parent relationship among the egg donor and the future child does not issue forth.

Moving from the gamete stage to the zygote and embryo stages, many articles discussed the merits, risks, and morality of pre-implantation and/or pre-natal genetic diagnosis and intervention. A central bioethical question related to the use of these technologies was do aspiring parents have the moral responsibility to produce the "best" children? This concept of procreative or reproductive beneficence was featured in multiple papers. Harris's aforementioned paper on reproductive cloning argues that cloning preserves the human genome more so than any other reproductive method, and should be considered in-line with reproductive beneficence since the ensuing clone would have a "tried and tested" genome (Harris 2014, 58). Other bioethicists carry forth the procreative beneficence argument to IVF and debate whether potential parents who use IVF to reproduce, or who discover abnormalities in their embryo based on prenatal genetic tests, have a moral obligation to insure that their future progeny are free of disability and/or to not select IVF-created embryos that carry disability-like traits (Soniewicka 2015, Weinberger and Greenbaum 2015). In the context of using genetic/genomic technologies to alter the genetic makeup of the to-be-born, Delaney argues that genetic engineering might be morally objectionable (2011). On the basis of Parfit's origin view and a grounds for complaint principle, he asserts that genetic engineering modifies a being that has already been "created" from a particular ova and sperm (Delaney 2011). Consequently that being has grounds to complain that they might have been better off in the pre-engineered state.

These ethical concepts were extended to the ethical implications of germline manipulation enabled by CRISPR-Cas 9 technology and MRTs (Harris 2015, Evitt, Mascharak, and Altman 2015). Such technologies change the DNA makeup of all future progeny of the fetus created using them. Accordingly, the genetic pool available to future generations is restricted, as some genes are not allowed to be passed onto future generation. The morality of thus constraining the reproductive "rights" of future generations remains a hot topic of debate. Moving from the clinic to the "free-market," Gynell and Douglas argued that the state has ethical grounds for regulating the use of genetic technology-based selection of traits on the basis of collective action problems that may ensue should the free-market be allowed to operate unrestricted (2015).

The bulk of the discussion about the ethical dimensions of genetic technologies that enable the creation of synthetic gametes and clones, introduce the possibility of different types of genetic linkages between donors of gametes and the embryo produced from such biological material, and allow for the selection of the genetic composition of progeny operates out of the view that members of the human species, in general, are entitled to reproduce. The implicit ontology of the human being, therefore, is one where the human is, in essence, a reproductive living organism. That is not to say that the entire bioethical discussion is focused on the moral duties that emerge from such reproductive rights, or that the assertion of such rights in universally accepted, rather the point here is that the undergirding ontology is one that needs to be unpacked and attended to in order to present a comprehensive moral vision for the use of such technologies.

The Human as a Biologically Evolving Entity

A third ontology that can help to explain the ethical concepts and questions embedded within the genethics literature is the human as an evolving biologic entity. Evolution, particularly in religious circles, is a particularly controversial topic. Much of the religious debate revolves around speciation, the formation of a new species from prior ones, over the course of time. This idea is not the focus of genethics literature, rather the evolutionary notion here is that the genetic composition of humankind is not static; it is dynamic. As noted above, sexual human reproduction involves the admixing of genetic material from both parents within the nucleus of the zygote. There are many different combinations that can occur and these combinations of parental DNA allow for novel genes to be present in the child that are not in the progenitors. Moreover, errors in DNA replication routinely introduce random changes in the cell's DNA, and environmental exposures can also alter one's DNA. Therefore the human genome is not a static entity but is always changing. As new genes can emerge within the human genome, human traits and features may also change. The questions at core of this set of articles revolve around the morality of deliberately changing the genetic profile of adults. If the human gene pool is always changing then using biomedicine to positively influence that change may be morally valid.

Illustratively, Glick discusses the morality of genetic enhancement from a Jewish perspective (2011). He notes that "man is commanded to be co-creator with his Creator in many areas of endeavor" and that Rabbinic authorities state that after the creation of humankind "there remained additional power to create anew, just like people create new animal species through inter-species breeding." (Glick 2011, 417). On the basis of scriptural texts and Rabbinic commentaries, he suggests that there is no "inherent banning of the use of [genetic] techniques for [human] enhancement" in the Jewish tradition (Glick 2011). Some bioethicists find genetic enhancement to be immoral and invoke the concept of human dignity to suggest that humans are part of a natural kind that demands non-interference (and optionally link this concept to a theistic worldview) (Greenbaum 2013, Chan 2015). Other bioethicists push back against such arguments. Chan argues that an Aristotlean view of humanity and ethics supports the usage of genetic technology to enhance human flourishing. He notes that Aristotle held that "a human being reaches its full potential through a combination of nature and nurture" and that "the development of human excellences [is] the goal of human existence."(Chan 2015, 280). One could extend this argument by saying that since the human project involves moral enhancement, therefore using genetic technology for moral enhancement should not be categorically prohibited. Indeed the debate around moral enhancement underlies Murphy's paper on biomoral modification. In it he outlines Persson and Savulescu's views on the moral obligation towards moral enhancement and how differences between individual and societal goals for human enhancement impact the moral assessment of such technologies (Murphy 2015). In another paper Murphy builds out an argument for genetic enhancement by noting that such modification can amplify choices, enrich lives, and consolidate identities (Murphy 2014a).The blurring of the morally important distinction between therapy and enhancement by genetic technologies is also discussed by several articles (Chan 2015, Holtug 2011).

Debates around the ethics of genetic enhancement of the individual or progeny (discussions of genetic intervention on progeny were outlined in the previous section) are based on the conceptualization of a human being undergoing constant change in his or her genetic composition. This biological fact is used by proponents of advancement to displace notions of humanity as a static creation, perfect in its nature, and thereby inviolable. Rather they view human nature as a project in development and that there are moral obligations to assist such that the development provided it accords with individual and societal conceptions of the good. An ontological perspective of the human being as an evolving entity informs the subsequent ethical argument for genetic enhancement. Other voices in the literature argue against enhancement but do not squarely challenge this ontological perspective.

Human Ontologies and Their Implications for Islamic Bioethical Perspectives on Genomic and Genetic Technologies

Our analysis of the genethics literature revealed at least three ontological perspectives on the human that are at the heart of the bioethical discussions- the human being (i) as a source of information about the past, present, and future, (ii) as a reproductive organism, and (iii) as an evolving entity. We hold that Islamic scholars wading into the bioethical debates over genetics and genomics must first assess how these ontologies compare with Islamic theological perspectives on the human being, and subsequently use the instruments of Islamic ethico-legal tradition to craft responses. Below we outline several issues relevant to the reception of these ontologies in the Islamic tradition.

The first ontology sees the human being as a repository of knowledge. Thus we must ask what is the relationship between knowledge and the human being in Islam? A Qur'anic worldview holds that God is the source of all knowledge and that He instructs humankind. Illustratively, the opening story of the Qur'an relates to teacher-learner relationship as God instructs the first human, Adam, about the "names of all things" and is instructed to obey God's commandments (Ali 1999 Verse 2:31). Thus while humans may hold onto knowledge, ultimately it is deemed to originate from God. If information contained in the genome originates from God's knowledge then how do we understand possession and ownership? For example, one might assert that the genomic/genetic data belongs to the individual it was derived from. Alternatively, Islamic scholars might contend that a stewardship responsibility emerges from the idea that humans are but custodians of God's knowledge encoded in the genome. Different legal analyses might therefore proceed from an ownership relationship when compared to a stewardship one. If a particular human "owned" that data then permission and consent might be required prior to any disclosure, however if the relation was of stewardship then God' rights (*ḥaqq Allah*) over the data can be asserted if the data are of public benefit (*maṣlaḥah*) (Emon 2006). A related question is whether genomic/genetic data is categorically beneficial. The Prophet Muhammad is recorded to have supplicated for beneficial knowledge and protection from knowledge of no benefit (Ibn Mājah Book 34: Hadith 17). How are these categories to be applied to probabilistic data from the genome about the past, present and future? As noted above, bioethicists debate the balancing of a biomedical scientist's responsibility and duty to warn people about disease risks with the individual's right not to know incidental findings. The ways in which the Islamic bioethicists classify genetic/genomic data in terms of its ownership and benefit will surely inform their view on the moral duties related to disclosing and protection these data.

Another related question is whether genetic and genomic data to be treated as certain knowledge in Islam? Science suggests that the genome conveys data about an individual's origin, helps to explain their present state, and forecast probabilities about their future. Islamic logicians define knowledge (*al-ʿilm*) to be propositional in nature, and consider knowledge to reflect the correspondence between one's understanding of, and true, reality. Accordingly certainty about correspondence between a particular truth claim and reality exists along a spectrum from absolute conviction, *yaqīn,* through predominant certainty (>50%), *al-ẓann al-rājiḥ* and equivocal certainty (50%), *al-shakk,* to an improbable conviction, *al-ẓann al-marjūḥ* (Qureshi and Padela 2016). The claims of genetic science regarding the past, present and future would need to be placed into this spectrum prior to making moral assessments. For example how does an Islamic perspective assess DNA evidence? One well-researched example of where genetic claims uneasily fit into Islamic law pertains to claims of paternity. As Shabana and others note even though genetic evidence might reach near certainty it does not negate traditional religious conventions that allow a father to claim paternity (Adlan and ten Have 2012, Shabana 2012, 2013).

In related fashion how do Islamic theologians contend with genetic determinism? And how does genetic knowledge matchup with Islamic theological views about fate, will, destiny and moral responsibility? For example, if the knowledge contained in one's genes reflects God's omniscience one could suggest that it is the "language of God" and contains knowledge about one's fate (Collins 2006). Indeed in our focus group interviews with diverse groups of Muslims, the idea that the genetic code reflects God's destined plan for a person with respect to disease was a dominant theme (Padela 2011). These public understandings merit recognition by scholars providing an Islamic moral vision for genetic and genomic technologies for they reflect the mindset of the technology's end-user.

An ontological perspective of the human as a reproductive organism by nature also bears upon Islamic bioethical positions of genomic and genetic technologies. As noted above, genetic technologies have a wide variety of uses in human reproduction. They can help in manufacturing offspring for those who cannot "naturally" reproduce, and can aid in diagnosing and repairing genetic diseases at the prenatal stage. As referenced above, some argue that members of the human species have the right to reproduce and thus the usage of technology to aid reproduction is inherently good. Would an Islamic worldview also contend that having offspring is part of what makes a human, human? A Qur'anic verse seems to suggest that one's biological capacity, as well as the eventual destiny to have children, is part of God's decree. The relevant verse reads "Or He bestows both males and females, and He leaves barren whom He will."(Ali 1999 Verse 42:50). On this basis might Islamic theologians interpret infertility among the human population as normative and thus not requiring genetic technologies to "fix"? While much of the Islamic bioethics literature is supportive of using IVF and similar methods to facilitate reproduction in the confines of marriage, do these judgements of permissibility also apply to manufacturing synthetic gametes, manipulating donor ova to accept DNA from other humans, and using reproductive cloning technologies? On one hand, Islamic legists deem having children to be a critically important human interest that Islamic law must protect. So much so that one of the overarching essential higher objectives of Islamic law, *maqāṣid al-Sharīʿah*, is the preservation of lineage (Raysūnī and International Institute of Islamic Thought. 2005, Shāṭibī et al. 2011). Indeed the permissibility of using IVF and other biomedical technologies for procreative assistance is often grounded in this objective. As the same time, it is beyond dispute that traditional views on lineage would need to be reimagined to cover the scenarios outlined above. For example, is the moral duty to preserve lineage a relevant construct through which to consider the morality of reproductive cloning? Does mitochondrial DNA create lineal relations according to Islam? Such questions can be better addressed once Islamic theologians reflect on whether reproduction is essential to the human being. If reproduction is deemed essential then perhaps reconceptualization of traditional constructs about lineage in light of new technologies, and the fashioning of detailed ethico-legal views on how to preserve lineage in the present genetic and genomic age is necessary. While Islamic jurists have prohibited gestational surrogacy due to concerns over disturbing traditional notions of lineage, would the ova donor whose cellular material is used in MRT have a claim to parenthood of the future embryo?

In a similar way, genetic and genomic technologies are reconfiguring views on what is means to be a parent, and the attendant moral obligations to the to-be-born. Does procreative beneficence cohere with an Islamic moral vision? While Islamic scriptural sources are replete with references to parents being responsible for the moral formation of children, do such responsibilities extend to using genetic selection and/or enhancement technologies on embryos pre- and post-implantation? As some have argued the selection of embryos to implant involves not only choosing who is born but also who is not. How would Islamic theology analyze the role of the human vis-à-vis God in selecting the to-be-born, and how is that reflected into Islamic ethics and law? Ideas such as accepting God's will and dominion versus competing with God's role in creation are often invoked by Islamic scholars discussing biomedical technologies, working out how these concepts relate to human reproduction in the context of genetic and genomic technologies that allow for the selection of traits in one's progeny and even in the future generation of progeny is much needed.

As our ability to modify the genetic composition of individuals and to make germline modifications improves, one wonders if whether we should? This ethical question moves the discussion to the realm of genetic enhancement and brings the notion of a human being as an evolving entity to the fore. Does Islam see the gradual process of human creation as having reached its end, and that the present human being is fully developed biologically? The Qur'an notes that the human is molded by God in the best of forms and Islamic theologians consider humankind to be at the pinnacle of God's creation (Ali 1999 Verse 40:64). Biomedical science and genetic technologies, however, give rise to the ability to modify genotype and the resulting phenotype of individuals and their progeny. Is this modification akin to altering God's creation? With the distinction between therapy and enhancement having been blurred by newfound capacities to intervene upon the human genome, can Islamic theology furnish a clear "red-line" for clinical practice based on this conventional framework? Arguably, an applied ethics must be rooted in theological conceptions of what the human being represents, and in judgements about what aspects of the human being's composition fall under God's sole purview. For example, Islamic legists offer that the Qur'anic prohibition of suicide, in part, stems from a moral condemnation of interfering with God's role as the originator and terminator of life (Yacoub 2001, Sachedina 2004). How are these views about God reconciled in an age when humans can select which embryos to gestate (and which ones not to) and can effect genetic modifications that allow for humanity to gain new, or improve upon existing, moral, physical and other capacities? Glick (2011) finds that Jewish scriptural sources do not categorically prohibit human enhancement, would a reading of the Islamic scriptures offer the same perspective? Moosa's exposition of the varied ways in which scholars have interpreted scriptural texts that appear to warn against the "changing of God's creation" but permit for the cross-pollination of date-palms to determine the illicitness of genetically modified foods, speaks to the need for context- dependent exegesis in order to delineate moral duties related genetic enhancements (Moosa, 2009).

Final Remarks

Bioethics sits at the intersection of many fields of knowledge, and has porous disciplinary boundaries. It certainly pertains to the biological sciences as ethical issues pertaining to living organisms are subsumed under the "bio" in the neologism; however conservation of the environment is viewed as part of bioethics as well. The "ethics" portion of the word is also expansive in that encompasses not only the traditional ethical disciplines of philosophy and law (and arguably theology) but politics and sociology as well. Accordingly, the field is marked with multidisciplinary, and a multitude of perspectives are offered on any given moral question.

This variety makes the bioethics discourse a complex one to navigate. For researchers an accurate description of the genre requires recognizing the particular disciplinary vantage points and ethical reasoning modes used by commentators to delineate and analyze the moral dimensions of a problem. For religious scholars seeking to provide moral guidance on a particular bioethical issue, their challenge is compounded by the fact that they need to not only understand the pertinent concepts and relevant ethical perspectives in the field, but also need to deconstruct those concepts and arguments in light of their own tradition prior to lending a religiously informed perspective.

By means of a systematic literature review and qualitative analytic methods this paper aims to help the deconstructive exercise by identifying several ontologies of the human that are implicit to the genethics discourse. We have demonstrated how the relevant bioethical questions as well as the bioethical concepts used to address those questions arise from such conceptualizations of the human being. While our literature search was not exhaustive, and our analytic methods introduce limits on the comprehensiveness with which we can detail specific ontologies, we hold that the developing field of Islamic bioethics must address the ontological claims of biomedicine in addition to the ethical and legal dimensions of the biomedical sciences and practices. Indeed it may be that Islamic theological perspectives on human ontology can supply visions for human flourishing that generate fresh, nuanced, and relevant Islamic bioethical guidance for the present era.

References

Adlan, Abdulla. A., and Henk. A. ten Have. 2012. "The dilemma of revealing sensitive information on paternity status in Arabian social and cultural contexts: telling the truth about paternity in Saudi Arabia." *J Bioeth Inq* 9 (4):403-9. doi: 10.1007/s11673-012-9390-y.

Ali, Abdullah Yusuf. 1999. *The Quran Translation*. New York: Tahrike Tarsile Quran.

Anastasova, Velizara, Alessandro Blasimme, Sophie Julia, and Anne Cambon-Thomsen. 2013. "Genomic incidental findings: reducing the burden to be fair." *Am J Bioeth* 13 (2):52-4. doi: 10.1080/15265161.2012.754066.

Appelbaum, Paul S., Erik Parens, Cameron R. Waldman, Robert Klitzman, Abby Fyer, Josue Martinez, W. Nicholson Price II, and Wendy K. Chung. 2014. "Models of consent to return of incidental findings in genomic research." *Hastings Cent Rep* 44 (4):22-32. doi: 10.1002/hast.328.

Berkman, Benjamin E., Sara. C. Hull, and Leslie G. Biesecker. 2015. "Scrutinizing the Right Not to Know." *Am J Bioeth* 15 (7):17-9. doi: 10.1080/15265161.2015.1039733.

Borgelt, Emily, James A. Anderson, and Judy Illes. 2013. "Managing incidental findings: lessons from neuroimaging." *Am J Bioeth* 13 (2):46-7. doi: 10.1080/15265161.2012.754069.

Bunnik, Eline. M. 2015. "Do genomic tests enhance autonomy?" *J Med Ethics* 41 (4):315-6. doi: 10.1136/medethics-2014-102171.

Chan, David. K. 2015. "The concept of human dignity in the ethics of genetic research." *Bioethics* 29 (4):274-82. doi: 10.1111/bioe.12102.

Collins, Francis S. 2006. *The language of God : a scientist presents evidence for belief*. New York: Free Press.

Corbin, Juliet M., Anselm L. Strauss, and Anselm L. Strauss. 2008. *Basics of qualitative research : techniques and procedures for developing grounded theory*. 3rd ed. Los Angeles, California: Sage Publications, Inc.

Costain, Gregory., and Anne. S. Bassett. 2013. "Incomplete knowledge of the clinical context as a barrier to interpreting incidental genetic research findings." *Am J Bioeth* 13 (2):58-60. doi: 10.1080/15265161.2012.754063.

Crabtree, Benjamin F., and William L. Miller. 1999. *Doing qualitative research*. 2nd ed. Thousand Oaks, Calif.: Sage Publications.

Delaney, James J. 2011. "Possible people, complaints, and the distinction between genetic planning and genetic engineering." *J Med Ethics* 37 (7):410-4. doi: 10.1136/jme.2010.039420.

Dheensa, Sandi, Angela Fenwick, and Anneke Lucassen. 2016. "'Is this knowledge mine and nobody else's? I don't feel that.' Patient views about consent, confidentiality and information-sharing in genetic medicine." *Journal of Medical Ethics* 42 (3):174-9.

Emon, Anver M. 2006. "Huquq Allah and Huquq Al-Ibad: A Legal Heuristic for a Natural Rights Regime." *Islamic Law and Society* 13 (3):325-391.

Evitt, Niklaus H., Shamik Mascharak, and Russ B. Altman. 2015. "Human Germline CRISPR-Cas Modification: Toward a Regulatory Framework." *Am J Bioeth* 15 (12):25-9. doi: 10.1080/15265161.2015.1104160.

Fairclough, Norman. 1995. *Critical discourse analysis : the critical study of language, Language in social life series*. London ; New York: Longman.

Fairclough, Norman. 2009. "A dialectical-relational approach to critical discourse analysis in social research." In *Methods of Critical Discourse Analysis*, edited by Ruth Wodack and Michael Meyer. London: Sage Publications.

Falagas, Matthew E., Eleni I. Pitsouni, George A. Malietzis, and Georgios Pappas. 2008. "Comparison of PubMed, Scopus, Web of Science, and Google Scholar: strengths and weaknesses." *The FASEB Journal* 22:338-342.

Fenwick, Angela, Sandi Dheensa, Gillian Crawford, Shiri Shkedi-Rafid, and Anneke Lucassen. 2015. "Rescue obligations and collective approaches: complexities in genomics." *Am J Bioeth* 15 (2):23-5. doi: 10.1080/15265161.2014.990763.

Fries, Melissa. 2016. "Analysis: OB/GYN-Genetics." *J Clin Ethics* 27 (1):59-60.

Garrett, J. R. 2013. "Reframing the ethical debate regarding incidental findings in genetic research." *Am J Bioeth* 13 (2):44-6. doi: 10.1080/15265161.2013.757972.

Garrett, Jeremy R. 2015. "Collectivizing rescue obligations in bioethics." *Am J Bioeth* 15 (2):3-11. doi: 10.1080/15265161.2014.990163.

Glick, Shimon M. 2011. "Some Jewish thoughts on genetic enhancement." *J Med Ethics* 37 (7):415-9. doi: 10.1136/jme.2009.034744.

Gliwa, Catherine, and Benjamin E. Berkman. 2013. "Do researchers have an obligation to actively look for genetic incidental findings?" *Am J Bioeth* 13 (2):32-42. doi: 10.1080/15265161.2012.754062.

Greenbaum, Dov. 2013. "If you can't walk the walk, do you have to talk the talk: ethical considerations for the emerging field of sports genomics." *Am J Bioeth* 13 (10):19-21. doi: 10.1080/15265161.2013.828121.

Greenbaum, Dov. 2014. "If you don't know where you are going, you might wind up someplace else: incidental findings in recreational personal genomics." *Am J Bioeth* 14 (3):12-4. doi: 10.1080/15265161.2013.879946.

Gyngell, Chris, and Thomas Douglas. 2015. "Stocking the genetic supermarket: reproductive genetic technologies and collective action problems." *Bioethics* 29 (4):241-50. doi: 10.1111/bioe.12098.

Harris, John. 2014. "Time to exorcise the cloning demon." *Camb Q Healthc Ethics* 23 (1):53-62. doi: 10.1017/S0963180113000443.

Harris, John. 2015. "Germline Manipulation and Our Future Worlds." *Am J Bioeth* 15 (12):30-4. doi: 10.1080/15265161.2015.1104163.

Hoffman, Sharona. 2016. "The Promise and Perils of Open Medical Data." *Hastings Cent Rep* 46 (1):6-7. doi: 10.1002/hast.529.

Hofmann, Bjørn. 2016. "Incidental findings of uncertain significance: To know or not to know--that is not the question." *BMC Med Ethics* 17:13. doi: 10.1186/s12910-016-0096-2.

Holtug, Nils. 2011. "Equality and the treatment-enhancement distinction." *Bioethics* 25 (3):137-44. doi: 10.1111/j.1467-8519.2009.01750.x.

Ibn Mājah, Abū 'Abdillāh Muḥammad ibn Yazīd. Sunan Ibn Majah. Sunnah.com

Jecker, Nancy S. 2015. "Rethinking rescue medicine." *Am J Bioeth* 15 (2):12-8. doi: 10.1080/15265161.2014.990169.

Kleiderman, Erika, Bartha M. Knoppers, Conrad V. Fernandez, Kym M. Boycott, Gail Ouellette, Durhane Wong-Rieger, Shelin Adam, Julie Richer, and Denise Avard. 2014. "Returning incidental findings from genetic research to children: views of parents of children affected by rare diseases." *J Med Ethics* 40 (10):691-6. doi: 10.1136/medethics-2013-101648.

Meagher, Karen M. 2015. "Seeking context for the duty to rescue: contractualism and trust in research institutions." *Am J Bioeth* 15 (2):18-20. doi: 10.1080/15265161.2014.990170.

Miles, Matthew B., and A. Michael Huberman. 1994. *Qualitative data analysis : an expanded sourcebook*. 2nd ed. Thousand Oaks: Sage Publications.

Milner, Lauren C., Emily Y. Liu, and Nanibaa' A. Garrison. 2013. "Relationships matter: ethical considerations for returning results to family members of deceased subjects." *Am J Bioeth* 13 (10):66-7. doi: 10.1080/15265161.2013.828533.

Moosa, Ebrahim. 2009. "Genetically Modified Foods and Muslim Ethics." In *Acceptable Genes? Religious Traditions and Genetically Modified Foods*, edited by Conrad G. Brunk and Harold Coward, 135-157. Albany: State University of New York Press.

Murphy, Timothy F. 2014a. "Genetic modifications for personal enhancement: a defence." *J Med Ethics* 40 (4):242-5. doi: 10.1136/medethics-2012-101026.

Murphy, Timothy F. 2014b. "The meaning of synthetic gametes for gay and lesbian people and bioethics too." *J Med Ethics* 40 (11):762-5. doi: 10.1136/medethics-2013-101699.

Murphy, Timothy F. 2015. "Preventing ultimate harm as the justification for biomoral modification." *Bioethics* 29 (5):369-77. doi: 10.1111/bioe.12108.

Padela, Aasim, Katie Gunter, and Amal Killawi. 2011. Meeting the Healthcare Needs of American Muslims: Challenges and Strategies for Healthcare Settings. Washington DC: Institute for Social Policy & Understanding.

Parens, Erik, Paul Appelbaum, and Wendy Chung. 2013. "Incidental findings in the era of whole genome sequencing?" *Hastings Cent Rep* 43 (4):16-9. doi: 10.1002/hast.189.

Parsi, Kayhan. 2015. "Rethinking the rescue paradigm." *Am J Bioeth* 15 (2):1-2. doi: 10.1080/15265161.2015.1006031.

Pellegrino, Edmund D., and David C. Thomasma. 1981. *A philosophical basis of medical practice : toward a philosophy and ethic of the healing professions*. New York: Oxford University Press.

Pennings, Guido. 2016. "Having a child together in lesbian families: combining gestation and genetics." *J Med Ethics* 42 (4):253-5. doi: 10.1136/medethics-2015-103007.

Price, W. Nicholson II. 2013. "Legal implications of an ethical duty to search for genetic incidental findings." *Am J Bioeth* 13 (2):48-9. doi: 10.1080/15265161.2012.754068.

Qureshi, Omar, and Aasim I. Padela. 2016. "When must a patient seek healthcare? Bringing the perspectives of islamic jurists and clinicians into dialogue." *Zygon* 51 (3):592-625. doi: 10.1111/zygo.12273.

Raysūnī, Aḥmad, and International Institute of Islamic Thought. 2005. *Imam al-Shatibi's theory of the higher objectives and intents of Islamic law*. London ; Washington: International Institute of Islamic Thought.

Ross, Kathryn M., and Marian Reiff. 2013. "A perspective from clinical providers and patients: researchers' duty to actively look for genetic incidental findings." *Am J Bioeth* 13 (2):56-8. doi: 10.1080/15265161.2012.754064.

Rothstein, Mark A. 2013. "Should researchers disclose results to descendants?" *Am J Bioeth* 13 (10):64-5. doi: 10.1080/15265161.2013.828531.

Sachedina, Abdulaziz. 2004. "Islamic Bioethics." *The Annals of Bioethics: Religious Perspectives in Bioethics*:153-171.

Shabana, Ayman. 2012. "Paternity between Law and Biology: The Reconstruction of the Islamic Law of Paternity in the wake of DNA Testing." *Zygon* 47 (1):214-239.

Shabana, Ayman. 2013. "Negation of Paternity in Islamic Law between Liʿān and DNA Fingerprinting." *Islamic Law and Society* 20 (3):157-201

Shah, Seema K., Sara C. Hull, Michael A. Spinner, Benjamin E. Berkman, Lauren A. Sanchez, Ruquyyah Abdul-Karim, Amy P. Hsu, Reginald Claypool, and Steven M. Holland. 2013. "What does the duty to warn require?" *Am J Bioeth* 13 (10):62-3. doi: 10.1080/15265161.2013.828528.

Shāṭibī, Ibrāhīm ibn Mūsá, Imran Ahsan Khan Nyazee, Raji M. Rammuny, Centre for Muslim Contribution to Civilization., and Qatar Foundation. 2011. *The reconciliation of the fundamentals of Islamic law*. 1st ed, *Great books of Islamic civilization*. Reading, UK: Garnet Pub.

Soniewicka, Marta. 2015. "Failures of Imagination: Disability and the Ethics of Selective Reproduction." *Bioethics* 29 (8):557-63. doi: 10.1111/bioe.12153.

Strauss, Anselm L., and Juliet M. Corbin. 1998. *Basics of qualitative research : techniques and procedures for developing grounded theory*. 2nd ed. Thousand Oaks: Sage Publications.

Taylor, Holly A., and Benjamin S. Wilfond. 2013. "The ethics of contacting family members of a subject in a genetic research study to return results for an autosomal dominant syndrome." *Am J Bioeth* 13 (10):61. doi: 10.1080/15265161.2013.828523.

Tiemersma, Douwe. 1987. "Ontology and ethics in the foundation of medicine and the relevance of Levina's view." *Theor Med* 8 (2):127-33.

Ulrich, Michael. 2013. "The duty to rescue in genomic research." *Am J Bioeth* 13 (2):50-1. doi: 10.1080/15265161.2012.754067.

Viberg, Jennifer, Pär Segerdahl, Sophie Langenskiold, and Mats G. Hansson. 2016. "Freedom of Choice About Incidental Findings Can Frustrate Participants' True Preferences." *Bioethics* 30 (3):203-9. doi: 10.1111/bioe.12160.

Wachbroit, Robert. 2015. "Rescue, strangers, and research participants." *Am J Bioeth* 15 (2):21-2. doi: 10.1080/15265161.2014.990759.

Weaver, Meaghann. 2016. "The Double Helix: Applying an Ethic of Care to the Duty to Warn Genetic Relatives of Genetic Information." *Bioethics* 30 (3):181-7.

Weinberger, Sara, and Dov Greenbaum. 2015. "Genetic technology to prevent disabilities: how popular culture informs our understanding of the use of genetics to define and prevent undesirable traits." *Am J Bioeth* 15 (6):32-4. doi: 10.1080/15265161.2015.1028665.

Wiesemann, Claudia. 2011. "Is there a right not to know one's sex? The ethics of 'gender verification' in women's sports competition." *J Med Ethics* 37 (4):216-20. doi: 10.1136/jme.2010.039081.

Wodack, Ruth, and Michael Meyer. 2009. "Critical Discourse Analysis: history, agenda, theory and methodology." In *Methods of Critical Discourse Analysis*, edited by Ruth Wodack and Michael Meyer. London: Sage Publications.

Wrigley, Anthony, Stephen Wilkinson, and John B. Appleby. 2015. "Mitochondrial Replacement: Ethics and Identity." *Bioethics* 29 (9):631-8. doi: 10.1111/bioe.12187.

Yacoub, Ahmed A. 2001. *The Fiqh of Medicine*. London: Ta-Ha Publishers Ltd.

Zabidi-Hussin, Z. A. 2012. "Does nondisclosure of genetic paternity status constitute a breach of ethics?: Commentary on "The dilemma of revealing sensitive information on paternity status in Arabian social and cultural contexts" by Abdallah A. Adlan and Henk A. M. J. ten Have." *J Bioeth Inq* 9 (4):413-4. doi: 10.1007/s11673-012-9401-z.

Żuradzki, Tomasz. 2015. «The preference toward identified victims and rescue duties.» *Am J Bioeth* 15 (2):25-7. doi: 10.1080/15265161.2014.990168.

CHAPTER 5

Islamic Perspectives on the Genome and the Human Person: Why the Soul Matters

Arzoo Ahmed[1] *and Mehrunisha Suleman*[2]

Introduction

A preoccupation with knowledge of the self and nature can be traced throughout human history. From writings of the Ancients to modern scientific inquiries, we find a collocation of ideas that range from the macrocosmic nature of the universe to microcosmic subcellular structures. Recent genomic advances, however, have not only added to a key facet of mankind's *raison d'etre*—seeking the truth about oneself and the world—it also offers a window into the repercussions of and consequences for manipulating matter on the nature of man, a nature that has hitherto been thought of as being immutable (Savulescu, ter Meulen, and Kahane 2011, XV). The myriad possibilities of genetically manipulating matter for human ends, raises critical questions about such endeavors and how they may influence our understanding of the human person (Savulescu, Ter Meulen, and Kahane 2011, XV). Central to the advances of mankind's ability to understand and manipulate matter is the unlocking of the cellular nucleus and the discovery of DNA with the identification of genes that code for particular phenotypes. This initial discovery has been rapidly applied to subsequent technological breakthroughs in genetic intervention. The latter offers mediation at the sub-cellular level and has the potential to alter the genetic constitution of individuals thereby offering them personalized therapies and unprecedented enhancement. The genetics revolution is thus profound-

1 Arzoo Ahmed is Director at the Centre for Islam and Medicine. She is reading for an MA in Philosophy at King's College London, and has a BA in Physics and an MPhil in Medieval Arabic Thought from the University of Oxford, arzoo.a@gmail.com

2 Dr. Mehrunisha Suleman is a researcher at the Centre of Islamic Studies, University of Cambridge. She has completed a DPhil in Population Health and a Medical Degree at the University of Oxford, ms520@cam.ac.uk

ly influencing our understanding of the human person, inheritance, genetic and species relatedness and distinctiveness. Concurrently, such advances have raised urgent moral questions about our ability to alter and influence the genetic makeup of individuals, and the potential adverse ethical impact these may have on future generations.

Recent genetic innovations have been studied through the prism of variegated fields including philosophy, theology and ethics. But very few studies have focused on the Islamic tradition's perspectives on the human genome and the human person. A constellation of Islamic philosophical, theological and spiritual narratives have contributed to the definitions for and understanding of the human person. Central to such accounts is the nature and role of the soul in defining and determining the human person. Yet very few scholarly contributions consider the relationship between the human genome, its association with the human person, and how these relate to Islamic considerations of the soul. The aim of this paper is to redress this imbalance.

This study seeks to extend the list of textual sources that contribute to ethical considerations of the genome question. We will show that these issues must, of necessity, include discussions of the human person and interrogate the interplay between them. Based on this hypothesis, this work will argue that, within the Islamic tradition, deliberations on the genome question should take into account knowledge of the soul and the body, since these are also believed to be constituents of the human person. Observations on these two primary determinants of the human person will be proffered from the fields of Qur'ānic exegesis, theology, and philosophy. The *modus operandi* of this investigation will be the exploration of key terms, how they relate to the approaches of different fields, and how they are conducive to offering a deeper understanding of the human person. The study will initially propound the relationship between the genome and the human person and how perspectives on the human genome are informing ideas about the human person and how such positions are stimulating ethical deliberations around speciesism, genetic determinism, enhancement, capacities, will and responsibility. This will be followed by an explication of the soul from within the Islamic tradition and how narratives from within this tradition provide novel perspectives to our understandings of the human person and the ethical considerations surrounding genomics. Finally, we will delineate areas for future engagement that this study has disinterred.

On the Relationship between the Genome and the Human Person: Definitions, Perspectives and Challenges

Genetic advancement, through the elucidation of the structure of DNA (Watson and Crick 1953) and the subsequent sequencing of the Human Genome,, has propelled us to a higher plateau of understanding our basic functions and the nature of human life (IHGSC 2004; Lander et al. 2001). Such developments are changing, not only what we can do, but also how we think and understand ourselves (Annas 1997, 157). George Annas explores how the mapping of the human genome has been a powerful "thought transformer" (Annas 1997, 157) similar to that of the astronomical mapping by Copernicus and the global mapping by Columbus. He argues that "maps model reality to help us understand it" (Annas 1997, 157). This proposition requires careful consideration vis-à-vis the mapping of the human genome in deriving understandings of the human person. The conceptualization of the human person from scientific, philosophical, ethical and legal perspectives, often refer to a composite – a person with distinct yet interacting parts (Taylor 1985). These fields oblige us to re-evaluate our understanding of the human person in light of genomics. This section will briefly explore perspectives on the human genome that are informing ideas about the human person and how such positions are stimulating ethical deliberations around speciesism, genetic determinism, enhancement, capacities, will and responsibility.

The Role of Genes in Determining Our Species Identity: Who Are We?

"Who are we? The question must be answered by each generation," suggests Archbishop Desmond Tutu reflecting on our need to reconcile recent scientific advancements with long held beliefs (Tutu 2015, ix). The discovery that our genes may define us as a species, *Homo sapiens*, brings into focus how pivotal scientific discoveries often involve "narcissistic offences" (Zwart 2009). Hub Zwart explains, for example, that scientific developments realized through Copernican heliocentrism demonstrate that earth and man do not occupy a central position in the organization of the universe; rather, we are deemed to be equals amongst other bodies in space (Zwart 2009). He adds that Darwin's seminal work on evolution elucidated how little we differ from other species (Zwart 2009), and that these estimations have been proven through comparative genomics, which reveal that humans may share up to 99.4% of their DNA with chimpanzees (Wildman et al. 2003).

Ethical considerations resulting from such findings are being realized through recent deliberations about and investigations into the moral status of *Homo sapiens*. Those who define being human along genetic lines argue that our species distinction confers certain privileges. This speciesism involves claims that humans should be able to manipulate other species for their own ends, such as killing them for food or using them for research (Steinbock 1978). The argument for employing the concept of a "genetic humanity" (Warren 2013, 308) as a way of defining human beings, is advocated by those who argue against abortion suggesting that it is wrong, not simply because fetuses are alive, but because they are human. However, this definition would require further evaluation when attempting to tackle questions around the best interests of the mother and fetus. If both are equal, since both are human, that is, each carries the human genome, then life-saving prioritization decisions cannot be made between mother and fetus on this basis.

Others, however, emphasize that comparative genomics, rather than conferring upon humans a genetic distinctiveness, demonstrate that, as a species, we are merely one entity on the species spectrum (Zwart 2009). How then do we relate to one another and comprehend, not only our identity as human beings, but also our identity as distinct persons? Peter Singer, who advocates for animal liberation, does, nevertheless, concede that human animals are distinct from non-human animals (Singer 1975; Steinbock 1978). If our genetic code alone does not confer such distinctness, then what accounts for it? And how can we understand intra-species differences, such as that of a mother and her fetus? Genomics may offer avenues for answering these questions, not through the Human Genome's application as a species map, but through it's conferring of individual phenotypic traits and therefore physical realities and capacities.

The Role of Genes in Determining Our Individual Identity: Who Am I?

James Watson, who co-discovered DNA with Francis Crick, famously claimed, "We used to think our fate was in our stars. Now we know, in large measure, our fate is in our genes" (Watson 1989). Richard Dawkins, in his pioneering work on evolution posits: "They (genes) are in you and in me; they created us, body and mind; and their preservation is the ultimate rationale for our existence. They have come a long way, those replicators. Now they go by the name of genes, and we are their survival machines" (Dawkins 2016). Such claims rely on reductionist models of the human person, which suggest a linear causal link between genes, the proteins for which they code, the subsequent cellular processes they initiate, their amalgamation at the level of individual tissues and organs and, finally, organ systems that are eventually realized as an entire human organism (Noble 2008). These models of biological systems find support in examples of so-called "genetic determinism" (Peters 2014) where genes determine phenotypic outcomes, illustrated by conditions such as Huntingdon's disease (HD) and cystic fibrosis (CF).

CF is a genetic disorder characterized by an autosomal recessive pattern of inheritance (Rommens et al. 1989). The disease is caused when an individual inherits mutations in both copies of the gene that code for the cystic fibrosis transmembrane conductance regulator (CFTR) protein. The condition affects multi-organ systems including the respiratory tract, reducing lung capacity, and impairing digestion resulting from suboptimal functioning of the pancreas; inheritance of the recessive alleles predict the presence of disease at birth. Currently, no treatment exists for CF and management comprises symptomatic relief for individual organ systems. It is a life-limiting disease (Tobias, Connor and Ferguson-Smith 2011) wherein the identification of the CF gene (Rommens et al. 1989) has raised the possibility of genetic intervention. Gene therapy targeting the faulty CFTR gene in individuals who carry the mutation has been hailed as a potential biomedical magic bullet for suffers of CF (Colledge and Evans 1995).

The identification of genes and their resulting phenotype, as well as the concurrent development of techniques to modify these genes hold great diagnostic and therapeutic potential. Nor are their uses limited to single gene disorders such as CF. Genetically predisposing factors such as the presence of BRCA1 and 2 in breast and ovarian cancer present subcellular level targets that may alter or ameliorate the risk of disease. The ability to alter our genetic makeup and in turn reduce risk of disease holds immense potential benefit. These advances and their proposed potential have also been welcomed as evidence for theories of genetic determinism. It also leads to claims that such genetic intervention ought to be pivotal in shifting our understanding of the fate of humans. For individuals carrying the genes that code for a potentially debilitating condition, like CF, gene therapy would undoubtedly have a tremendous impact on their sense of self and quality of life. It would thus support the notion that genetic intervention through gene therapy plays a critical role for genes in determining the human person by offering an unprecedented ability to alter the fate of humans at the level of the germ line.

Such genetic interventions also raise ethical concerns, however, around the human person in terms of health, illness and quality of life. Few would argue that to intervene at the genetic level to prevent or treat a debilitating condition, like CF is ethical. But what can be said of gene therapy induced cosmetic changes or the potential to extend life through telomere repair? How do we distinguish between therapeutic benefit and enhancement when both aim to improve quality of life? A genotypic emphasis for understanding normal traits (Buchanan et al. 2001) and the human person raise questions, not only about our existing attributes, but also about our aspirational endeavours. Who do we want to become? Some authors argue that it is not only desirable but also a moral obligation to enhance ourselves (Savulescu, ter Meulen, and Kahane 2011). If such enhancement enables people to lead better lives, the argument goes, then we have a responsibility to promote such an endeavor. Human engineering or "transhumanism" raises hopes in many spheres about how the human species can be advanced in fields of physical and intellectual endeavor.

Nevertheless, such views raise critical questions about what it means to lead a good life or better life. These discussions highlight the need for societies to gain a deeper understand of the meaning of health, disease and wellbeing in the wake of genetic intervention and enhancement. Modern eugenics or selective breeding to "optimize" the population gene pool is commonly practiced, with genetic testing and subsequent selection against disorders such as Down's syndrome. This practice is considered acceptable as it is predicated on parental consent and has the overall objective of producing a child with the best opportunities for a good life (Savulescu 2009). However, are people the best guardians of their own interests and capable of making appropriate choices, not just for themselves, but also for their offspring and future generations? The potential uses of genetic technologies have revolutionized our understanding of biomedical therapeutics and enhancement. However, such innovations have raised not only bioethical questions about personal capabilities, but also potential diachronic implications for future generations.

When biomedical interventions seek to be both lifesaving and life enhancing, how do we decide the boundary between enhancement and therapy? Consideration of genetics alone does not enable us to answer such questions. Instead, they require a deeper evaluation of how we individually, and collectively, understand concepts such as quality of life and suffering, and alleviation thereof. It is also important to consider that few diseases confer such a strong causal link, such as is seen in CF, between genotype and phenotype. The human phenotype comprises, not just physical characteristics such as eye color or a faulty protein, it also constitutes features such as behavior and general disposition. How far do genes account for these other aspects of our phenotype? Studies of twins reveal that they can share identical genotypes, yet display distinct phenotypes. Individual genotypic definitions of the human person are therefore limited and require us to consider broader phenotypic models that incorporate features such as our behavior and general disposition.

The Role of Genes in Determining Human Psychology: Am I My Thoughts and Actions?

Crick believes the role of genes determine thoughts and actions. He remarks, "You, your joys and your sorrows, your memories and your ambitions, your sense of personal identity and free will, are in fact no more than the behavior of a vast assembly of nerve cells and their associated molecules" (Crick 1994, 3). Other scientists, on the other hand, caution against such genetic determinism by emphasizing that "stretches of DNA that we now call genes do nothing on their own. They are simply databases used by the organism as a whole" (Noble 2008, 18). Denis Noble goes on to argue that genes are "captured entities, no longer having a life of their own, independent of the organism" (Noble 2008, 18). Dawkins also recently clarified that in his thesis, "The Selfish Gene," the "anthropomorphic personification" (Dawkins 2016, xi) of DNA that was initially employed ought to be clarified. He explains that "no sane person thinks DNA molecules have conscious personalities" (Dawkins 2016, xii) and that the personification of molecular structures simply offers scientists a "didactic device" (Dawkins 2016, xii). The disambiguation of Dawkins notwithstanding, there have been increasing misconceptions about the role of genes in determining behaviors and psychologies. Studies show growing cognitive biases in terms of genetic essentialism where people consider that their lives, thoughts and behaviors are an inevitable reflection of their genotype (Dar-Nimrod and Heine 2011).

The discussion hitherto has been on the role of genes in determining the human person with regards to our species identity as *Homo sapiens*, and to their lesser role in determining our individual biological composition through physical phenotypic determination. As persons, however, our identity goes beyond belonging to a species and having a physical or bodily identity. Some authors have suggested that "persons are separate entities to human beings" (Savulescu 2009, 220). Charles Taylor, in his ground-breaking account of "Human Agency and Language," elucidates the concept of a person. He writes:

> Where it is more than simply a synonym for 'human being,' 'person' figures primarily in moral and legal discourse. A person is a being with *a certain moral status*, or a bearer of rights. But underlying the moral status, as its condition, are *certain capacities*. A person is a being who has a sense of self, has a notion of the future and the past, can hold values, make choices; in short, can adopt life-plans. At least, a person must be the kind of being who is in principle capable of all this, however damaged these capacities may be in practice. (Taylor 1985, 97)

Moral philosophers, such as Singer, argue that what characterizes us as persons and confers upon us certain moral implications, is not membership within the species: *Homo Sapien*, but rather, the particular properties of rationality and self-consciousness (Singer 2011, 83-87). Both of these accounts indicate that a person is someone who is able to retain a sense of self over time and hold values and preferences into the future. Additionally, Julian Savulescu comments that a necessary, but not sufficient, component of persons is their "capacity to act from normative reasons, including moral reasons" (Savulescu 2009, 243). He also explains that "animals have desires and wants about what to do. Humans alone have beliefs about what they should do" and that they "sometimes act on the basis of these" (Savulescu 2009, 244).

Such philosophical narratives imply a complex interplay between genetic predispositions, rationality and how we relate to our past and future selves. Yet there are also claims that particular genetic traits predispose to certain actions that challenge notions of individual reasoning and will, and subsequent notions of responsibility. For instance, the XYY or super-male karyotype, which has been linked with criminal behavior (Farrell 1969), has been employed as a legal defense. Individuals arrested for a crime who were found to have this genetic makeup argued that, due to a genetic predisposition, they could not be held responsible and should thus not be considered criminally accountable for their actions. This so-called "XYY defence" has broadly been rejected and, in cases where it was accepted, it was on the grounds of diminished mental capacity, where the defendant was confined to a mental institution (Annas 1997, 158). Be that as it may, the "XYY defence," and genetic myths (Fox 1971) implying genes are the "metaphorical locus of our fate" (Wolpe 1997), bespeak not only a nascent understanding of how and if genes impact our ability to reason and act, they exhibit our willingness to believe that such faculties are beyond our conscious control (Dar-Nimrod and Heine 2011). Sigmund Freud, for example, famously posited that we are not rationally driven beings and are entirely subject to our subconscious (Zwart 2009). Freud's psychoanalysis thesis forces a further reconfiguration of our understanding of the human person and the role of the human genome in the latter's determination. Current and future developments within science, psychology, philosophy and ethics will continue to inform such discussions.

It is important to reflect here that recent research has also elucidated the role of genes in determining non-physical characteristics. Studies have illustrated the genetic heritability of a range of cognitive abilities, including intelligence (Plomin and Spinath 2004). However, these links are posited currently as possibilities and not accurate predictions. Such advances also raise ethical concerns about future possibilities of artificially selecting for such genes. The ethical implications of "designer descendants" are vast; including how such interventions may alter our capacity for moral reasoning. If these designer descendants are genetically selected and programmed to be morally infallible, then are such entities still considered persons? Such interventions would have an unprecedented bearing on our understandings of free will, consciousness, determinism, responsibility and our aspirations for and obligations towards future generations of "persons."

The foregoing displays how genetic advancements, which have offered new perspectives on our understanding of the human person, have also stimulated novel ethical deliberations. Accounts presented above indicate that there are many gaps in our construction of the human person when relying solely on the genome. Ethical deliberations around speciesism, genetic determinism, enhancement, capacities, will and responsibility that are stimulated by the human genome also require broader examination through additional sources of knowledge. The scientific, ethical, philosophical and legal accounts of the human person presented above suggest that more work needs to be done to characterize the human person, in terms of personal and collective traits, before the associated ethical considerations can be fully addressed and scrutinized. Now we offer perspectives from the Islamic tradition and, in particular, accounts on the human soul, which offer novel perspectives on aspects of the human person, and the associated ethical considerations surrounding the genome question.

On the Relationship between the Soul and the Human Person within the Islamic Tradition: Definitions, Perspectives and Implications for the Genome Question

Knowledge about the concept of the human person is informed, not only through biological and scientific advances as they relate to the genome, and other fields of inquiry mentioned earlier, but also through religious and spiritual traditions which offer specific insights into the non-physical dimensions of the human person.

The concept of the human person within the Islamic tradition commonly features in bioethical discussions concerning abortion and the specific events marking the beginning and end of life (Brockopp 2008; Ghaly 2012; Shaw 2014). In these contexts, the human person is defined in accordance with the status conferred upon it by the movement of the soul as it enters or leaves the human body. A broader consideration of what constitutes the human person, and the nature of the soul, according to the Islamic tradition, is not only helpful, but also necessary before assessing the impact and implications for the genomics project.

Key Qur'ānic terms that relate to the physical and spiritual aspects of the human person are identified and explored within various disciplines of Islamic thought. Here we will present a narrative on the human person that combines beliefs about the body, soul and spirit - their origins, existence, fate, and purpose of creation. The primary focus is on highlighting the complexity that the existence of a soul adds to a conceptualisation of the human person. Where possible, the implications for the genomics project are touched upon, insofar as they may impact the soul's propensity for reflecting on and returning to its pure and natural (*fiṭrah*) primordial state, as originated by God.

Defining Key Terms Related to the Human Person: Qur'ānic, Theological and Philosophical Perspectives

This section surveys Qur'ānic terms related to the human person, which are further elaborated in exegetical, theological and philosophical sources. Table 1 displays these terms, their associated translations and their frequency of occurrence in the Qur'ān.

TABLE 1 *Key terms from the qur'ān relating to the human person*

Arabic term in Qur'ān	Translation	Frequency of term in Qur'ān	Note	References of Qur'ānic verses mentioning term
Insān	Man, mankind, human beings	71*		(2:60) (4:28) (7:82) (10:12) (11:9) (12:5) (14:34) (15:26) (16:4) (19:67) (21:37) (22:66) (23:12) (82:6) (89:23) (96:5) (100:6)
Ins	Men, mankind	18		(6:112) (6:128) (6:130) (7:38) (7:179) (17:88) (27:17) (41:25) (41:29) (46:18) (51:56) (55:33) (55:39) (55:56) (55:74) (72:5) (72:6)
Bashar	Man, human, human beings	37*		(3:47) (3:79) (5:18) (6:91) (11:27) (12:31) (14:10) (14:11) (15:28) (15:33) (16:103) (17:93)
Arḍ	Earth	2	Arḍ occurs 461 times, but only on two occasions does it refer to the human being as produced from earth (arḍ)**	(11:61) (53:32)
Ṭīn	Clay	8	Material out of which the human being was fashioned	(6:2) (7:12) (17:61) (23:12) (32:7) (37:11) (38:71) (38:76)
Turāb	Dust	14	Half of the verses refer to the human being's creation from dust, other half refer to the human being's return to dust	(3:59) (13:5) (18:37) (22:5) (30:20) (35:11) (40:67) (23:35) (23:82) (27:67) (37:16) (37:53) (50:3) (56:47)
ṣalṣāl	Clay	4	Three of the references describe the clay as being from an altered black mud	(15:26) (15:28) (15:33) (55:14)
asad	Body	1	Refers to the prophets' forms	(21:8)
Badan	Body	1	A purely physical bodily form after the spirit has departed	(10:92)

TABLE 1 *Key terms from the qur'ān relating to the human person* (cont.)

Arabic term in Qur'ān	Translation	Frequency of term in Qur'ān	Note	References of Qur'ānic verses mentioning term
Qalb	Heart	132*		(2:7) (2:10) (2:74) (2:88) (2:93) (2:204) (2:283) (3:7) (3:126) (3:156) (4:155) (5:41) (6:25) (7:100) (7:179)(9:93) (9:127) (10:88) (26:89) (33:26) (41:5) (49:7) (50:33) (57:16) (61:5) (66:4) (74:31)
Fu'ād	Heart	16		(6:110) (6:113) (11:120) (14:37) (14:43) (16:78) (17:36) (23:78) (25:32) (28:10) (32:9) (46:26) (53:11) (67:23) (104:7)
ṣūra	Form	3	The perfected form of human beings	(40:64)(64:3) (82:8)
Taqwīm	Mould	1	The stature or mould of human beings	(95:4)
Nafs	Soul	290	In an additional 5 verses, God refers to 'Himself' using the same root word nafs. Occurs as a noun denoting 'soul,' 'self,' 'person,' 'mind,' dependent upon context	(2:9) (2:44) (2:48) (2:54) (2:72) (2:123) (2:155) (3:93) (5:32) (5:70) (5:80) (5:116) (6:152) (7:42) (7:189) (9:81) (9:118) (11:21) (12:53) (13:42) (17:15) (21:35) (39:6) (41:31) (50:21) (82:19) (91:7)
Rūḥ	Spirit	21	References the spirit of God, the Holy spirit, the spirit in human beings, inspiration, as well as the angel Gabriel	(2:87) (2:253) (4:171) (5:110) (15:29) (16:2) (16:10) (17:85) (19:17) (21:91) (26:193) (32:9) (38:72) (40:15) (42:52) (58:22) (66:12) (70:4) (78:38) (97:4)
'Aql	Intellect, reason	49*	Not mentioned as a noun. Occurs as form I verb 'aqalu meaning to understand, reason, intellect	(2:44) (2:73) (2:76) (2:164) (3:65) (3:118) (5:58) (5:103) (10:16) (12:2) (29:43) (30:24) (36:68)
Fiṭrah	Nature	1	Original, natural state upon which God created human beings	(30:30)

* Terms with a large frequency where only select references are mentioned.

** Nine Qur'ānic verses mention God making humans 'vicegerents' (*khalīfah*) on earth (*arḍ*). Four of those mentions (Q. 2:30; 6:165; 27:62; 35:39) refer to vicegerents on earth in the general sense. It appears that being a vicegerent on earth is a property of humans that is intrinsically tied to their physical existence on earth. Humans, produced from earth, sent through the Fall to be vicegerents on earth, must inevitably return to earth through death, and be resurrected again from earth, before continuing their journey to the hereafter.

The term *nās* (translated as people, or mankind) occurs 241 times as a noun in its plural form. This has been omitted from the table because it is employed as a general address to humans, which is extraneous to the specific physical and spiritual constitution of the human person that this paper addresses.

In the Qur'ānic narrative, man (*insān/bashar*) was created from clay (*ṭīn/ṣalṣāl*) and dust (*turāb*) (Q. 32:7; 15:28; 35:11), and fashioned (*ṣawwara*) into the best of forms (*ṣuwar*) (Q. 64:3), before God blew into him, of His spirit (*rūḥ*) (Q. 38:72). Elsewhere in the Qur'ān, God states that "He created you from a single soul (*nafs*)" (Q. 4:1). Man is thus constituted of three key elements: the form (encompassing the body), the soul, and the spirit, and has been endowed with a physical and spiritual existence, both of which are central to a conception of the human person.

Yet, as Table 1. clearly demonstrates, these terms are not equally significant in the Qur'ānic narrative. Of all the terms outlined, one stands out as being the most important: *nafs*. This term features in the Qur'an more than fourteen times its potential synonym, *rūḥ*. It is used more than twice as many times as its nearest related term (*qalb*, 132), and it is cited with a significantly higher frequency as compared to all the other terms under consideration combined (290: 379). There can be little doubt, then, as to its centrality from a Qur'ānic perspective. Any study that makes bold to address the notion of the human person in the Islamic tradition, therefore, must place *nafs* at the centre of that endeavour. This is the reason *nafs* shall assume pride of place in this paper, with all other terms being examined as they are connected to or affiliated with it.

Nafs and *rūḥ*: Distinct Entities or Synonyms?

A person's metaphysical and spiritual existence is predicated on, and determined by, the existence of *nafs* (soul, self) and *rūḥ* (spirit). From Table 1, it can be seen that the Qur'ān mentions *nafs* far more frequently than *rūḥ* (290:21, respectively). The vast array of verses that mention *nafs* provide copious information about its nature, its different levels and functions, and how it sometimes inclines towards good and, at other times, away from it. On the other hand, the Qur'ān is far less expansive in its elucidation of *rūḥ*: "They question you about the Spirit (*rūḥ*). Say: the Spirit is from the command of my Lord. And you have not been given knowledge about it, save a little." This verse bespeaks an essential numinosity, which is why many scholars chose not to elaborate on matters pertaining to the spirit.

The renowned exegete, Abū 'Abd Allah al-Qurṭubī (d. 671/1273), writes "it [the *rūḥ*] is a great matter and a significant affair from the matters of God that He has obscured for us and not explained, so that man realizes his apodictic incapacity to know the reality of his self, whilst knowing that [his reality] exists. And if man is so incapable of knowing his own reality, then he is, *a fortiori*, incapable of knowing the reality of God" (Qurṭubī 2004, 10: 210).

Some scholars hold that *nafs* and *rūḥ* are one and the same, thus the terms for each can be used interchangeably to indicate the masculine and feminine forms of the soul. Others, like Saʿīd Ḥawwa, distinguish superficially between the two entities, and posit that *rūḥ* is the form of the spirit's independent existence prior to being affiliated with a body, and that *rūḥ* resides as *nafs* after it is "clothed in the body" (Saʿīd Ḥawwa 1995). Thus, *nafs* is a term for the body being fertilized with the spirit.

Not all orthodox scholars accepted the synonymy of *nafs* and *rūḥ;* the Sufi exegete, 'Abd al-Karīm al-Qushayrī (d. 465/1072), for instance, comments that, "*rūḥ* is the locus of subtle states (*aḥwāl laṭīfah*) and praiseworthy actions, just as sight is the locus of visions… the *rūḥ* is [thus] the locus of all praiseworthy attributes, just as *nafs* is the locus of all blameworthy ones" (Qushayrī, 4: 304).

The Qur'ān assigns three states to *nafs*: the satisfied soul (*nafs al-muṭma-innah*) (Q. 89:27); the self-accusing soul (*nafs al-lawwāmah*) (Q. 75:2); and the soul inclined to evil (*nafs al-ammārah bi'l-sū'*) (Q. 12:53). That *nafs* is associated with states inclined towards actions, both good and bad, means that it offers a tangible link between the physical and metaphysical dimensions of a person.

Others were more succinct, and offered a deeper distinction between *nafs* and *rūḥ*, stating that "the latter is that whereby is life; and the former, that whereby is intellect (*ʿaql*), or reason; so that when one sleeps, God takes away the *nafs*, but not the *rūḥ*, which is not taken save at death" (Lane, Arabic Lexicon, 1111). Opinions of scholars citing a significant difference between *nafs* and *rūḥ* present a perspective that is more coherent with the Qur'ānic characterization of both aspects of the human person. In the Qur'ān, God attributes *rūḥ* to Himself, as part of His command, whereas *nafs* is more directly associated with humans. Were *nafs* and *rūḥ* similar in their origin, existence and function, the Qur'ān would not have distinguished between the two so clearly, by specifying that *rūḥ* is from the matter of God, of which we know little, whilst offering far greater insight into *nafs*.

A consideration of *nafs,* therefore, is more conducive to constructing a framework connecting the physical and metaphysical aspects of the human person, particularly through the lens of human action. That there is an intimate connection between the body (*jasad, jism, badan*) and the *nafs,* it is impossible to deny. But how are they connected, and how, and to what extent, do actions of the body influence the soul?

Nafs and Body

The body and soul are intimately connected. The mystic Muhammad b. al-Ḥusayn al-Sulamī (d. 412/1021), writes that 'the soul is a band of rays of the Reality whose traces differ in bodies' (al-Sulamī 2001, 1:395). The soul is the receptor of the Divine ray, whose output is through the body. The soul thus percolates through the actions of the body, and so, would it be too outlandish to assert that the actions of the body, in mutualistic fashion, affect the soul? Before we consider this specific conundrum, we must look at the general relationship between the body and the soul, and whether the human person is the body alone, or the body—soul amalgam.

The Qur'ān asserts that 'every soul (*nafs*) shall taste death' (Q. 3:185). The term employed in this verse, *nafs,* is significant because there is unified opinion that the soul (*nafs*) does not die. It is the body that dies, its death marked by the departure of the soul. Thus, in this verse, the term *nafs* more literally represents the body or person, who experiences death. It is indeed the case that the Qur'ānic usage of the term *nafs* oscillates between 'soul' and 'self' and the 'human person' emphasizing the inextricable association of the body and soul. The Qur'ān's deliberate application of the term *nafs* to indicate both physical and metaphysical realities supports the body-soul amalgam of the human person, addressing what is visible: the body, and that which isn't visible: the soul.

The nature of the interaction between the body and the soul was of central concern to Muslim philosophers and theologians. Their debates about the nature of the human person took place within wider discussions about human agency and theories of action, as they sought to understand nature, their role as humans, and the role of God (Wolfson 1976; Calverley and Pollock 2002). Their discussions and the questions they raised resonate within our current scientific milieu, as we tread the boundaries of what is natural and question again what it means to be a human person.

Many opinions circulated among theologians, as al-Ashʿarī (d. 324/935) in his doxographical work, *Maqālāt al-Islāmiyyīn*, catalogues. The discussions ranged from whether the body alone could define man, that is, "the individual that you see," or whether accidents (*aʿrāḍ*), too, formed part of the body. A group opined that the body itself is man and its accidents are not part of it. It was also debated whether the soul was regarded as a separate substance (*jawhar*) to man, in addition to the body. Ḍirār b. ʿAmr (d. 200/815?), rejecting the idea of a spirit or an immaterial soul, remarked "man is made up of many things, namely: colour, taste, smell, the ability to touch and so forth. They constitute man whenever they are combined and there is no separate substance other than these." Others, like the dualist Bishr b. al-Muʿtamir (d. 210/825), disagreed, "man is body and spirit, and these two together constitute man" (Ritter 1929, 329). Writing a couple of centuries later, Al-Rāghib al-Iṣfahānī (d. 502/1108) conveyed his understanding of the role of the body and soul, respectively: "man (*insān*) is composed of a body (*jism*) with the faculty of sight (*baṣar*), and a soul (*nafs*) with the faculty of insight (*baṣīra*)'... the soul is the locus of spiritual accidents, and the body the locus of bodily accidents" (Mohamed 2006, 456).

In his capacity as a celebrated physician, Avicenna (d. 429/1037) wrote about the human body, most notably in his Canon of Medicine, whilst as a distinguished philosopher, he wrote on the human soul, and how intellectual perfection allows the soul to attain ultimate happiness. Avicenna, adopting the Aristotelian concept of the soul, defines it in his *magnum opus*, *al-Najāt* as "the first entelechy (perfection) of a natural body possessing organs that potentially has life" (Rahman 1952, 25). The soul, therefore, is the defining feature that perfects a body to make it part of a species and distinct from other species. Avicenna's conception of soul as "entelechy is wider than that of form" because forms necessarily subsist in matter, whereas entelechy allows the soul to be associated with the body without exclusively being a form inherent in the body (Rahman 1952, 9).

Avicenna postulates a tripartite division of the souls (vegetative, animal, and human). These are bifurcated into the physical (vegetative and animal) and spiritual (human) souls, with the former "passing away upon the death of the body" and the latter (human, spiritual) being classified as an "independent and immaterial substance" (Rahman 1952).

Although the foundations of an immaterial soul were laid by the Ancient Greeks, with Plotinus asserting that the 'soul is not in the body as in space, it is not related to the body as form to matter, as a whole to a part or a part to a whole,' it was Avicenna, who, through his unique thought experiment of 'the flying man,' helped explain the immaterial nature of the soul as a form of consciousness and self-awareness. He thus opined that the soul, by its very nature, has self-knowledge (Reisman 2004).

The immortality of the soul is another unique aspect of Avicenna's definition of the soul. In chapter 13 of his Najāt, the philosopher writes, "we say that the soul does not die with the death of the body and is absolutely incorruptible. As for the former proposition, this is because everything, which is corrupted of something else, is in some way attached to it. And anything, which is in some way attached to something else, is either coexistent with it or posterior to it in existence or prior to it, this priority being essential and not temporal (Rahman 1952, 58).

Avicenna's theory of the substantiality of the soul and the assertion of the separability of the human rational soul from the body grants the soul an independence from the body. Goodman comments on the idea that the soul does not exist in the body as mere form in a substrate. He states that "human actions are not to be conceived solely in terms of the behavior of the body, and are not reducible to physical terms or explicable wholly and solely by reference to mechanical events… souls can affect bodies; it is not always a case of bodies affecting souls… this thesis is crucial to our ability to maintain or restore the idea that a person is an agent, that human thought, action and experience are not adequately described or explained in mechanistic terms" (Goodman 1992, 161).

The soul's experience of life on earth is intrinsically tied to the body. al-Attas writes that "the human body and the world of sense and sensible experiences provide the soul with a school for its training to know God" (al-Attas 1995, 175). Further questions can also be posed about whether the body and soul are independent in the spiritual sense, since physical actions have been shown in the Qur'ān and other writings, to have an effect on the human person's spiritual existence, and eternal fate. To disinter the full complexity of this relationship, we must first look at the point when the body and soul became united.

How Old is the Association between the Body and the Soul? The First Covenant

The story of creation, which details different stages and events in the origin of the human person and its purpose for creation, is useful in providing further insights into the nature of the soul and how it relates to physical, psychological and spiritual aspects of the human person, in the space-time realm which precedes that of the human being's earthly existence. In its primordial existence, as described in the Qur'ān, the human being took his first covenant: "And (remember) when your Lord brought forth from the Children of Ādam, from their loins, their seed, and made them testify about themselves, (saying): Am I not your Lord? They said: Yes, verily. We testify" (Q. 7:172).

Scholars, who have commented on this first covenant and the nature of the extraction of human beings from the loins of Ādam, differ concerning the form of human existence during the event. Some assert that the covenant was taken in a corporeal form, whereas others hold that it was just the souls that were gathered for this moment and testified. Ibn Jarīr al-Tabarī (d. 310/923) cites ʿAbd Allāh ibn ʿAbbās (d. 74/687), the exegete *par excellence*, as having said, 'When God created Ādam, He extracted his progeny from his back [and they were] like atoms (*dharr*).' Al-Tabarī is more categorical as to the constitution of Ādam's progeny, when he states that upon extraction from Ādam, his progeny was 'in the form (*hay'a*) of atoms (*dharr*)' (al-Tabarī 2005, 6:111). The lesser-known exegete, Abū Isḥāq al-Thaʿālibī (d. 427/1035), avers that the souls were either like atoms (*dharr*) or mustard seeds (*khardal*). He then quotes Muhammad ibn Kaʿb (d. 108-120/726-738?) who remarks, "they were souls (*arwāḥ*) to whom a task was given' (al-Thaʿālibī 1996, 1:586). Al-Qurṭubī also records the viewpoint that the covenant was taken pre-phenomenally. He writes, "God extracted the souls (*arwāh*) [from Ādam] before their bodies (*ajsām*). He also mentions, a somewhat cryptic notion that "He, be He exalted, extracted simulacra (*ashbāh*) [of bodies] in which were souls from the back of Ādam, … and God made them cognisant (*ʿuqūl*)….' (Al-Qurṭubī 2004, 7:200). Precisely what state the pre-existent souls or bodies existed in, and their respective modes of testifying are unclear. Furthermore, it is not known whether it was a testimony through speech, or through their very being before God. However, it is clear that created beings possessed a knowledge of God and themselves, in the earliest moments of the creation story.

The timing of the first covenant is also disputed, with opinions existing on both sides claiming that the covenant took place either before, or after, earthly existence. Ismāʿīl ibn ʿAbd al-Raḥmān al-Suddī (d. 127/745?) believes that it occurred on the lowest heaven (*al-samāʾ al-dunyā*), which would indicate it was before earthly existence, but, significantly, still after the fall. Al-Ḥasan al-Baṣrī (d. 110/728) believes that after the progeny of Ādam was extracted from his back and the covenant was taken, it was returned to him (Al-Qurṭubī 2004, 7:201). Accounting for the different opinions, a potential sequence of events could then be: the creation of Ādam and Ḥawāʾ; the fall of man to the lowest heaven; the progeny extracted from Ādam's back; covenant taken; progeny returned to Ādam's back; Ādam lowered to earth. When addressing a conception of the human person, it may be helpful to consider the implications of a pre-earthly existence, and give further thought as to how aspects of the unity and uniqueness of creation contribute to an understanding of the human person.

Notwithstanding the differences of opinion vis-à-vis when the body and soul came to be united, it is clear that the association of the body and soul, at least in the opinion of many scholars, predates our corporeal existence in the world. We have not yet considered the specific implications of actions on the soul. For this, we must look at the first act of human transgression, and whether it had an effect on the soul.

Fall to Earth: a Case Study of the Interaction between the Body and *nafs*

One of the most significant events in the creation story is the Fall of Ādam and Ḥawāʾ from heaven to earth, after they consumed fruit from the forbidden tree. The most detailed account of this oft-mentioned narrative in the Qurʾān is found in The Heights (*al-Aʿrāf*):

> "O Ādam, dwell, you and your wife, in Paradise and eat from wherever you will but do not approach this tree, lest you be among the wrongdoers." But Satan whispered to them to make apparent to them that which was concealed from them of their private parts. He said, "Your Lord did not forbid you this tree except that you become angels or become of the immortal." And he swore [by God] to them, "Indeed, I am to you from among the sincere advisors."
>
> So, he made them fall, through deception. And when they tasted of the tree, their private parts became apparent to them, and they began to fasten together over themselves from the leaves of Paradise. And their Lord called to them, "Did I not forbid you from that tree and tell you that Satan is to you a clear enemy?"
>
> They said, "Our Lord, we have wronged ourselves, and if You do not forgive us and have mercy upon us, we will surely be among the losers." God said, "Descend, being to one another enemies. And for you on the earth is a place of settlement and enjoyment for a time. Therein you will live, and therein you will die, and from it you will be brought forth" (Q. 7:19-25).

In this account, it is unclear precisely what form characterized the existence of Ādam and Ḥawāʾ in paradise: was it as corporeal or incorporeal beings? Or perhaps it was an existence in between or neither. Further, what role did the form, whether it existed corporeally or ethereally, play in the decision to consume the fruit and what effects manifested in their form as a ramification of that act? Finally, what is one to infer from the notion that their private parts were exposed? Though there is a farrago of opinions on each of these issues, what can be concluded is that their physical act had both physical and spiritual consequences for their existence, which caused them to repent and be descended to the earth.

Al-Ṭabarī, in his exegesis, mentions that there was "a light over their private parts" that dissipated when they ate the forbidden fruit (Al-Ṭabarī 2005, 5:449). Al-Thaʿālibī comments that a number of exegetes view the manifestation of Ādam and Ḥawā's private parts (*saw'a*) as denoting folly; thus ingestion of the forbidden effected an alteration in their consciousness and awareness of self. He writes, 'this phrase means only that there are imperfections (*maʿā'ib*) and what diminished their state was exposed to them, it does not denote "private parts"' (Al-Thaʿālibī,1996, 1:534). Al-Rāzī is even more explicit that the manifestation of *saw'a* could connote deterioration in their spiritual state. He writes, 'the manifestation of their *saw'a* is a metonym (*kināya*) for their loss of sanctity (*ḥurma*) and dissipation of dignity (*jāh*)' (Al-Rāzī 2004, 14:39).

Al-Qurṭubī declares that there was an actual alteration, though not in the physical constitution of Ādam and Ḥawā', but in the light surrounding their private parts, which still reflected a visible change in them. He writes, "their private parts only became manifest unto both of them, not to anyone besides them, as there was [theretofore] a light (*nūr*) over them, such that their private parts could not be seen, so the light disappeared [when they ate the forbidden fruit]" (Al-Qurṭubī 2004, 7:115).

It is evident from this, and other accounts, that the physical act chosen by Ādam and Ḥawā', effected a change to their consciousness and psyche, and contributed to how they perceived themselves and their bodies. This creates a psychological and physical link between the body and *nafs*, and raises the question of how physical acts may impact our psychological and physical states, as well as the physical environment around us. The actions of Ādam and Ḥawā' with their physical, psychological and spiritual consequences, demonstrate that inasmuch as human beings act and produce actions, they and their physical and spiritual fate are also impacted by their actions. In the context of the Islamic tradition, one could go further; in questioning what effect this early event in creation may have had on the collective psyche of humans, and whether this could affect individual notions pertaining to conceptions of the human person. The human genome project which opens up the potential for new knowledge and actions, should consider how it is that this knowledge and subsequent actions could indeed shape what it means to be a human person in its multifaceted dimensions.

Returning to the Fall, whether or not there was a physical change in Ādam and Ḥawā' following their act of transgression, most scholars agree that there was a metamorphosis in perception and cognizance. This would suggest an intimate relationship between *nafs* and the intellect (*ʿaql*). Indeed, Ibn Sīnā believes the two to be so inextricably intertwined that he predicates happiness of the *nafs* upon achieving intellectual perfection. It thus behooves us to investigate this link further.

Nafs and *ʿaql*

The ability to reason and engage in rational thought is common to several conceptions of the human person. In the Islamic tradition, the intellect (*ʿaql*), and the act of reasoning, located in the soul, is a central step in the journey of human beings fulfilling their purpose for creation. The capacity to reason is called upon in order to establish and strengthen faith (*īmān*), and for the performance of virtuous actions. Linguistically, "*ʿaql*, is derived from the expression for the strapping of a camel in order to prevent it from running away...the intellect is thus imagined as a cause for man being restrained from practicing that which is not ethically beautiful"[3] (ʿAjamī 1985).

The Qur'an repeatedly calls humans to think, reflect and reason, using the root word *ʿa-q-l*. Not restricting itself to this term alone, it also employs a number of other terms, such as *fahm* (understanding) (Q. 21:79), *nuhā* (intelligence) (Q. 20:54), *ḥijr* (intelligence) (Q. 89:5), *ʿilm* (knowledge), *tafakkur* (reflection) (Q. 13:3), *tadabbur* (contemplation) (Q. 38:29), *lubb* (inner heart) (Q. 3:190) and *ḥikma* (wisdom) (Q. 2:269). The strenuous emphasis on reason in the Qur'an may have inspired the preoccupation of Muslim scholars and philosophers to speculate about the intellect, as evidenced by their numerous epistles[4] on its definitions, loci, nature and functions, as well as ontologically placing the intellect within their respective cosmic schemes of the universe (Davidson 1992).

Islamic philosophers assiduously studied the Greek tradition wherein one of the most widely discussed topics was the development of the theoretical intellect (*nous*) in man, as elaborated by Aristotle and his commentators. This resulted in a more prominent focus on knowledge of the soul (as a synonym of the intellect), such as we find in Aristotle's *De Anima*. Some philosophers distinguished between the intellect and the soul, with the former being envisaged

3 Ibn Taymiyya (d. 728/1328) cites Abū l-Barakāt al-Baghdādī (d. 560/1164?) in his book *al-muʿtabar fī'l-ḥikmah* who mentions: "In Arabic, *ʿaql* means the thing which controls man's whims and desires... This thinking element which controls whims is called *ʿaql* because it prevents man from carrying out his intentions in the same way that the rope called *ʿuqāl* binds the camel, preventing it from moving to any place it wants.' (Salim 1979)

4 Examples of such works include Abu Nasr Fārābī's (d. 339/950), *Risālat fī'l-ʿaql* and Miskawayh's (d. 421/1030) *Min kitāb al-ʿaql wa al-Maʿqūl*.

as a spiritual and incorporeal substance, while the latter being conceived of as something linked to the body as a potential intellect.

For Avicenna, the rational soul has two components, namely, a practical intellect and a theoretical intellect. The practical intellect is the lower, downward facing, active component of the soul that is responsible for movement, controls bodily appetites and governs all other faculties of the body. Through the act of deliberation, and the help of the theoretical intellect, it forms the ordinary and commonly accepted opinions, concerning actions and other premises. Good ethical behavior is a result of a successfully functioning practical intellect.

The theoretical intellect, on the other hand, is on a higher plane, and this is the soul's upward facing component, it is passive and receives information, through contemplation from the celestial intelligences. This is the faculty responsible for the pure cognition of truth, receiving and acquiring intelligibles, and impressions (imprinted on the mind) of universal forms (ideas) abstracted from matter, through a connection with the Active Intellect (*ʿaql faʿʿāl*).

The Active Intellect is an intelligence that is always in actuality; it bestows intelligible forms on the potential intellect. In Avicenna's cosmological model, the Active Intellect arises from the emanation scheme, in a top-down hierarchical structure for the flow of existence and thought in the universe, starting with God. Avicenna includes three stages of potentiality before the intellect becomes an actual intellect. "The first is the material intellect (*ʿaql hayulānī*) which is a potentiality for thinking, the second which is possible potentiality (*ʿaql bi'l malaka*), and possesses principles of knowledge; while the third is the perfection of this potentiality, the actual intellect (*ʿaql bi'l-fiʿl*)' (Rahman 1952, 89).

Thus, despite the crucial role of the body, human intellection is not purely a bodily function, nor is it entirely individual. Islamic schemes of intelligence rely on the soul, and a cosmic intelligence, "which broadcasts an undifferentiated range of forms," in the wider cosmological schemes of the universe. Islamic conceptions of the human person may be enriched by accounting for the individual's ability, in accordance with its soul-body amalgam, of being able to reason and acquire knowledge. This knowledge, in the Islamic tradition, includes knowledge of the self and God, for "He who knows himself, knows God."

The acquisition of knowledge, nevertheless, is not the only conduit to knowing oneself and one's Lord. Indeed, inspired by the Qurʾān, Abū Hāmid al-Ghazālī (d. 505/1111) clearly advocates a path to God through repristination of the heart, and the soul. In so doing, he forges a connection between the two most significant terms, from a Qurʾānic perspective, that are here considered: *nafs* and *qalb*. Tying the diverse threads of intellect, *nafs*, and heart (*qalb*) together, and considering these in light of the human genome project - which has the capacity to affect and influence some, if not all, of these loci of intellection - remains a challenge that has to date, not only not been met, but also has not been considered.

Nafs and *qalb*

The Qurʾān predicates the human being's successful return to God upon a sound heart (*qalb salīm*) (Q. 26:89; 37:84) and a purified soul (*nafs zakiyyah*) (Q. 91:7-9). As the human person is described in physical and metaphysical terms, so do Qurʾānic references of the heart that similarly signify both physical and spiritual dimensions. Although Qurʾānic descriptions of the heart include words loaded with physical connotations such as diseased (Q. 2:10), sealed (Q. 2:7), and hardened (Q. 5:13), the overriding emphasis remains very much on the non-physical heart. *Nafs* and *qalb* share traits that are described in the Qurʾān in similar ways: both have the capacity to "earn" deeds (*'nafsin mā kasabat'* (Q. 2:281) and *'kasabat qulūbukum'* (Q. 2:225)) and as a consequence (of earning good ones) have the potential to reach a state that is *mutmaʾin* (*'an-nafs al-muṭma-innah'* (Q. 89:27) and *'taṭma-inn al-qulūb'* (Q. 13.28)). It can thus, be supposed that *nafs* and *qalb* each contribute significantly to the physically manifest metaphysical conceptualization of the human person.

Contrary to Islamic philosophical schema wherein intellect is localized in the *nafs*, the Qurʾānic paradigm designates the heart as the locus of reflection (*yatadabbaru*) (Q. 47:24), reasoning (*yaʿqilūna*) (Q. 22:46), and understanding (*yafqahūna*) (Q. 7:179). The heart deliberates, makes judgements, and becomes resolute on decisions for which it will be held to account, and in cases deemed sinful (Q. 2:225; 2:283). On several occasions the Qurʾān makes mention of *qalb* alongside hearing (*samʿ*) and sight (*baṣar*) (Q. 2:7; 6:46; 22:46), thus connecting the heart directly with sensory input. It calls human beings to use their hearts to reflect upon information garnered by their senses to arrive at faith. Thus, the heart is made the seat of faith (*fi qulūbihim al-īmān*) (Q. 58:22). This interaction of the heart with sensory input offers another consideration for the human genome project, where it ought to take into account the potential impact of the physical on other dimensions of the human person's existence and experience. In the Islamic narrative, the actions of a person influence the level of faith (*īmān*) in the heart, which can increase or decrease. Given that the heart is the locus of faith, the body therefore has the capacity to affect the state of the heart. *Qalb*, as a nexus between the sensory and spiritual worlds, may thus be the connective tissue that links the body—in its outward facing, physical role-- with the *nafs*—in its inward facing, spiritual one— thereby forging a body-*qalb-nafs* composite.

In his chef d'oeuvre, *Iḥyāʾ ʿulūm al-dīn*, under the chapter of "'ajā'ib al-qalb", al-Ghazālī deals with the terms *qalb*, *nafs* and *rūh*, with the focus being on the heart, the actions it inspires and the knowledge to which it may be privy. Purity of the heart, attained through acts of worship and remembrance of God, is a prerequisite for authentic knowledge of God, says the theologian. He draws a connection between the physical heart and the spiritual realms, between the limbs and the heart, and how the heart is influenced by the actions of the limbs (al-Ghazālī 2007, 1:858-86).

The foregoing views of theologians, philosophers and exegetes on the nature of the human person demonstrate that a range of positions was held regarding the physical and metaphysical dimensions that constitute the human person. These include the soul, the spirit, body, intellect, and heart, and their respective functions in human person's journey. Notwithstanding differences in the nature, method and comprehensiveness of these opinions, the essence is that an understanding of the human person cannot be limited to the physical, and that the spiritual and metaphysical dimensions must also be considered. This necessitates that an ethical evaluation of the human genome project should examine how it facilitates or hampers the journey of the human person towards God.

On the Interrelationship of the Soul, Genome and Human Person: a Primer for Future Research

The above discussions elucidate the inextricable connection of the soul to the human person, and how conceptions of the latter are being further informed by recent developments in genomics. Earlier sections of this paper presented the complex interplay between genes and the human psyche and how genes may confer aspects of our mental capacities. One mechanism of genetic traits determining cognitive states is through the phenotypic determination of our sensory receptors and neural circuitry. Recent neuropsychological studies have investigated the role that external stimuli play on emotions and the subconscious (Winkielman and Berridge 2011). As we rely on our sensory receptors to first collate information from external stimuli and our cortical and subcortical structures to process this data through internal modification, the embryonic formation and subsequent development of these devices all rely on precise genetic coding and programming. Yet, we cannot precisely determine how far this internal modification is carried out through our hard-wired neural circuitry. Despite this, we may reasonably assert that sensory input, at least to some degree, affects the intellect (*'aql*) in ways which we are neither aware, nor have taken cognizance of. Since the metaphysical relationship between the *'aql* and *nafs* and *qalb* has already been established, we can speculate a link between genes, their role in sensory reception and the subsequent influence on the soul. The soul plays an integral role in the performance of actions, and reciprocally, the actions of a person impact the soul. These attributes of the soul interconnect and overlap with the aforementioned descriptions of genetic traits and their probable correlation with physical attributes that confer the ability to perform actions. In the context of a discussion relating to faith, this relationship allows us to tentatively consider how it is that a potential alteration of the human genome may impact the soul - not in its constitution, but in its propensity for particular types of actions. Given that the actions of a person are considered to have an impact on the soul (*nafs*) and heart (*qalb*), we may cautiously posit that genomic alterations of the body, which can impact a person's actions, could in turn have metaphysical and spiritual consequences, by impacting the non-physical dimensions of a human person.

This paradigm can be further elaborated in terms of health, illness and suffering. Illness and disease in Islam are suffering incurred by believers that may act as a means of spiritual cleansing where religious transgressions are manifest as ailments, or as a means of elevating the devotee (Al-Shahri 2005, 432-6). Patiently endured, such suffering is considered a means of expiation of sins and thus a vehicle for increasing one's level of faith. Although the Islamic tradition urges the prevention of disease and the seeking of cures, if genetic alteration leads to a sterile or disease-free human genome then the question can be raised of how such advancements would impact existing understandings of suffering, spiritual cleansing and subsequent faith. Despite such possibilities, the role and interaction of genes with the environment in determining suffering remain. It may be argued that although genes can undergo "disease-free" modification, there are myriad environmental risks that leave man's fate prone to disease.

Furthermore, could a change in the capacity of persons, conferred through genomic alteration of organs and limbs, influence their accountability? Given that "God does not charge a soul except [with that within] its capacity" (Q. 2:286) how would manifestations of these altered capacities impact the soul? Additional to disease and therapy, how would concerns related to enhancement map onto Islamic understandings of human capacity and accountability? Elaboration of such questions, through future research, may further elucidate the interrelatedness of the genome, soul and human person that this paper seeks to initiate.

Conclusion

This paper explores the connection between the physical and non-physical dimensions of the human person to broaden the scope of ethical discussions concerning the human genome project. A multifaceted conception of the human person in the Islamic tradition is informed by differing beliefs about the body, soul and spirit - their origin and existence, nature of interaction, and purpose for creation. Through surveying key terms, and subsequent theories on the nature and interaction of the non-physical aspects of a human person with the body, the centrality of the soul for a person's experience and existence is established. The non-physical dimension of a human person, identified here as *nafs*, influences the body, and the body, as the physical dimension of a human person influences the *nafs*. A more detailed exploration is required to determine the extent to which the body influences the non-physical dimension of human existence, and the role of *nafs* in inclining towards or away from good actions. Yet, this narrative opens up a pathway to question the metaphysical repercussions of physical changes in the body through the human genome project, which may facilitate the perpetuation of good or bad deeds.

In the Islamic narrative, the purpose of human existence is associated with a higher spiritual journey and a return to *fiṭrah;* thus any project impacting this existence ought to consider the nature and extent to which the soul's propensity for returning to its primordial state of purity and obedience may be affected. The human person's striving in this life to reenact the first covenant—the moment at which the realization of faith was at its greatest—is a physical endeavor with spiritual manifestations. This striving is undertaken by the human person through actions carried out by the body, which have a purifying potential and the possibility of increasing a person's faith. Given that the locus of faith is the soul, the body is therefore capable of influencing, through its actions, the state of the soul. It can be reasonably questioned what impact the human genome project could have on a person's ability to recollect the first moment of witnessing and being before God.

Although this paper is constructed on the views of Islamic scholars who inherited the bifurcation of body and spirit from the Hellenistic tradition, the Qur'ānic conception and categorization of man does not gainsay the possibility, even probability, of spirituality being resident in, and an intrinsic property of the human body. This sacralised conception of the body demands closer scrutiny through the lens of the Qur'ānic discourse, and may have ethical implications for genomics.

This is a preliminary survey, engaging two fields that have hitherto been presumed to be disparate, if not entirely antipodal. It has forged an interstice between them and carved a landscape wherein they cohabit. This paper lays the groundwork for subsequent analyses that can take up the multitudinous strands of relations delineated, and interactions defined here, to arrive at a higher plateau of knowledge about the spiritual dimension of human existence, and our understandings of the human person.

References

al-Ashʿarī, Abū al-Ḥasan ʿAlī ibn Ismāʿīl. 1963. *Maqālāt al-Islāmīyīn wa ikhtilāf al-muṣallīn.* Edited by Helmut Ritter. Vīsbādin : Dār al-Nashr Frānz Shtāynir.

al-Attas, Syed Mohammad N. 1995. *Prolegomena to the Metaphysics of Islam.* Kuala Lumpur : International Institute of Islamic Thought and Civilization.

al-Ghazālī, Abū Hamid. 1963. *Maʿārij al-quds fī madārij maʿrifat al-nafs.* al-Maktabah al-Tijārīyah al- Kubrá, Cairo.

al-Ghazālī, Abū Ḥāmid. 2007. *Ihyāʾ ʿulūm al-dīn.* Cairo: Dār al-Salām. Vol. 1:858-908

al-Iṣfahānī, Rāghib Abū al-Qāsim al-Ḥusayn ibn Muḥammad. 2001. *Kitāb al-dharīʿah ilá makārim al- sharīʿah.* Dimashq: Dār Iqraʾ.

al-Qurṭubī, Abū ʿAbd Allāh. 2004. *Tafsir al-Qurṭubī.* Beirut: Dār al-Kutub al-ʿilmiyya,

al-Qushayrī, ʿAbd al-Karīm. *Laṭāʾif al-Ishārāt.* 4: 304. Online version available at Maktabah Shamilah – Islamic Library.

al-Rāzī, Fakhr al-Dīn. 2004. *Al-Tafsīr al-kabīr.* Beirut: Dār al-kutub al-ʿilmiyya.

al-Shahri, Mohammad Z., and Abdullah Al-Khenaizan. 2005. “Palliative care for Muslim patients.” *J Support Oncol* 3.6 432-6.

al-Sulamī, Abū ʿAbd al-Rahmān. 2001. *Ḥaqāʾiq al-tafsīr.* Beirut: Dār al-kutub al-ʿilmiyyah.

al-Ṭabarī, Muhammad ibn Jarīr. 2005. *Jāmiʿ al-bayān fī taʾwīl al-Qurʾan.* Beirut: Dār al-kutub al-ʿilmiyyah.

al-Thaʿālibī, ʿAbd al-Rahmān. 1996. *Al-Jawāhir al-ḥisān fī Tafsīr al-Qurʾān.* Beirut: Dār al-kutub al-ʿilmiyyah.

Brockopp, Jonathan E. 2008. “Islam and bioethics.” Journal of Religious Ethics, 36 (1):3-12.

Buchanan, Allen, Dan W. Brock, Norman Daniels, and Daniel Wikler. 2001. *From Chance to Choice: Genetics and justice.* Cambridge University Press.

Calverley, Edwin Elliott, James W. Pollock. 2002. *Nature, man and God in medieval Islam : ʿAbd Allāh Bayḍāwī's text, Tawaliʿ al-anwar min mataliʿ al-anzar, along with Mahmud Isfahani's commentary, Mataliʿ al-anzar, sharh Tawaliʿ al-anwar.* Leiden: Brill V1.

Colledge, W. H., and M. J. Evans. 1995. “Cystic fibrosis gene therapy.” *British medical bulletin* 51, no. 1: 82-90.

Crick, Francis. 1994. *The Astonishing Hypothesis: The Scientific Search for the Soul.* Si-

mon and Schuster, London.
Dar-Nimrod, Ilan, and Steven J. Heine. 2011. "Genetic essentialism: on the deceptive determinism of DNA." *Psychological bulletin* 137, no. 5: 800.
Davidson, Herbert A. 1992. *Alfarabi, Avicenna, and Averroes, on intellect: their cosmologies, theories of the active intellect, and theories of human intellect.* Oxford University Press
Dawkins, Richard. 2016. *The selfish gene*. Oxford: Oxford University Press.
Fārābī. 1938. *Risālat fi'l-ʿaql.* Edited by Maurice Bouyges. Beyrouth : Imprimerie Catholique
Farrell, Peter T. 1969. "The XYY syndrome in criminal law: an introduction." *John's L. Rev.* 44: 217.
Fox, Richard G. 1971. "The XYY offender: a modern myth?" The Journal of Criminal Law, Criminology, and Police Science 62, no. 1: 59-73.
George J. Annas. 1997. *Standard of care: the law of American bioethics*. Oxford University Press, USA.
Ghaly, Mohammed. 2012. "The beginning of human life: Islamic bioethical perspectives." *Zygon* 47 (1):175-213.
Goodman, Lenn Evan. 1992. *Avicenna.* London: Routledge.
Ḥawwā, Saʿīd. 1995. *al-Asās fi'l-sunnah wa fiqhihā.* Vol 1. Online version available at Maktabah Shamilah – Islamic Library.
Ibn Taymīyah, Aḥmad ibn ʿAbd al-Ḥalīm, 1979. *Darʾ taʿāruḍ al-ʿaql wa-al-naql* ed. Muḥammad R.S. Jāmiʿat al-Imām Muḥammad ibn Suʿūd al-Islāmīyah.
International Human Genome Sequencing Consortium (IHGSC). 2004. "Finishing the euchromatic sequence of the human genome." *Nature* 431, no. 7011: 931-945.
Izutsu, Toshihiko. 2002. *Ethico-religious Concepts in the Qur'ān.* Montreal: McGill-Queen's University Press
Lander, Eric S., Lauren M. Linton, Bruce Birren, Chad Nusbaum, Michael C. Zody, Jennifer Baldwin, Keri Devon et al. 2001. "Initial sequencing and analysis of the human genome." *Nature* 409, no. 6822: 860-921.
Lane, Edward W., and Stanley Lane-Poole. 1984. *Arabic-English lexicon.* Cambridge: Islamic Texts Society
Mohamed, Yasien. 2006. *The Path to Virtue: The Ethical Philosophy of al-Rāghib al-Iṣfahānī : An Annotated Translation, with Critical Introduction, of kitāb al-dharīʿah ilā makārim al-Sharīʿah.* Kuala Lumpur: ISTAC
Noble, Denis. 2008. "Claude Bernard, the first systems biologist, and the future of physiology." *Experimental Physiology* 93, no. 1: 16-26.
Peters, Ted. 2014. *Playing God? Genetic Determinism and Human Freedom.* London: Routledge.
Plomin, Robert, and Frank M. Spinath. 2014. "Intelligence: genetics, genes, and genomics." *Journal of personality and social psychology* 86, no. 1: 112.

Rahman, Fazlur. 1952. *Avicenna's Psychology: an English translation of Kitāb al-Najāt, Book II, Chapter VI with historico- philosophical notes and textual improvements on the Cairo edition.* Oxford: Oxford University Press.

Reisman, David, and Jon McGinnis. 2004. *Interpreting Avicenna: science and philosophy in medieval Islam : proceedings of the Second Conference of the Avicenna Study Group.* Leiden, Boston: Brill.

Rommens, Johanna M., Michael C. Iannuzzi, Bat-sheva Kerem, Mitchell L. Drumm, Georg Melmer, Michhael Dean, Richard Rozmahel et al. 1989. "Identification of the cystic fibrosis gene: chromosome walking and jumping." *Science* 245, no. 4922: 1059-1065.

Savulescu, Julian, Ruud ter Meulen, and Guy Kahane, eds. 2011. *Enhancing human capacities.* New Jersey: John Wiley & Sons.

Savulescu, Julian. 2009. "Genetic interventions and the ethics of enhancement of human beings." *Read Philosoph of Tech* 16, no. 1: 417-430.

Savulescu, Julian. 2009. "Moral Status of Enhanced Beings: What Do We Owe the Gods?" In: *Human Enhancement,* edited by Savulescu, Julian, and Nick Bostrom, 211-250. Oxford University Press on Demand.

Shaw, Alison. 2014. "Rituals of Infant Death: Defining Life and Islamic Personhood." *Bioethics* 28 (2): 84-95.

Singer, Peter. 1995. *Animal liberation.* New York: Random House.

Singer, Peter. 2011. *Practical ethics.* Cambridge: Cambridge University Press.

Steinbock, Bonnie. 1978. "Speciesism and the Idea of Equality." *Philosophy* 53, no. 204: 247-256.

Taylor, Charles. 1985. Philosophical papers: Volume 1, Human agency and language. Cambridge: Cambridge University Press.

Tobias, Edward S., Michael Connor, and Malcolm Ferguson-Smith. 2011. *Essential medical genetics.* Vol. 22. New Jersey: John Wiley & Sons.

Tutu, Desmond. 2015. In *The Emergence of Personhood: A Quantum Leap?* Edited by Jeeves, Malcolm, ix. Michigan: William B. Eerdmans Publishing.

Warren A. Mary. 2013. "Abortion". In *A companion to ethics.* Edited by Peter Singer. New Jersey: John Wiley & Sons. 303-315.

Watson, J, quoted in Leon Jaroff. 1989. The Gene Hunt, TIME, Mar. 20, at 62, 67.

Watson, James D., and Francis H. C. Crick. 1953. "Molecular structure of deoxypentose nucleic acids." *Nature* 171, no. 4356: 737-738.

Wildman, Derek E., Monica Uddin, Guozhen Liu, Lawrence I. Grossman, and Morris Goodman. 2003. "Implications of natural selection in shaping 99.4% nonsynonymous DNA identity between humans and chimpanzees: enlarging genus Homo." *Proceedings of the national Academy of Sciences* 100, no. 12: 7181-7188.

Wolfson, Harry Austryn. 1976. *The philosophy of the Kalam.* Cambridge, Massachusetts: Harvard University Press.

Wolpe, Paul Root. 1997. "If I am only my genes, what am I? Genetic essentialism and a Jewish response." *Kennedy Institute of Ethics Journal* 7, no. 3: 213-230.

Zwart, Hub. 2009. "Genomics and identity: the bioinformatisation of human life." *Medicine, Health Care and Philosophy* 12, no. 2: 125-136.

CHAPTER 6

The Ethical Limits of Genetic Intervention: Genethics in Philosophical and Fiqhi Discourses

Mutaz al-Khatib[1]

The world has witnessed three important transformations. The first is the Darwinian theory of evolution, which subverted the perception of the human anthropological configuration and its origin.[2] It was through this image that man occupied an exceptional position among the species and acquired his superiority and sacredness. The second transformation is the discovery of the movement of the earth whereby Copernicus destroyed the perception of geographical centrality or the way the geography of our world was known at that time. The third transformation lies in sophisticated biotechnology, which could be representative of the third decentering of our worldview[3]. Subjugation of our body and our life to biotechnology results in philosophical, ethical and religious issues related to human life and reflective of a specific vision of man and his nature, as well as the limits of dealing with, or disposing of, his body. These problematic issues include questions like: when does life begin? What criteria should be used to support the belief that the human being is human? Who should require ethical rights? Why do we consider life sacred? What effect does this have on the way we deal with the human being at various life stages?

Historically, the abovementioned questions have been linked to the specific issue of abortion, although biotechnological developments have broadened the familiar potential of work to include reproduction and procreation. These

---

1 Assistant professor of Methodologies and history of Islamic Ethics at the Research Center for Islamic Legislation and Ethics (CILE), College of Islamic Studies, Hamad Bin Khalifa University in Qatar, malkhatib@hbku.edu.qa

2 This research was made possible by the NPRP grant "Indigenizing Genomics in the Gulf Region (IGGR): The Missing Islamic Bioethical Discourse", no. NPRP8-1620-6-057 from the Qatar National Research Fund (a member of The Qatar Foundation). The statements made herein are solely the responsibility of the author. This chapter is based on an earlier Arabic version.

3 Habermas 2003, 54.

technical developments open the door to vast possibilities, allowing for new patterns of reproductive and procreative interventions whose consequences are difficult to predict, while they were purely human acts or practices in the past. This has led to numerous debates and raised problematic issues at the religious, legal, political, philosophical, ethical and social levels. As well as, issues with regards to emergence of the term "liberal eugenics."[4]

In view of the previous questions and the many possibilities provided by the technical revolution, the subject of "genetic engineering" is dealt with in transdisciplinary studies and consequently, specialists in philosophy, ethics, medical ethics, biology and the sociology of medicine all engaged in it. Moreover, in the field of medical ethics, biotechnology has provoked a serious debate about the use of prenatal genetic engineering and other modern genetic techniques. There is also the question of whether we have to impose ethical constraints on this field to limit its possibilities and to keep the technology under human control and not the other way around. In philosophy, the applications of genetic engineering raise several problematic issues that pertain to the field of applied ethics and thus preoccupy philosophers.

Despite all these ramifications, the central question posed by genetic engineering is: what should a human being do to avoid compromising his life? Or what should man do in the course of the life he is destined to live? In other words, what is an exemplary life worthy of following, which is also reflective of the perception of Jürgen Habermas in the context of his treatment of human nature, which itself is threatened by genetic intervention?[5] If we use the religious formula, however, the question will be about the good deed and the good life that relate to the mission incumbent on the human beings on earth as "*khulafāʾ fī al-arḍ*" (Allah's vicegerents), which includes concepts of worship and the promotion of growth and prosperity on earth. This implies that we are faced with a philosophical and religious question simultaneously, a question that deals with the "binding force" relating to the personal and communal life, as well as the "doctrine of life" and the way it should be lived. This fundamental question relates to ethics and meta-ethics, and the relationship between morality at its ethical level of universalistic deontology and ethics at the level of critical self-clarification of values concerning societies and individuals.

The central question is a metaphysical one, while the prevailing trend in philosophy and ethics is secular and liberal, thus posing new challenges that

4 In the German version of his book, *The Future of Human Nature*, Jürgen Habermas uses the subtitle: "auf dem weg zu einer liberalen eugenik" ("Towards liberal eugenics"). However, this subtitle does not figure in the English version.

5 See Habermas 2003, 2.

add to the preceding challenge. In dealing with this question, there are two main directions: the first is the post-metaphysical trend, which is espoused by Jürgen Habermas who argues that liberal societies must seek metaphysical answers to the metaphysical questions about existence, its aim and end. The second trend is the metaphysical one and is represented by the Christian and Islamic theological visions, which engages in the study of the issue of genetic engineering and its effects, passing judgments according to perceptions emanating from jurisprudence, *kalām* (Islamic scholastic theology), and ethics.

Since biomedical technology raises all these questions, debates, and stimulates this influence we should address it by first examining the position of this technique itself and understanding each party's perceptions of the possibilities it offers, as well as by understanding how it operates and considering its implications. This is because any debate or position will be the result of two stands: the first deals with the form and extent of awareness of this technology and its possibilities. The second is concerned with the philosophical and religious perceptions of man and his existence. The third lies in identifying the problematic issues raised by this technology, which are the corollary of the two preceding issues. This deductive and sequential method informs this study.

1 Genetic Engineering and the Limits of Awareness

Generally, thinking about technology as such is always problematic in itself, and one's attitude towards it depends on how one defines it. In trying to answer the question "what is technology?" We face two answers. The first sees technology as a means to achieve some objectives, while the second sees technology as an ability which is specific to human beings. The two answers are mutually supportive. The pursuit of goals and the utilization of means is a human act. In addition, the manufacturing of tools and machines is part of the nature of the technique, and the perception that makes of it both a means and an ability characterizing human beings is an "instrumental and anthropological" conception. It is a vision that directs every effort to put man in an appropriate relationship with technology, thus enabling man to control and guide it for the sake of spiritual purposes (i.e., bringing technology under human control). Thus, the discussion revolves around the means that are being used and the ends that are being researched.

Martin Heidegger, on the other hand, argues that "when we consider technology neutral we surrender to it in the worst form, because this perception makes us lose sight of what technology is," which refers, according to him, to a pattern of exposure to any form of truth. Technology reveals that which does not self-create (i.e. outside the causal relationship), which is not yet in existence but which can take different forms. Technology is production, in the sense of disclosure and not manufacture. Disclosure then takes the form of provocation and incitement of natural energies and of reality as a supply that is ready for use.[6]

With regard to biotechnology, in particular, we can distinguish between two views: first, a simple or instrumental view that deals with the question of means and ends, which is the jurisprudential view that determines its positions and judgments based on this perspective. The second is a complex philosophical view that explores the scope of this technique and its effects on human kind and its ethical dimensions.

1.1 *The Philosophical Analysis*

Aristotle's philosophy offers the possibility of distinguishing between three positions: the theoretical position that observes nature in a non-advantageous way, the technical position that works with the aim of producing, thus intervening in nature, through the development of means and the use of tools, and the practical position which works according to the rules of upholding customs. Habermas adds the assertive position which works through a communicative manner that requires communication with someone else on something (Habermas 2003, 45-46). Experimental sciences, however, have combined the theoretical position of the "neutral observer" and the technical position of the "intervening observer" who seeks the experimental effects. Thus, the revolution in the technical practice of genetics has developed from the simple to the complex level. It has taken the human being from the expansion of the possibilities of familiar actions to the creation of a new limitless type of intervention. In other words, by eluding to Habermas's arguments,[7] on the principle that development is comprehensive and can unfold into considerable features.

The first feature reveals that in adapting to the human "special nature dynamics," the biological technique has transcended the limits of "therapy" to the search for what is "precautionary," which is excluded or avoided. Therefore, this led to the gradual dissolution of the boundaries separating "justified intervention" (negative) from "unjustified intervention" (positive), which was previously transparent, especially with the emergence of "liberal eugenics." The latter does not recognize the boundaries between therapeutic interventions and those which aim at development and are subject to market dynamics.

---

6 See Heidegger 1995, 44.

7 See Habermas 2003, 16-74, and for a critical view see Christiansen 2009, 147-156.

The second feature is that the biological technique constructs man as a body, and here there is a difference, according to Helmut Plessner, between being a body and having a body (Plessner 1969, 9-10). Hence, turning man into a body eliminates the boundary between nature (us) and the bodily organ where the self resides. Thus, the technique creates a new relationship with the self which is immersed in the depths of the organic carrier. Consequently, influencing how the self understands itself and how the self is used in the new reality, independently or arbitrarily, according to self-choices through the market. In other words, technical progress affects our understanding of ourselves as having responsibility over our actions (the normative understanding of self).

The third feature is that the new possibilities offered by the biological techniques are an expansion of freedom, thus posing new challenges to the modern understanding of freedom. Given the impact of this freedom on the future of the human species, in its search for "limits" to be imposed on procreation (the improvement of progeny) so as to prevent serious deformities. Meanwhile, these possibilities can help us acquire a new understanding of ethical freedom—that is the freedom of each person to possess the management and be responsible for his own "conduct," and the right for each newborn to a genetic formation which is unhindered by any deliberate programming or manipulation. As an adult, he will be able to submit his own personal history to critical judgment and revision, as Kierkegaard argues in the context of the individual's possession of his own life history or the ethical self-understanding.[8]

The fourth feature demonstrates that the development of biological techniques is a dynamic that threatens to compromise the normative clarification process, which will affect our self-understanding as beings of a qualitative essence.

The fifth feature shows how interventions in genetics can turn into a hegemonic act that affects the self and causes present-day people to exert domination over future generations by turning them into subjects, to the extent that the other side of today's authority will be reflected in the subsequent subjugation of the living by the dead.

The sixth feature shows that biological genetic research is embodied in what increases the investor's profit and the pressures of national governments whose success depends on these developments and achievements.

All these considerations lead to a complex vision of the technique and its effects, thus considering genetic intervention an infringement upon the physical foundations of the spontaneous relationship with the self and on the ethical freedom of another person. This intervention raises problematic issues of an

---

8 For further details, see Stack 1973, 108-125; Holmer 1953, 157-170.

ethical nature, on the one hand, and other types of problematic issues, on the other hand. Some aspects of these controversies will be tackled below.

1.2 *The Contemporary Jurisprudential Analysis*

Technology has imposed important challenges on contemporary jurists in many respects, both in terms of the changes and the possibilities it creates, which will contribute to the resolution of some serious differences of opinion that exist in the jurisprudential tradition (menstruation, moonsighting, pregnancy, proof of parentage, etc.). Yet, the understanding of the dimensions of the techniques has been confined within the limits of the simple and intuitive view. Besides, some attempts to study the "impact of modern technology" have failed to formulate a systematic and mature attitude and have, thus, been characterized by superficiality and hesitancy in using technology in many instances, which is the result of an ambiguous vision and a weak method.[9] The result is an acceptance of some of its impacts due to "necessity," despite the homogeneous acknowledgment among jurists that changes in jurisprudential traditions may occur with the advancement of the technical means.

In terms of genetic technology, the jurisprudential consideration is constantly governed by practical attitudes. The jurist's view is focused on the examination of the means used, the discussion of the goals and the differentiation between them based on a latent conception of technology and its limits, on the one hand, and on a specific representation of the meaning as intended by the lawgiver, in whole or in part, on the other hand.

The issue of "genetic intervention" encompasses all the applications affecting human genes: genetic testing, genetic engineering, diagnosis, treatment, cloning and research. These can generally be classified into three types: diagnosis, treatment, improvement or enhancement. While the intervention is apparent in the treatment and improvement cases, much of the examination or diagnosis is usually but a precursor to most types of interventions. For instance, the intervention aimed at diagnosing existing or anticipated genetic diseases, the detection of anticipated genetic traits in both the sex cells or the fetus for the genetic improvement of the offspring, knowledge of the structure of the genetic fingerprint as evidence for the detection of crime and the establishment of lineage, and genetic testing for the purpose of executing various types of civil contracts (marriage, insurance, employment, etc.)[10]. In other words, genetic technology refers to a network of possibilities interrelated to changes and effects, sometimes making it difficult to differentiate amongst them or bring parts of them under control without the reciprocal interacting

---

9 For example: Al-Sheikh 2010.

10 See al-Lūdaʿmī 2006, 141; al-Lūdaʿmī 2011, 55-67.

effects. Thus, the distance between the means and the results may become narrow, complicated, and overlap as a result of the intervention of several parties, and the gene's owner would be merely one of the parties. Therefore, I argue that the neutral observer merges with the intervening observer.

The instrumental consideration provided in the legal decree issued by *Dār al-Iftāʾ al-Miṣriyyah*,[11] focuses on the great therapeutic potential offered by genetic technology and that "a large part of it is in the interest of the human being and serves to preserve his health," arguing that reservations are due to the possibility of some rare harm. This, however, does not explain the concept of "harm" and its dimensions. It neither defines the concept of "interest" or its dimensions, especially when it seems that the Mufti's view does not exceed the duality of body and disease.

It is true that the International Islamic Fiqh Academy provides a more comprehensive and detailed perception, since it has brought together jurists and experts in this regard,[12] but it does not transcend the simple consideration of genetic technology. The decision revolves around two things: defining the concept of genetic technology and its uses. Learning about the human genome "is part of man's discovery of himself and of God's ways in His creation and, as a means to identify some genetic diseases and the possibility of infection, it has become an added value to health and medical sciences in their pursuit to cure diseases, thus being considered part of the collective obligations (*furūḍ al-kifāya*)." (International Islamic Fiqh Academy 2013). In other words, it is a neutral technical view and the relevant judgment is geared to how the technique is used, including genetic therapy and genetic engineering. Therefore, words such as "use" and "procedure" are repeated to express this practical technical view.

The discourse here addresses the field of practice (the physician-practitioner and the subject-person), which is restricted to the scope of "beneficial areas" and that there should be no "harmful use of the genome." What is meant by "useful" here is the "treatment and prevention" purpose. It is then clear that the positive perception has prevailed in the decision of the International Islamic Fiqh Academy by determining the genome's technique and dimensions. Yet, it seems that the International Islamic Fiqh Academy is still held captive by the experts' technical medical outlook which favors technology as a means of knowledge over the treatment and control it offers, without considering the

11 See the legal decree "Al-handasa al-wirāthiyya wa-istikhdāmuhā fī majāl al-ʿilāj" on *Dār al-Iftāʾ al-Miṣriyya* (website 09-09-2014).

12 It is based on a previous symposium organized in Jeddah in cooperation with the Islamic Organization for Medical Sciences under the title "Genetics, Genetic Engineering and the Human Genome," February 2013.

overall philosophical perception that governs this development and the possibilities it promises. The concern was focused on the technique's uses only as a commodity. Hence, one of the recommendations following the decision was to appeal to Islamic countries "to adopt genetic engineering in all its legitimate fields and applications" and "enact laws and regulations necessary to protect the citizens from being used for experimental purposes," calling upon states "to provide these services to their citizens who require them." The Makkah Council, on the other hand, recommends at the end of its resolution that "doctors and laboratories fear God who is watching them and stay away from actions that harm the individual, society, and the environment" (Islamic Fiqh Council 1998).

The jurisprudential view herein is dominated by instrumental considerations, and therefore it focuses on the restriction and control of the practices of technology. Consequently, it is directed at individuals and doctors. Since the International Islamic Fiqh Academy is an affiliate of the Organization of the Islamic Conference, it appeals to states, while the Makkah-based Council addresses only the individual conscience.

Some theologians became aware that "technology is power and power is never neutral."[13] Habermas talks about the investor's profits and the pressure of governments to implement technology. If we consider the fact that the Islamic world is a mere consumer to such techniques, we will discover that jurisprudential decisions respond to this particular case without thinking about what lies outside its boundaries and questioning the development and its dimensions. This could also be attributed to the function of the jurist himself who is not preoccupied with the totality of the developments in the field and the principle of possibilities, but he is concerned with practical positions *vis-à-vis* specific issues. He may not be able to achieve these without the use of means and objectives in the light of a principled and legislative vision, be it a text, a general, or an all-inclusive rule. Generally, genetic technology and biological developments may impose the transition from jurisprudence to *kalām* theology, linking the practical to the ontological so as to avoid the consideration of the practical or the partial in isolation from the overall perceptions.

2 Genetic Interventions as an Ethical Dilemma

We have already pointed out that genetic technology raises problematic issues of a multidimensional nature, but, we want to focus here on the ethical and jurisprudential dimensions. Ethical problems belong to the fields of philoso-

13 See Simmons 1983, 211.

phy and Islamic jurisprudence, and issues related to reproduction are at the heart of religion which considers the safeguarding of lineage as one of its major objectives, and this is linked to the pattern of human life which has preoccupied philosophers and religious scholars alike.

2.1 *The Philosophical Debate*

A genetic test is allowed at the pre-transplant phase for the benefit of the patient's family members, for the sake of having an estimated examination, overcoming organ deficiency or, for intervention to repair the genome. Although this intervention offers positive and therapeutic possibilities, it is not limited to that. Therefore, the philosophical discussion initiated by Jürgen Habermas revolves around global technical development, a matter which raises three main problematic issues. The first one is that it undermines the concept of autonomy, which in turn threatens human morality, because all human actions are linked to the human being as an ethical creature. If his autonomy is undermined, then his effectiveness is undermined as well. Second, it affects the ability to understand the self/identity, and self-understanding is a sine qua non for engagement in liberal democratic culture and in secular ethics. The third problematic issue is the attitude towards pre-personal life and human dignity.

2.1.a Genetic Interventions and the Concept of Autonomy

The decision to intervene in the formulation of another person's identity imperils the Kantian concept of autonomy which is based on freedom and equality for all persons as a birthright. It also compromises the ethics of a person who is responsible for his actions under his own free-will and to act without influence. This cannot happen in genetically programmed humans whose genes have been manipulated by virtue of another's will. Genetic intervention threatens Habermas' communicative action theory which seeks to understand human existence through the assumption that human beings recognize themselves in terms of Kantian individuals, who possess freedom and reason. Genetic intervention is, accordingly, achieved by controlling the ethics of personal relationships and the ethical orientation of people as human beings who should be able to determine—without interference—who they are and what they want to be, without one acting on their behalf. Intervention in the natural formation of another person is a non-reciprocal predilection, and a relationship with an unknown personality evolves out of it. This represents a strange form in the mutual recognition of relationships at the judicial and institutional levels in modern societies. Such an intervention necessarily curbs equality, that is, the parallel responsibility initially held among free and equal people. In normal situations, one can establish a rational relationship as per his formational process, carving his own self-understanding through revision and possessing a self-critical stand on his own maturation. None of this will be possible in the case of genetic intervention.

Elizabeth Fenton[14] raised the issue that Habermas exaggerated his concern with the idea that genetic modification would distort human relations, especially between the begetter and the begotten, because this kind of relationship is already lacking in equality, and the parents have the authority to exert control over their children during their growth and development. It is, therefore, a relationship which disrupts Habermas's presumptive thesis that "all people receive the same normative status, and that it is part of their duty to mutually and simultaneously recognize each other, based on the principle of exchange in human relations." Habermas, however, argues that there is a difference between social dependency and liberal eugenics. He states that "the irreversible decision of a person to regulate someone else's genome, according to his wishes, will generate a new pattern of relationships between these two individuals," which undermines the ethical understanding of the self that acts and emits judgments in an independent manner without the existence of a primary obstacle that conflicts with the system of equal relations between people. Genetic programming generates a non-paralleled relationship, which is "a special kind of parental," because the child's dependency on parents, although non-reversible, dissolves when children reach adulthood without affecting the children's existence or any specific determinants of their future lives. As for genetic dependency, it is focused on a single act attributed to the programmer and it eliminates the "normal exchange between equals by birth." The programmed person can explain the parental intention of the programmer but he cannot modify it , prevent it from happening, or reverse its occurrence. In other words, this genetic parenting intention takes in this case an in-kind form through a genetic program that creates an effect without any social practice or social exchange based on communication.

2.1.b Genetic Interventions and the Ethical Understanding of Self/Species

Genealogy relates to the central discussion about the concept of "human nature" and the effects that genetic technology can have in this case. The ethical question "what should I do?" is linked to the other fundamental question "what can I do?" Both questions lead to self-knowledge, which is the foundation on which ethical theories have been based since Socrates, who said: "Know yourself." Kierkegaard believes that knowing one's self is an important requirement

---

14 See Prusak and Malmqvist 2007, 4-6.

for man to become an ethical being.[15] He and Kant attached importance to the inner voice as a reliable source of information and guide in the control of one's actions. The ethical understanding of human kind is based on the understanding of one's self in the same way one writes *alone* his own autobiography. This ethical understanding also recognizes the self as an individual working independently. Genetic technology, however, undermines confidence in this inner voice, which becomes the product of genetic modification that parents can choose a month before their child's birth. Genetic interventions challenge the ethical field that Kant established, the idea of objective morality. However, since the beginning of the nineteenth century, this has shifted toward self-ethics, so much so that it became difficult to sustain Kierkegaard's relativist vision. Kierkegaard argues that even though we do not share the same values with our neighbors, at least we can all be honest with ourselves (being ourselves), because of the control of individualism over the field of ethics.[16]

Habermas calls "moral" such "issues as deal with the just way of living together. Actors who may come into conflict with one another address these issues when they are confronted with social interactions in need of normative regulation" (Habermas 2003, 38). What the codification of human nature does is a different understanding of ourselves to the extent that we no longer understand ourselves as human beings who are able to act freely, ethically equal, and guided by norms and proofs. The manipulation of the genome, with the aim of decoding it and changing it, destabilizes the stark differences that exist between the "grown" and the "made", the subjective and the objective; whereas the body is the medium in which the person's existence is embodied. Thus, we can distinguish between the actor and the acted upon, between creation and process, between the mind and revelation, between acts that are attributed to the self and those attributable to others. Any change in this respect would imply transformation of our understanding of ourselves, affect the moral conscience and alter the conditions of natural growth that are necessary for our understanding of our ourselves as the creators of our own personal lives and as equal members with respect to the communal ethical rights. Thus, this kind of change threatens the freedom and the constant equality of all people with regards to rights granted to people at birth, and is equally a form of domination over the ethics of personal relationships and the ethical orientation of human beings as species who know themselves and what they want to be, i.e., it obfuscates our perception of who we are as created beings with a body. This can result in a new pattern of unequal relationships.

---

15 See Stack 1973, 108-125.

16 See Guilfoyle 2004, 483.

Genetic intervention leads to "fetal reification" and impacts the ethical understanding acquired by humanity as a whole. The ethics of one's will is going to become but one among other possibilities, and this will affect the philosophical question about life as it should be lived. Since, it will disclose a general debate about the understanding we may have on the cultural forms of life as such, and thus normativity will be lost. Also, the expansion of reproductive techniques and its transformation into something "natural" will contribute to a change in the intellectual outlook of human life, and the elimination of our moral sensitivity to the deference of "cost and profit" calculations. "The eugenic practice may cause damage to the status of the potential human, as a member of a universal group composed of ethical creatures." The debate about the commodification of fetuses for research purposes or about the conditional creation of embryos is "a violation of the boundaries of the species that we consider firm and it is the beginning of the uncertainty associated with the identity of the human race," on the one hand, and it is the context in which our legal and ethical representations are organized, on the other hand (Karmein 2012, 106-107).

Elizabeth Fenton has argued in this context against the belief that there is a decisive boundary between the natural and the artificial, and therefore in her view there are no criteria to follow in this differentiation. Habermas, however, asked serious critical questions about the seriousness of the absence of these limits, because this undermines the required ethical clarity. If we consider the totality of this trundling evolution, we will end up "commodifying" the human being and turning him into a device. Hence, we must talk about the right to genetic inheritance that is free of manipulation so as to prevent any interference or control of our physical existence.

The philosophical debate here is based on the concept of man and human nature. The German and European debate focuses on a normative concept of man which is centered on principles, it adopts a metaphysical perception of human nature and seeks to bring the future developments of the genetic technique under control. The American debate, however, seeks to find an appropriate way to accept the technological developments. In other words, it is based on the unshakeable confidence in science and technological development and relies on the liberal legacy—from John Locke onwards. It is mainly concerned with the protection of the freedom of choice enjoyed by a "person" according to the law. This approach to new challenges focuses on the vertical dimension of the relations established by the social partner with the social force surrounding him; consequently, the decisions here are left for the parents' own estimations. The technical possibilities, however, are no more than an extension of the freedom of reproduction and the rights of the parents (basic individual rights).[17]

2.1.c Genetic Interventions and the Pre-Personal Nature

The third problematic issue is based on the following central question: is the act of intervention (creation with reservation) compatible with human dignity? Do we have the right to expose human life to selection and alteration? This view concerns two stages: The pre-embryonic stage and the embryonic stage. In either case, we are talking about the pre-personal life and the debate around its nature, either as a described or determined nature, and whether it acquires the status of a being with moral rights or not.

The ethical status of the fetus, or the fertilized egg, is determined through a central concept of human sacredness based on the common characteristics that give it value. Defining the concept of human sacredness is the key to all issues relating to medicine and biology. Since the ancient times, the religious concept has given man a unique position because he is God-made, this status has been shaken with the development of science which has deprived him of his unique position. Kant, who established his understanding of man on a purely rational basis away from methods of faith, considered man a purpose in himself, and developed principles that would analyze man's ethical behavior as it should be. Thus, Kant restored man's natural position as no longer merely a means to achieve others' purposes. Hegel as well explored this path by arguing that the absolute right lies in the principle: "be a person and respect others as persons." Hegel derived all the political rights of man (ownership, acquisition, contracting, ...etc.) from this Kantian premise (Hegel 2007, 14-16; 146). Given that all of this relates to personal life, the problem here is in regard to the nature of pre-personal life.

Several definitions were offered to define the meaning of the sacredness of human life, including "appreciation of life," in the sense that whenever there is life, intervention to terminate it is contrary to its sacredness; the "quality of life," which means that life deserves to be lived. But, determining it remains ambiguous, since based on this, one can argue that in order to preserve the quality of life, other lives can be sacrificed in the process and experiments can also be carried out on embryos if the aim is to preserve the general public life. Other ideas concerning the sacredness of life include the characteristic of the "distinctness of life" in the sense that life itself is a characteristic related to man and his existence, and it is immutable before the acquisition of any experience. The problem, however, is that this sacredness does not tell us what to do

17 See Habermas 2003, 77; J. Karnein 2012, 93-116.

about procreation and morphogenesis. This also includes the perception of life in the sense that the sacredness here is internal and that our feelings towards life are imbued with sanctification and respect. Another relevant idea is that the sacredness of life reflects the general direction of life by being a comprehensive view that affects our attitudes towards it both in terms of its normative and applied aspects. A human being has the right to live (life), and life should not be wasted without a justifiable reason, which makes of it an ethical law that allows for exception, according to logical justification. Life includes a clear and accurate judgment that depends on care for one's self and for others, and concern for exceptions as well as for the original principle.[18]

The main issue, however, remains the definition of the concept of "sanctified life." Does it begin after a certain stage of fertilization, or does it do so in line with the Aristotelian view that "the human being becomes human during his first movements in the mother's womb and when the mother starts feeling this movement"? Does it start from the moment of the fusion of the sperm with the egg, as biologists believe? The ethical debates focus on identifying a constant criterion through which the human being can be evaluated; these debates propose a number of possibilities, such as self-awareness that distinguishes the human being from other beings and enables him to exercise his independence and communicate with others; the ability to evaluate life as an internal perception; responsibility over one's actions and behaviors and the ability to account for this.[19] Yet, these criteria do not help us evaluate the pre-personal life stage; they do not even include the fetus, while other criteria do not include the child after birth.

This issue seems to have prompted Teresa Iglesias, a Roman Catholic philosopher, to defend the idea that the human being is a person throughout the stages of his growth, and that what makes us people as such is the kind of beings we are, as well as the nature we possess; all phases in our existence carry this identity. That is, the concept of a person cannot be determined by a particular phase or be restricted to it. Accordingly, the respect due to the fetus should be the same as the one due to the person without any distinction. Therefore, it should not be killed or subjected to other uses or means of exploitation (Iglesias 1984, 32-37).

Although personal nature is only proven in the case of the born neonate, Anja J. Karnein argues that our ethical obligations *vis-à-vis* the "person" means towards the developing fetus that is going to take the shape of a fully developed human being, i.e. the respect shown for the fetus should be accredited

18 See al-Baqṣamī 1993, 111-120.

19 See al-Baqṣamī 1993, 121-137.

to what it will subsequently be. True, it will be difficult to distinguish between what will develop and become a full-fledged person and what will not, but the assumption is based on the principle of precaution, which requires us to treat all embryos as potential people. Once the living organism becomes a person, it means that the stages it went through before should be taken into consideration. Another reason to show respect to embryos comes from the difficulty of establishing hierarchical levels during the course of human life; that is, embryos must be protected from harm, and harm is anything that is liable to cause any change in the unique characteristics that embryos bear in an absolute sense (hold in trust) (J. Karnein 2012, 26-34).

Habermas acknowledges that, given the multifarious views, it is difficult to give the embryo—from inception— "complete protection" in life since this is attributed to those who enjoy basic rights. Yet, Habermas believes that human dignity dictates that life be preserved at the pre-personality stage and not be subjected to speculation or entreaty. Hence, the difference between the debate on genetic intervention and that on abortion can be summarized in terms of the "commodified" research on embryos and the pre-transplant diagnosis. Monitoring the intended or desirable nature turns human life into a "mechanical one" created in certain conditions in accordance with third-party preferences. It is true that the diagnosis can warn against the likelihood of abortion, but it leads to the conflict between the protection of the child's right to life and the right of parents who put him through the research testing as though he were property. This conflict, however, occurs when allowing the fetus to undergo genetic testing, thus showing that the parents have been part of this contradiction from the beginning. Nonetheless, the debate on unintended pregnancy concerns the woman's right to determine abortion and the need to protect the fetus, meaning that the life decision to terminate the pregnancy has nothing to do with this readiness for a consumer type use. There is also a difference between abortion and pre-personal life. The discussion on abortion is a discussion on embryonic life, and there are two different views on this matter. The first describes human life in neutral terms (free from a prior judgments), such as describing the fetus as a "clump of cells," unlike the "neonate" which represents the first stage in which the person is described as having acquired human dignity. The second view describes the fertilization of cells as the actual start of a sophisticated process which enjoys uniqueness and organizes as well as regulates itself, and this can be biologically described as perforce a human prototype and, therefore, this prototype should be entitled to fundamental rights.

Habermas endeavored to find the criterion that would allow pre-personal life to be worthy of protection by arguing that the world of ethical rights and duties lies in determining the "basis of ethics" provided by the group of ethical beings who decide their own laws. This group is concerned with all the relations that require organization in accordance to given criteria. It imposes ethical obligations on itself, and the members expect each other to behave in conformity with the criteria. Even animals can benefit from the ethical duties of our relationships, including all creatures who are sensitive to feeling pain. Human dignity, therefore, is one that is equal to this parallel in relations and resides in the concept of "inalienability" which is meaningless unless it is conducted within relations among persons who mutually recognize each other within the framework of parallel and equal exchanges (Habermas 2003, 33).

2.2 *The Jurisprudential Debate*

Contemporary jurisprudence debates have shifted from genetic technology towards research on its "uses" and identifying the legitimate vs. illegitimate uses. Some researchers have limited the framework of discussion to three areas: medication and therapy in respect to the principle of inviolability of body and soul; the legitimate concept of mating and procreation in order to preserve offspring and lineage; and control as well as follow-ups of these techniques.[20] The uses of genetic technology, however, relate to several central concepts, such as the beginning of life, the sacredness of the body, the system of rights; the rights of God and the rights of people, and the preservation of self and lineage. These concepts intersect with the abovementioned philosophical problematic issues, because they constitute the issues of "justified intervention" (negative), "unjustified intervention" (positive), and the "boundaries" to be imposed on "eugenics" (the improvement of lineage) with the intent of preventing serious malformations. Other intersections are reflected in the distinction between what is "clinical/therapeutic" and what is "precautionary." As will be explained below, these concepts are concerned with other dimensions, including the human body and the limits of intervention in the human body, whether carried out by the person himself or by his custodian, in order to avoid "domination over self."

2.2.a The Beginning of Human Life

Determining the beginning of human life remains a controversial matter among jurists. The old jurisprudential discussions revolved around the fetus in its three phases: sperm drop, the clinging substance, and then the embryo, followed by breathing of the soul. The duration of each stage is forty days. The debate about the fertilized sperm outside the uterus (before implantation) was

---

20 See ʿAbd al-Raḥīm 2002.

not part of the debates or even imagined before the emergence of modern technology. In other words, the debate is about two issues: the nature of the fertilized egg and the fetus, especially in its early stages.

Disagreement regarding the fertilized egg is but an extension of the disagreement about its nature: is it considered a fetus so that it can be judged accordingly? Most of the participants in the symposium held by the Islamic Organization for Medical Sciences[21] held the viewpoint that the fertilized egg has no legitimate sacredness of any kind, nor consideration for before it is implanted in the uterine wall, for it is not yet called an embryo. In Arabic, the word embryo is derived from "*ijtinān*" which means "concealing", which is "the name for the baby as long as it is in the womb" (al-Qalyūbī 1998, 4:160). In surveying jurisprudence and its different domains, we find that Ḥanbalī scholars, among others, base a number of rulings on the embryo as it is implanted in the uterus, which means that the estimation of the age of the embryo begins on the day the sperm drop sticks to the uterus.[22] Therefore, the majority of jurists do not ascribe sacredness to that which lies outside the womb. If life deserving of respect begins in the uterus, according to the recommendations of the abovementioned symposium, this means that testing on un-implanted ova is permissible. But under two restrictions: that the nature of God's creation should not be altered and that the exploitation of science for evil and corrupt practices should not be pursued.

There is a second view which argues that the location is not a criterion in establishing the ruling.[23] This meaning can be elicited from the words of Imam Abū Ḥāmid al-Ghazālī, who considered that existence is founded on levels, the first of which is "that the sperm drop falls into the uterus and mixes with the woman's liquid to become prepared to accept life." Al-Ghazālī says: "I argue that human life starts when semen drops inside the uterus, not in terms of its exit from the female urethra, and given that the sperm in the vertebrate does not create the baby, the same applies to the sperm's exit from the female urethra, unless it mixes with the woman's liquid and blood, because this should be a tangible criterion" (n.d. 2:51). Accordingly, the semen drop, when mixed

21 See the third symposium of the Islamic Organization for Medical Sciences in Kuwait 1987; al-Ashqar 2001, 305-310.

22 A reference was found later in some of the fatwas of the Permanent Committee for Scholarly Research and Iftā' in Saudi Arabia. See al-Buhūtī 1993,1:646; 2:619; Wizārat al-Awqāf (n.d.), 30:295; and fatwas of the Permanent Committee for Scholarly Research and Iftā' (No. 17576) see Āl al-Shaykh and Bin Bāz 2004.

23 This view is expressed by Sheikh Muḥammed al-Mukhtār al-Sallāmī in the discussions that took place during the symposium held by the Islamic Organization for Medical Sciences. See: (http://islamset.net/arabic/aioms/). See also: Abā al-Khayl (n.d.), 28-29.

with the woman's liquid, regardless of the site of this mixture, is affected by the development of technical means and capabilities. This is consistent with medical knowledge which has concluded that life begins at conception in a gradual manner and that the characteristics of the human being are completed in the fertilized embryo, and the subsequent stages are those of growth and development.

Holders of the first opinion consider only the concept of the "fetus" and neglect the other provisions relating to this fertilized egg, such as the rights of either or both parents once they initiate the fertilizing process but then one of them retracts from the completion of the conception process. This applies to the disagreement regarding the ruling on coitus interruptus, for those who forbid this process confer on the sperm some sort of "respect," as the principle should be its dissemination on the part of the man inside the woman's uterus and not waste it. From this perspective, and in view of the confirmation of the child's lineage, Shāfiʿī jurists stipulate that the sperm be "respected," from the moment it is exuded from the male body to the moment it is injected into the female body. This can only take place during sexual intercourse between a legally married couple[24]. Therefore, one cannot say that the fertilized egg is wasted and not respected at all!

As for the issue of the fetus, there are divergent and elaborate views on it. In the context of this chapter, the dispute between the jurists revolves around the process of determining the occurrence of the inception and the stages of life. All jurists agree on the prohibition of abortion once the soul has been breathed into the fetus. This act of breathing of the soul only occurs four months after the inception, but the jurists differ on the ruling concerning abortion in the three stages preceding this one. Mālikī jurists argue for total prohibition even at the semen-drop stage. The Ḥanafīs and Shāfiʿīs, on the other hand, argue for the permissibility of abortion before the absolute breathing of the soul into the fetus, while the Ḥanbalīs have adopted an intermediate position between prohibition and permissibility, advocating permissibility at the semen-drop stage and prohibition at the stages of the clinging substance and the embryo. There are also overlapping views across the doctrines, but in the present study I have only focused on the statements adopted in each doctrine.[25]

The legitimate life agreed upon begins with breathing of the soul. The soul is what makes the living being a human being. Therefore, jurists have agreed that it is forbidden to abort the fetus at this stage and have issued a number of rulings, both worldly and holy, such as the performance of *Janāzah* prayer

---

24 See al-Khaṭīb al-Shirbīnī 1994, 6: 516.

25 See the different views on abortion in Wizārat al-Awqāf (n.d.), 2: 57-59; Yāsīn 2008.

on it, wrapping it in a shroud and burying it, arguing for its resurrection on the Day of Judgment, etc. This, however, does not imply that the fetus's life should be tampered with before the soul is breathed into it, for there is a "legal life" involved here. Even though the embryo, before the soul is breathed into it, is like an inanimate thing,[26] many jurists have forbidden its abortion because they see in it a legal life though they differ on the timing of its inception. Does it start with the "clinging substance" and the settlement of the semen drop in the womb or from "clotting" and the beginning of its transformation into pregnancy in the formation of the clinging substance stage? Those who argue that it starts with the "settlement" forbid abortion altogether, and those who see that it starts at the clotting stage conclude that the sperm drop does not require consideration or concern and it may be wasted, because it has not yet with certitude turned into life.

The question of legal existence (*al-ḥayāt al-iʿtibāriyya*) is based on the principle of precaution in religion (*al-iḥtiyāṭ fī-l-dīn*), whether in terms of the rulings of pregnancy or respect for the life of the fetus. Aspects of precaution that determine the rulings show that Ḥanbalī jurists have not issued rulings for the clinging substance the way they have for pregnancy (such as the postpartum period, the prescribed legal period, etc.), because they do not consider the clinging substance to represent a real state of pregnancy even if the life of the embryo is considered to be a legal one. Part of the precautions for the life of the fetus itself is what some Shāfiʿī jurists refer to as the "sanctuary of the soul," (*ḥarīm al-rūḥ*) i.e. the period preceding the breathing of the soul,[27] in order to take precautions concerning the actual life agreed upon, ensuring no injustice is inflicted upon it. Some jurists have issued rulings concerning the embryo, for they see it as the locus of the formation of the lump of flesh. Others, however, have exercised reservation concerning the time span of the sperm. Hence, Ibn al-Jawzī, of the Ḥanbalī school, prohibits abortion starting from the stage of the semen drop, because "the pregnancy is in the process of development and leading towards completion and perfection. Therefore, abortion is a violation of the divine will" (1981, 374). Ibn ʿĀbidīn, of the Ḥanafī school, reports that some fellow Ḥanafī jurists abhor this practice because "after the sperm falls

26 The Ḥanbalīs differentiate between the soul and life and argue that they do not correlate. Before the soul is breathed into it, the embryo experiences movement, growth and nourishment just like plants, but with no sensation or will. When the soul is breathed into it, then sensation and will join the process of growth and nourishment. The movement of the fetus is of two types: a volitional self-movement that is made possible thanks to the soul, and a casual movement caused by membranes and moisture. See: Ibn Qayyim Ibn al-Qayyim (2008), 509.

27 See al-Ramlī 1984, 8: 442.

into the womb, it is destined to live; therefore, it falls under the ruling applicable to life" (1992, 3:176). This view is the most recognized in the Shāfiʿī doctrine, because "the sperm drop, after becoming settled, is destined for formation and prepared for the breathing of the soul" (al-Shabrāmallisī 1984, 6: 182; al-Sharwānī 1983, 7:186).

However, this precaution does not eliminate the difference between the ethical legal aspect and the aspect of rights, although a number of provisions have been established to protect the fetus, they do not confer the aspect of "personality" (*shakhṣiyya*) upon it until it is born. This means that its existence in any form is only recognized when it leaves the uterus and as far as it bears characteristics of a human being. Hence, the rulings have been based on birth and the perception of a human being, even if this humanness itself is concealed.

2.2.b The Attribute of Humanness (*al-ʾādamiyya*) and the Sacredness of the Body

The ruling principle in this matter is that souls and bodies have sanctity (*ḥurma*) (al-Bukhārī 1997, 3: 147), and pregnancy is considered as part of the mother's body since the embryo relies on it.[28] It is also related to the fact that humanness is defined in terms of a number of concepts observed by jurists, who have placed obligational provisions, such as the concept of "clotting," (*al-inʿiqād*) which is the principle of turning the semen into a clinging substance. The latter, then, is the semen turned into a congealed blood clot. Ḥanbalī texts and rulings consider the clinging substance as the first indication that the pregnancy is a conceived baby; therefore, they use the principle of sacredness to forbid the abortion of the clinging substance because it is a "solidified embryo" (Ibn Mufliḥ 1997, 7:74; Ibn Rajab 2004, 1:161; Al-Buhūtī (1993), 3:193; Al-Ruḥaybānī 1994, 1:267&5:561). Ibn al-ʿUthaymīn forbids the abortion of the clinging substance on the principle that it is "blood, and blood is the substance of life" (2007, 13:342). This is also true of the concept of "formation" (*al-takhalluq*) or "the principle of human creation" (i.e. its beginning), and "the hidden image of humanity," among other expressions.

Humanness (*al-ʾādamiyya*), or the principle of its formation, requires respect as it is developing to completion to become a full-fledged body that is ready to receive the soul. On this basis, one should distinguish between "humanness" (*al-ʾādamiyya*) and "human life (*al-ḥayāt al-ʾādamiyya*)." Human life is established by virtue of breathing of the soul, which is commonly agreed upon, but humanness is inherent to the creation of man before and after breathing of

28 See al-Kāsānī 1986, 4:94.

the soul and is recurrent in the debate on abortion of the early fetus. There are special obligational provisions not to compromise humanness, denigrate it, or destroy it in discretionary punishment (projected penalties include imprisonment, etc.). Corpses should not be subjected to torture or denigration. These provisions also exist in dealing with slavery, as in ancient times, whereby the jurists decided that the "characteristics of humanness in a slave are not compromised as a result his status as a slave" despite him not being free, and that "humanness in a slave is more elevated than money."[29]

These concepts and considerations refer to the relationship between body and self, discussed previously under the philosophical debate, especially that the human body is valued from the beginning of its formation: the clotting and the perception, and even after the departure of the soul. In a Prophetic tradition, it is stated, "when the soul of a believer goes out (of his body) it will be received by two angels who will take it to the sky," and it will be told: "let the blessings of Allah be upon the body in which you used to reside."[30]

Genetic intervention in the body's cells is considered an intervention in an actually existing person whose own characteristics have been identified and established; that is, the intervention takes place in the components of the human body itself. Therefore, the intervention must abide by the considerations of its nature, motive and implications, in case harm is expected to result from this. This situation may, therefore, be subject to case-by-case considerations.

2.2.c The System of Rights

Islamic law identifies three types of rights, those owed to God, to people, and to both.[31] The fetus, on the other hand, is endowed with several rights according to different considerations that govern the issuance of the ruling in this matter. These are: God's right, the fetus's right and the parents' right. These aspects constitute the basis of the following discussion.

The embryo is primordially a divine creation, and this is a religious issue that distinguishes between parents as the means of procreation and God as the owner and creator of the fetus. The formation and development of the embryo is part of the system of creation, from the beginning to the end. Thus, all kinds of abortion are acts of transgression against God's design and creation. There are some Ḥadīths that refer to the existence of an angel in charge of the semen; yet, some jurists disagree on the timing of the angel's pledge to take charge of it depending on different Ḥadīths. However, some of these jurists conclude

---

29 See Wizārat al-Awqāf (n.d.), 23:13&73.

30 Muslim 1998, No. 2872.

31 See Mutaz al-Khatib 2013, 27.

that "the angels are committed to and caring for the condition of the semen at its different periods" (Al-ʿAynī 2001, 3:435). Some late Ḥanbalīs argue that the angel designated for the affairs of the fetus starts recording its destiny in its first forty seconds (Ibn al-Qayyim 1988, 173; Ibn Rajab 2004, 1:173). In view of this religious explanation, there is a distinction between the sin of committing abortion and the determination of the obligatory expiation and "blood money" related to the destruction of the fetus. The religious sin is established as a crime against the clinging substance and the unformed lump of flesh, though the jurists do not require that the offender pay blood money or perform acts of expiation in this situation. The required expiation in case of a crime against the fetus who bears human traits is a punishment imposed in the name of Almighty God's right. In this case, the punishment carries the meaning of deterrence as well as worship, because it is carried out through the act of fasting. This punishment is proof that the sinful act represents a transgression against the Creator's wisdom and design, hence the imposition of expiation as an obligation towards Almighty God. Genetic interventions can be included in this rubric, and the permission of the legislator is necessary in this regard, because it represents an act of disposition that concerns God's Kingdom and system of creation.

In respect to the fetus and its entitlement to rights, the jurisprudential view examines and determines the nature of the fetus: is it an independent being? If it is looked at as being part of the mother and is nourished by her nourishment, the ruling then is that it is not independent, and no obligation is required of it. If, however, it is considered an independent body with a life of its own, the ruling of obligation applies to it, and by virtue of this, it becomes entitled to rights and obligations.[32] Nonetheless, "given that either way cannot be decisively confirmed, Muslim jurists treat the fetus as part of its mother because it is not fit to assume obligations, but it has also been treated as an independent soul with its own life, which makes it fit to undertake obligations. Accordingly, the embryo acquires incomplete obligations."[33] In this regard, Hanafi jurists define the fetus as "a faceless body" (Ibn Qudāma 1968, 8:406; Ibn Mufliḥ 1997, 7:295; Al-Kāsānī 1986, 7:325). On this basis, it requires rights that do not need acceptance, such as the confirmation of lineage, inheritance and eligibility to endowment. Since the fetus is potentially capable of future separation and independence, only rights that do not require acceptance are recognized. The genetic intervention that affects its formation cannot be performed without its permission, which is actually impossible, or its guardian's, which raises the

32 See Ibn Rajab 1998, 2:225-251.

33 See Wizārat al-Awqāf (n.d.), 16:118-19.

question of the limits of guardianship in such cases. Old jurists' views concerning «guardianship of the soul» (*al-wilāya ʿalā al-nafs*) involve medicating for therapeutic purposes and educating. Therefore, in this case non-therapeutic genetic interventions are not involved and the actions of the guardian are conditioned by the permission of the legislator as well.

Concerning the right of parents, it is confirmed as early as the right to procreation, even though there is a dispute on whether this is the husband's right or it is a joint right between the spouses. This comes to surface especially when tackling the question of abortion of the sperm drop after it reaches the uterus or the case of coitus interruptus to prevent the sperm from reaching the womb (*al-ʿazl*). Shāfiʿī jurists see that it is the husband's right, while the majority of jurists argue that the child is a common right between the spouses (Ibn ʿĀbidīn 1992, 3:175-176; Al-Buhūtī 1993, 1:122&3:44; Al-Mirdāwī 1956, 1:383).

Based on the above, the right of each party must be taken into account, and the rule governing genetic intervention stipulates that "no one should grant himself the liberty of using that which belongs to others without their permission" (Al-Zarqā 1989, 461). With regard to rights, the resolution of the International Islamic Fiqh Academy on the human genome emphasizes, "clear and legally valid permission is mandatory and should be solicited from the person himself or his legal guardian for the examination of his genetic map, ensuring the interests of the person concerned." The Academy declares, "everyone has the right to decide whether he wishes to be kept apprised of the results or implications of any genetic test he undergoes." This resolution also applies to the requirement that "all archived genetic diagnoses or those prepared for other purposes like research should be subject to full confidentiality." In addition, "no person should be made subject to any form of discrimination because of his genetic attributes, if it proves that the purpose is to compromise his freedom and fundamental rights and violate his dignity" (IIFA 2013, 21).[34] The International Islamic Fiqh Academy also ruled that genetic intervention is legal when carried out for therapeutic purposes and in compliance with specific conditions; for example, the type of treatment does not lead to greater harm than the harm the person is inflicted with, that the preponderant intent should be to heal or alleviate pain, that alternative treatment does not exist, that the proper and legal conditions of the organ transfer from the donor to the recipient should be observed, and finally that the operation is carried out by specialized and highly experienced staff known for their skillfulness and trust.

34 A similar resolution was adopted by the Islamic Fiqh Council, affiliated to the Muslim World League in Makkah 1998,15th Session.

2.2.d Preservation of Self (*ḥifẓ al-nafs*)

"Preservation of self" is considered a higher objective of Islamic law and one of its most fundamental pillars. The concept of "preservation of self" transcends the modern philosophical criticism of the meaning of "sacredness of life." It differs from the expressions "appreciation of life," "quality of life," and "distinctiveness of life" which is referred to earlier. It refers to a general moral law agreed upon by universal proclamations, including the precise and clear meaning of general obligation, also their genera: care for self and for others. Preservation of self is based on two principles: "the first one is that which builds its cornerstones, establishes the rules and observes existence. The second concerns all that can ward off disruption, existing or expected, and ensure its preservation" (Al-Shāṭibī 1997, 2:18). It also includes two views: preservation in whole and in parts. Retribution, although it entails killing a single person, aims for the preservation of the whole. In other words, preserving the self is an absolute law that takes into account the part, and brings together the normative and applied levels. The expression "preservation of self" helps us assimilate modern techniques and genetic engineering. Genetic manipulation, for example, is not an act of killing. Thus, the rule of "do not kill without justification" is inadequate to accommodate such developments, while it is covered by the law of "preservation of self" from both sides of existence and nonexistence. This also includes the real self and the legal self, as previously explained. Some contemporary studies[35] have underrated these meanings which concern "human interests" and their ethical dimensions.

The concept of "preservation" (*ḥifẓ*), therefore, helps us assess genetic intervention by identifying three things: the form and nature of the intervention, its motives, objectives and expected outcomes. This will be based on a case-by-case study. The International Islamic Fiqh Academy and the Makkah-based Islamic Fiqh Council stipulate "a prior and accurate assessment of the potential risks and benefits associated with these activities" before conducting any research, treatment, or diagnosis related to the genome.[36]

Preservation of self also includes treatment, which was debated by ancient jurists and considered a human rights issue in the way it is perceived in the modern context. The evolution of biological techniques expanded the concept of medicine beyond the traditional perception of "treatment" with its six principles. Ibn al-Qayyim explained its six principles as: "the framework of the doctor's authority is to make treatment and procedure revolve around six pillars: preservation of existing health, restoration of lost health as far as possible,

35 For example, the study by ʿAbd al-Nūr Bazā 2008.

36 See the International Islamic Fiqh Academy 2013, 21; the Islamic Fiqh Council 1998, 15.

removal or alleviation of illness to the extent possible, identification of the lesser harm to remove the more harmful, and dismissal of the lower interest to achieve the greater one. Treatment should be carried out based on these six fundamentals" (Ibn al-Qayyim 1994, 4:132-133). However, the significant development in genetic technology has revealed the need to address three issues: first, all genetic diseases and malformations should be treated to meet the demands of a society that is illness-free but, whose configuration is hard to predict. Second, the genetic disease that must be treated still lacks precise identification: what should be considered a disease and what should not? Does the variation relate to customs and cultures? Third, the human race should be developed and enhanced, which means that the color of the skin or eyes, or the person's height, will be subject to cultural influences and preferences, as well as the other potential and desirable qualities related to mental and physical abilities.[37]

The concept of treatment cannot be extended to encompass all these developments; thus, these matters should be accommodated under the concept of "preservation of self." On the contrary, preservation of self entails the restriction of interventionist practices in the human's body and life, especially when some of these interventions concern the individual and his fate (somatic cells), while others affect him and his progeny (sex cells). The International Islamic Fiqh Academy has authorized the genetic treatment of somatic cells under certain conditions, as stated previously in this chapter, and has also approved genetic surveys provided that the means are permissible and safe, while protecting the confidentiality of information. What oversteps therapy, such as enhancement (engineering) purposes, are often practiced before fertilization or on the fetus at an early stage. This transcends the purpose of preservation of self and is rather closer to the purpose of preservation of lineage.

2.2.e Preservation of Lineage (*ḥifẓ al-nasl*)

The preservation of lineage is one of the five objectives of Islamic Law. The reproductive cells relate to sexual organs, and the ruling about sexual organs is prohibition. Therefore, intervention is conditioned here by the principle of lineage preservation, but therapeutic intervention is often dominated by the introduction of foreign elements that lead to confusion in lineage. Accordingly, the International Islamic Fiqh Academy decided to allow the examination of the sexual cells to identify diseases, but prohibited their treatment in its

37 See Mussa al-Khalaf 2003, 84-85.

"current form," because this act does not take into consideration Islamic legal provisions, leads to the mixing of lineage and, therefore, is risky and harmful.[38]

On the other hand, intervention that serves an enhancement purpose when performed on sexual cells is somewhat problematic. The International Islamic Fiqh Academy and the Makkah-based Islamic Fiqh Council agreed to prohibit it because it interferes with the origin of the human being, tampers with the creation, transgresses human dignity (through the process of human engineering and experimentation), abuses the human character and his individual responsibility, and it does not respond to lawful needs and necessities.[39]

This kind of intervention may be viewed from the perspective of "enhancement" of the body (*tazyīn al-jasad*), which has evolved into plastic surgery and then genetic improvement. Yet, genetic enhancement is different in that it deals with the origin of the human being and his formation, and this is subject to the same difference as in the case of the "sacredness" of the sperm drop and the embryo. Those who argue against their "independent" features will draw on protracted views regarding therapeutic and enhancive interventions. The reason put forth by the International Islamic Fiqh Academy and the Islamic Fiqh Council that this matter involves an act of "changing God's creation" (*taghyīr khalq Allāh*) is subject to several textual interpretations: does this process include absolute physical transformation or is it specific to moral change (God's religion), as reported by Ibn ʿAbbās and others? Is the change specific to the postpartum stage or does it include the prenatal one as well? Physical change (tattooing, removal of eyelash extensions and attachment of hair extensions, etc.) is a matter of disagreement among jurists, which reflects their different views about the cause of the prohibition of these practices. Is it a change in God's creation, an act of fraud, etc.? Thus, the analogical measurement of judgment on eugenics on the basis of enhancement and beautification would involve the same discord. Unless, we argue that the improvement of lineage is different from the enhancement of the body for the same considerations, as previously stated, in terms of time, impact, the type of intervention and the decision-maker in the intervention. Moreover, an enhancement-based intervention is analyzed based on the degree of enhancement, the part of the body that will be subject to such an intervention, and whether this will transgress the necessities or needs of the part (the special case), or of the whole (for the human race), if the modification slips out of control and becomes an act of interference in the system of creation.

38 See the resolution of the International Islamic Fiqh Academy referred to earlier.

39 See the two resolutions made by the International Islamic Fiqh Academy and the Makkah-based Islamic Fiqh Council on genetic engineering referred to earlier.

The above considerations problematize the conclusions reached by some researchers. The ruling stipulates that, in origin, the enhancement intervention is authorized based on the generalities of the texts that convey the need for strong lineage, and that the characteristics familiar to human beings are part of the permissible when there is no text that either proscribes it or orders it.[40] Here, we should question the concept of strength, its nature and source, and whether it includes the energy desired by athletes. For example, athletes who should attribute their achievements not to themselves and their abilities, but to the modifications made onto them. Differentiation should also be made between the self-propelling (intrinsic) human nature and the contrived one (extrinsic). Al-Shāṭibī talks about the innate human characteristics, for which there are no provisions of order or proscription. However, the case here is that of a person who is in control of the nature of another person who is making progress as a human being and is in the process of complete development.

The elements of lineage preservation authorized by the International Islamic Fiqh Academy concern premarital genetic examination, provided the instrument is lawful and safe, diagnosis of the fertilized egg before implantation, provided that samples are not mixed and examination during the pregnancy. So, if a hereditary disease is detected, then the abortive procedure is permitted. The International Islamic Fiqh Academy and the Islamic Fiqh Council have even required the genetic screening of newborns for early intervention in curable cases. All these practices are included in modern applications concerning the preservation of lineage from genetic diseases.

From the discussion above, we can identify the following axes as the subject matter of the contemporary jurisprudential debate with reference to genetic technology: first, there are interests and harms which result from the uses of this technology, and there should be restrictions on how to use this technology in "useful areas." In this respect, there are several observations: 1) the benefit should initially serve to prevent, treat and improve. Much of the discussion actually revolves around the physical or material benefit and harm, and there are hardly any traces of the impact on moral considerations and the influences that these interventions can have on the human personality, psychologically and mentally, in terms of the connectedness of self-awareness and physical awareness. The discussion does not refer either to the ethical individual or to the social responsibilities resulting from the intervention, especially if the latter concerns development or enhancement. 2) The interests and harms represent different ranks; therefore, they must be brought under control based on a number of considerations, such as the analogical evaluation of the motive be-

40 See Tamam al-Lūdaʿmī 2006, 168-169.

hind the intervention and the actual need for it, as well as, the expected benefit and the distinction between the need for (correction of defects for instance) and the enhancive (preferential purposes).

Second, the main question of medical treatment is foundational in the field of medicine. The contemporary jurisprudential discussions, however, hardly discuss the evolution of the concept of medicine, the need to grasp the notion of treatment, and its criteria. Further, a contemporary jurisprudential discussion on whether it is influenced by culture or if it is a normative matter.

Third, the objectives of Islamic law, especially in relation to the preservation of self, preservation of lineage, and that the means used to achieve these should be lawful and safe.

Fourth, the outcomes should be well considered by requiring the pre-evaluation of uses, interventions, and impacts of genetic engineering, be they physical or social. This includes evaluation of outcomes which lead to discrimination or harm in contracts, such as employment, marriage, etc.

Conclusion

In general, the discussion in the context of this chapter does not concern the genetic technique per se, but, it does concern its uses, aims and impacts. Therefore, only through the position that we adopt *vis-à-vis* the power of genetic engineering and how we monitor the possibilities it offers can we ensure that no ethical harm will be inflicted upon the lineage of the potential human being. The extent of the agreement that can be reached is the need to prevent diseases that are ineluctably dangerous. The philosophical debate is preoccupied with the idea of elaborating convincing criteria that determine the health and disease of the physical body in the context of a non-mechanical relationship, and the establishment of a distinct line between "lineage" whose aims are therapeutic and "eugenics" whose aims are developmental.

This vital and current issue has posed challenges to philosophical theories like Kant's on the ethical being and Habermas on the communicative act. The issue at stake is urgent and affects the future of coming generations. The main debate has been about the ethics of self-understanding as applicable to the human species, but, the critical question that can be raised here is that it is not clear how this type of ethics can engender individual obligations that are threatened by genetic intervention.

The discussion on genetic intervention evokes the divide between science, ethics and jurisprudence in the way they look at the future. While science tries to understand physical reality, it rarely examines the greater questions concerning existence itself. Science seeks to push for development to the farthest extent possible and to turn the possible into reality, which in turn opens an existing space for the emergence of other possibilities, and so on and so forth. This incremental process involves movement from the simple to the compound, the results of which should be dealt with by science, ethics and jurisprudence. Thus, Habermas tried to push philosophical thinking to anticipate the developments by reflecting on the principle of total development and predicting the course of the possible. This implies that ethical thinking should transcend science, keep its movement in check, and reduce its attempts to dominate and manipulate human nature. Instead of merely tracking it down and finding solutions and outcomes for the present reality it imposes. Few contemporary jurists show more interest in the jurisprudence of projections versus that of the present-day reality. Moreover, few contemporary jurists consider the outcomes at both levels, the partial (individual and specific practice) and global (human nature).

Genetic technique is the product of Western modernity and its perceptions of man and the world. It raises all these philosophical, ethical, and jurisprudential discussions, which intersect as well as diverge. The discussion here revolves around issues which have long been considered central to religion, such as lineage and reproduction. The developments, however, have generalized the debate and made the discussions appeal to different branches of knowledge, thus involving the philosopher in discussions about biological applications and making him justify his participation therein, instead of, leaving the terrain only to biologists and genetic engineers. These developments have equally prompted the jurist to interact with these same issues, even though the use of these techniques is still limited in the Muslim world.[41] However, the significant legal and ethical legacies available to the jurist, as well as, the nature of his job and specialization, require that he be at the heart of these philosophical and ethical debates.

Paradoxically, the notion of "autonomy" raised by Kant and advocated by other philosophers, although, it was received negatively as an attempt to "destroy the traditional view that defines man in terms of progeny created by God," poses a serious challenge to genetic technology. It also makes us wonder about the concept of "individual freedom," because the available and expanding possibilities widen the extents of this freedom. Thus, Habermas found himself compelled to engage in this kind of debate and defend what he called a "justified reservation" *vis-à-vis* genetic technology.

---

41 There are several projects in the GCC countries, namely in Saudi Arabia, Qatar and the UAE, which aim to indigenize research studies about the genome, its uses and techniques.

On the other hand, some concepts like autonomy itself still need to be thoroughly examined by jurists. The Islamic debates around it have not yet been exhausted, especially with regards to the notion of autonomy of the fetus who is about to become a full-fledged being, and the extent of parental authority over him. It is also important to rise above the practical and partial jurisprudential discussion to embrace theological discourse (*kalām*). This is due to the fact that the challenges posed by genetic technology cannot be addressed in isolation from the perceptions about existence and its aims, the relationship of self and body, and the essence of human life as divine creation. This also concerns the view of the body in relation to the soul, the impact of the possibilities of genetic technology on the concept of "trials and tribulations," which is a central concept in the Qur'ān, in the perceptions of creation and life, in the study of the relationship between expanding horizons and limited acts, or between the possible and the obligatory within the relationship between parts and wholes.

References

Abā al-Khayl, Sulaymān. n.d. "Isqāṭ al-ʿadad al-Zāʾid min al-Ajinnah al-Mulaqqaḥah Ṣināʿiyyan". Paper presented at the 2nd Fiqh Conference in Imām Muḥammad Ibn Saʿūd Univesity, Saudi Arabia, 28-29.

ʿAbd al-Raḥīm, Maḥmūd. 2002. "Al-Aḥkām Al-Sharʿiyyah wa-l- Qānūniyyah li-l-Tadakhkhul fī ʿAwāmil al-Wirāthah wa-l- Takāthur". PhD dissertation: ., Cairo: Al-Azhar University.

Āl al-Shaykh, Hishām. 2010. *Athar al-Tiqaniyyah al-Ḥadīthah fi-l-Khilāf al-Fiqhī*. Riyadh: Al-Rushd Library

Āl al-Shaykh, Muḥammad and ʿAbdul ʿAzīz Bin Bāz. 2004. *Al-Fatāwā al-Mutʿalliqah bi a-l-Ṭib wa Aḥkām al-Marḍā*. Cairo: Ulī al-Nuhā li-l-Intāj al-ʿIlmī.

Al-Ashqar, ʿUmar. 2001. "Al-Istifādah min al-Ajinnah al-Mujhaḍah aw al-Zāʾidah ʿan al-Ḥajah fī al-Tajārib al-ʿIlmiyyah wa Zirāʿat al-Aʿḍāʾ," In *Dirāsāt Fiqhiyya fī Qaḍāya Ṭibbya Muʿāṣirah*, edited by ʿUmar Al-Ashqar et al., 1:305-310. Amman: Dār Al-Nafāʾis.

Al-ʿAynī, Bdr al-Dīn. 2001. *Umdat al-Qārī: Sharḥ Saḥīḥ al-Bukhārī*. Beirut: Dār al-Kutub al-ʿIlmiyyah.

Al-Baqṣamī, Nahida. 1993. *Al-Handasah al-Wirāthiyyah wa-l-akhlāq*. Kuwait: National Council for Culture, Arts and Literature.

Bazā, ʿAbd al-Nūr. 2008. *Maṣāliḥ al-Insān: Muqārabah Maqāṣidiyyah*. Herndon: The International Institute of Islamic Thought.

Al-Buhūtī, Yūnus. 1993. *Daqāʾiq Ulī al-Nuhā li-Sharḥ al-Muntahā*. Beirut: ʿĀlam al-Kutub.

Al-Bukhārī, ʿAbdul ʿAzīz. 1997. *Kashf al-Asrār ʿan Uṣūl Fakhr al-Islām al-Bazdawī.* 3:147. Beirut: Dār al-Kutub al-ʿIlmiyyah.

Christiansen, Karin. 2009. "The Silencing of Kierkegaard in Habermas' Critique of Genetic Enhancement" *Medicine, Health Care and Philosophy*, 12: 147-156.

Dār al-Iftāʾ al-Miṣriyya. 2014. *Al-Handasa al-Wirāthiyyah wa Istikhdāmuhā fī Majāl al-ʿilāj*. al-Miṣriyyah website.

Al-Ghazālī, Abū Ḥāmid. n.d. *Iḥyāʾ ʿUlūm al-Dīn*. Beirut: Dār al-Maʿrifah.

Guilfoyle, Ciaran. 2004. "The Future of Human Nature by Jurgen Habermas" *Philosophy* 79:483.

Habermas, Jürgen. 2003. *The Future of Human Nature*. Cambridge: Polity Press.

Hegel. 2007. *Elements of the Philosophy of Right*. Trans. Imām ʿAbduel Fattaḥ. Beirut: Dār al-Tanwīr.

Heidegger, Martin. 1995. *The Question Concerning Technology, and Other Essays*. Trans. Sabīla, Muḥammad and Miftāḥ, ʿAbduel Hādī. Beirut: The Arab Cultural Center.

Holmer, Paul. 1953. "Kierkegaard and Ethical Theory". *Ethics* 63: 157-170.

Ibn ʿĀbidīn. 1992. *Radd al-Muḥtār ʿalā al-Durr al-Mukhtār.* Beirut: Dār al-Fikr.

Ibn al-Jawzī, Abū al-Faraj. 1981. *Aḥkām al-Nisāʾ.* Beirut: Manshūrāt al-Kutub al-ʿAṣriyyah.

Ibn Mufliḥ, Shams al-Dīn. 1997. *Al-Mubdiʾ fī Sharḥ al-Muqniʾī.* Beirut: Dār al-Kutub al-ʿIlmiyyah.

Ibn al-Qayyim. 1994. *Zād al-Maʿād fī Hady Khayr al-ʿIbād.* 4:132-133. Beirut: Muʾassasatu al-Risālah.

Ibn al-Qayyim, Abū ʿAbdullāh. 1988. *Tuḥfat al-Mawdūd bi Aḥkām al- Mawlūd.* Cairo: Maktabat al-Qurʾān.

Ibn al-Qayyim, Abū ʿAbdullāh. 2008. *Al-Tibyān fī Aymān al Qurʾān*. Edited by ʿAbdullāh Bin Sālim al-Baṭāṭī, Saudi Arabia: ʿĀlam al-Fawāʾid

Ibn Qudāma, Muwaffaq. 1968. *Al-Mughnī.* Cairo: Maktabat al-Qāhirah.

Ibn Rajab, ʿAbdul Raḥmān. 1998. *Taqrīr al-Qawāʿid wa Taḥrīr al-Fawāʾid.* Saudi Arabia: Dār Ibn ʿAffān.

Ibn Rajab, ʿAbdul Raḥmān. 2004. *Jāmiʿ al-ʿUlūm wa-l-Ḥikam.* Cairo: Dār al-Salām.

Ibn al-ʿUthaymīn. 2007. *Al-Sharḥ al-Mumtiʿ ʿalā Zād al-Mustaqniʿ*. Saudi Arabia: Dār Ibn al-Jawzī.

Iglesias, Teresa. 1984. "In Vitro Fertilization: The Major Issues". *Journal of Medical Ethics* 32-37.

J. Karnein, Anja. 2012. *A Theory of Unborn Life*. New York: Oxford University Press.

Al-Kāsānī, ʿAlāʾ. 1986. *Badāʾiʿ Al-Ṣanāʾiʿ fī Tartīb al-Sharāʾiʿ.* Beirut: Dār al-Kutub al-ʿIlmiyyah.

Al-Khalaf, Mūssā. 2003. *Al-ʿAṣr al-Jīnūmī: Istrātījīyyāt al-Mustaqbal al-Basharī*. Kuwait: ʿAlam al-Maʿrifah.

Al-Khatib, Mutaz. 2013. "Manẓūmat al- Ḥuqūq wa Maqāṣid al-Sharīʿah". *Al-Tafāhum* 39:27.

Al-Khatīb al-Shirbīnī, Shams al-Dīn. 1994. *Mughnī al-Muḥtāj ilā Maʿrifati Maʿānī Alfāth-dh al-Minhāj*. Beirut: Dār al-Kutub al-ʿIlmiyyah.

Al-Lūdaʿmī, Tammām. 2006. "Al-Tadakhkhul fi-l- Jīnūm al-Basharī fī al-Sharīʿah wa-l-Qānūn". *ʿĀlam al-Fikr Journal* 35:141.

Al-Lūdaʿmī, Tammām. 2011. *Al-Jīnāt al-Bashariyyah wa Taṭbīqātuhā: Dirāsah Fiqhiyyah Muqāranah*. Hirenden: International Institute of Islamic Thought.

Al-Mirdāwī. 1956. *Al-Inṣāf fī Maʿrifat al-Rājiḥ min al-Khilāf.* Beirut: Iḥyāʾ al-Turāth al-ʿArabī.

Muslim Ibn al-Ḥajjāj. 1998. *Saḥīḥ Muslim*. Riyadh: Bayt al-Afkār al-Adawliyyah.

Plessner, Helmuth. 1969. "The Living Human Being." Trans. by Magdi Youssef. *Fikr wa Fann* 9-10.

Prusak, Bernard G. and Erik Malmqvist. 2007. "Back to the Future: Habermas's 'The Future of Human Nature'". *The Hastings Center Report* 37:4-6.

Al-Qalyūbī, Shihāb al-Dīn. 1998. *Ḥāshiyat Qalyūbī ʿala Sharḥ Jalāl al-Ddīn al-Maḥallī ʿala Minhāj al-Ṭālibīn*. Beirut: Dār al-Fikr.

Al-Ramlī. 1984. *Nihāyat al-Muḥtāj ila Sharḥ al-Minhāj.* 8:442. Beirut: Dār al-Fikr.

Al-Ruḥaybānī, Muṣtafā. 1994. *Maṭālib Ulī al-Nuhā fī Sharḥ Ghāyat al-Muntahā*. Beirut: al-Maktab al-Islāmī.

Al-Sallāmī, Muḥammad. "'Symposium held by the Islamic Organization for Medical Sciences'". http://islamset.net/arabic/aioms/.

Al-Shabrāmallisī, Nūr al-Dīn. 1984. *Ḥashiyyatun ʿalā Nihāyat al-Muḥtāj ilā sharḥ al-Minhāj li Al-Ramlī.* Beirut: Dār al-Fikr.

Al-Shāṭibī, Abū Isḥāq. 1997. *Al-Muwāfaqāṭ fī Uṣūl Al-Sharīʿah*. Saudi Arabia: Dār Ibn ʿAffān.

Al-Sharawānī, ʿAbdul Ḥamīd. 1983. *Ḥāshiyya ʿalā Tuḥfat al-Muḥtāj fī Sharḥ al-Minhāj li Ibn Ḥajar al-Haytamī.* Egypt: Al-Maktabah al-Tijāriyyah al-Kubrā.

Simmons, Paul D. 1983. *Birth and Death: Bioethical Decision Making*. Westminster Press.

Stack, George J. 1973. "Kierkegaard: "The Self and Ethical Existence". *Ethics* 83:108-125.

Wizārat al-Awqāf wa al-Shuʾūn al-Islāmiyya bi al-Kuwayt. n.d. *Al-Mawsūʿa al-Fiqhiyya al-Kuwaytiyya*. Kuwait: Wizārat al-Awqāf wa-l-Shuʾūn al-Islāmiyya.

Yāsīn, Muḥammad Naʿīm. 2008. *Abḥāth Fiqhiyyah fī Qaḍāyā Ṭibbiyya*. Jordan: Dār al-Nafāʾīs.

Al-Zarqā, Aḥmad. 1989. *Sharḥ al-Qawāʿid al-Fiqhiyyah*. Revision and Commentary by Muṣṭafā al-Zarqā. Damascus: Dār al-Qalam.

PART 3

Widening the Scope of Ethical Deliberations

CHAPTER 7

In the Beginning Was the Genome: Genomics and the Bi-Textuality of Human Existence

Hub Zwart[1]

Introduction[2]

The *Human Genome Project* (HGP) has been hailed as an important milestone, not only for the history of the life sciences, but even for humanity as such (Collins 2006). It was presented as an endeavour that would transform the practice of medicine but also change the course of human history (Davies 2001, cf. Zwart 2015). And yet, although the HGP undoubtedly altered the way in which biomedical research is conducted (Collins 1999), the actual benefits for human society (notably in terms of novel treatments for diseases, for instance) have been limited so far (Collins 2011), so that the great expectations initially associated with classical genomics have now been displaced to newer hype-prone areas of research, such as next generation sequencing (NGS), personalised genomics, precision medicine and gene editing. In this contribution I will argue, however, that the cultural and spiritual relevance of the HGP, has been quite substantial. Genomics, I will contend, has affected our self-understanding as 'rational animals' and as stewards of creation (Zwart 2009). More specifically, the HGP revivified the (allegedly outdated) question of the soul (Ahmed and Suleman 2017), a key issue not only in Christian, but also in Islamic thinking; two intellectual traditions of global significance for which Aristotle has been a major source of inspiration, in combination with the Bible and the Quran respectively. The aim of this contribution is to assess the broader, cultural rel-

1 Professor of Philosophy at the Faculty of Science and Director of the Institute for Science in Society, Radboud University Nijmegen, the Netherlands, h.zwart@science.ru.nl

2 I profited significantly from the comments and discussions which evolved during the seminar "Islamic Ethics and the Genome Question". A previous version of this paper was published in *The New Bioethics: A Multidisciplinary Journal of Biotechnology and the Body*, 24 (1), pp. 26-43. DOI: 10.1080/20502877.2018.1438776

evance of human genomics and its philosophical, spiritual and ethical ramifications by staging a mutual learning dialogue (or triangulation) between genomics research, continental philosophy and religious (notably Christian and Islamic) anthropology.

First of all I will consider the way in which the cultural and spiritual relevance of the HGP was addressed by Francis Collins, at the time Director of the *International Human Genome Sequencing Consortium* during the famous Press Conference in June 2000 when the human genome sequence was proudly presented to a global audience, but even more elaborately in his autobiographical retrospect (Collins 2006). Subsequently, opting for a continental philosophical perspective (notably building on the work of key authors such as Hegel, Teilhard de Chardin and Lacan), I will address the question whether and to what extent the human genome can be regarded as the "language of God" (Collins 2006), or as a molecular update of the Aristotelian concept of the soul. Starting from the claim made by Max Delbrück that Aristotle must be credited with having predicted DNA, I will reread *De Anima* to explore whether insights coming from genomics indeed concur with Aristotle's understanding of the relationship between soul and life. My conclusion will be that human existence results from a dialectical interplay between two types of text: on the one hand the molecular language of DNA, on the other hand the languages of our socio-cultural environments. As living beings we are susceptible to the language of the genome, but as cultural and spiritual beings humans are also susceptible and answerable to the "language of the Other", providing a symbolic scaffold for moral responsibility and ethics.

The Adoration of a Genome

Eighteen years ago, on June 26, 2000, President Bill Clinton, together with scientists Francis Collins and Craig Venter, solemnly announced, from the East Room of the White House (*urbi et orbi*, so to speak), that the *Human Genome Project* (HGP) was rapidly nearing its completion. This carefully orchestrated, widely broadcasted press conference resembled a religious ceremony in various ways. As if Clinton, Collins and Venter conducted a spiritual service before an international gathering of top scientists, journalists and politicians, congregated in solemn adoration. The near-religious atmosphere was underscored by the fact that the addresses delivered on that occasion were punctuated by "blatantly religious references", as Collins himself phrased it (2006, 2), such as the statement by Clinton: "Today we are learning the language in which God created life"; or the statement by Collins: "Today we celebrate the revelation of the first draft of the human book of life" (National Human Genome Research Institute, 2000).

In his autobiographical retrospect, published six years after the event, Francis Collins confessed that for him, as a Christian scientist, the sequencing of the human genome was indeed "an occasion of worship" (2006, 3). He confessed to be "in awe of this molecule" (102), this "wondrous" map, this "miraculous" code, "previously known only to God". The process of "uncovering this most remarkable of all texts" held a special significance, since the human genome was "written in the DNA language by which God spoke life into being" (123). Therefore, the human sequence invoked in him "an overwhelming sense of awe" (123).

Interestingly, however, besides an electronic screen claiming that the decoding of the book of life represented a milestone for humanity, there was nothing to be seen during the press conference, nothing visibly on display. The mysterious centrepiece of the whole event, the focal point of attention, was emphatically absent, like a spectral Lacanian "thing". The dramaturgic *mise-en-scène* revolved around a void. Not only because the sequencing process was still ongoing (the celebration, for various complicated strategic reasons, was organised somewhat prematurely), but first and foremost because a strand of nucleotide code can only be made visible through highly technical means. It is a molecularised, computerised version of what we are, rather than a portrait or mirror lay audiences can relate to. At best, the typical output of automated sequencing machines resembles modernistic (decidedly non-figurative) art. Similar to Holy Mass, one could argue, the presence of the object of worship (the human genome) had to be presupposed or envisioned by a congregation of committed believers. But this concurred with Collins' conviction that the human code is the molecular equivalent of sacred Scripture, something that would be trivialised and desecrated by direct exposure to a public gaze.

The June 2000 event has been compared with the *Adoration of the Mystic Lamb* (Zwart 2010), a famous medieval polyptych altarpiece on display in Ghent's Saint Bavo Cathedral, created almost six centuries ago (between 1430 and 1432) by the Limburgian artists Hubert and Jan van Eyck: a highlight of late medieval religious art. Its central panel assembles knights, martyrs, hermits, pilgrims, saints, priests, burghers and nobility in a joint celebration, comparable to how the White House press conference brought together scientists, journalists, policymakers and heads of state in a similar gathering, as representatives (the front row as it were) of humankind, beholding a sublime, ethereal object. The Van Eyck altarpiece stages a culmination point, a final station in a collective pilgrimage. Like the lamb on the central panel, the human sequence is expected to deliver humanity from all kinds of evil: erasing cancer and producing cures for degenerative congenital diseases such as Alzheimer's: the HGP as a "soteriological" project (Song 2003). Indeed, it was claimed that, due to this "most wondrous map… Our children's children will know the term cancer only as a constellation of stars" (National Human Genome Research Institute, 2000).

But whereas human beings, trees, flowers and buildings are represented by the Van Eyck brothers with dexterous craftsmanship and exacting realism, the central figure (the mystic lamb) is an iconic, formulaic image, a screen or semblance covering up an empty spot, an unconceivable, un-representable "something", and the same applies to the dove hovering above (representing the Holy Spirit, the third component of the Holy Trinity). In Catholic liturgy, the Lamb (Jesus) is not literally visible, but present via a mystic event known as transubstantiation. The lamb image is inserted at the focal point of convergence to conceal a void, for the divine object is only spiritually perceivable (for true believers). Also in this respect, the late medieval artwork and the HGP press conference resemble one another. The focus of attention is a spectral, absent, imaginary entity, an adulated iconic screen covering a void, indicating the advent of something which is keenly anticipated, but not yet tangibly there.

Genomics and Self-knowledge

The spiritual aura radiating from the HGP (as one the highlights of contemporary technoscience) seems at odds with the decidedly "secular" profile of modern scientific research. Is this a coincidence, an oddity resulting from the fact that one of the key players in the room (Collins) happened to be a Christian? Or should we rather see it as a symptomatic feature which points to a more fundamental dimension of contemporary science: an unconscious aspiration, obfuscated and disavowed perhaps in normal every-day research, but resurging on such prominent occasions? At the Press Conference, the HGP was framed as a crucial station on a long journey of exploration, which began with the famous motto inscribed on the temple of Apollo's at Delphi more than twenty-five centuries ago: "Know thyself" (γνῶθι σεαυτόν). Self-knowledge remains the ultimate goal of our *cupido sciendi*, our "will to know", and the HGP entailed the promise that we will now finally be able to know ourselves (Zwart 2007). Venter for instance, both at the press conference and in his autobiography (Venter 2007), describes the human genome as "our own instruction book" and as "the draft of the human book of life". The human genome was regarded by many as our "blueprint", and as the HGP came off ground, the twenty-five century old quest for human self-knowledge seemed to be entering a decisive phase.

But also in terms of self-knowledge, the HGP resulted in a disappointment, for in the course of the project something remarkable happened. Initially, estimates of the number of genes on the human genome tended to vary greatly. Walter Gilbert (1992) had suggested that the human genome contained something like 100.000 genes a figure widely quoted and adopted (IHGSC 2001, 898), but James Watson (2002) even mentioned 248.000 genes as a probable estimate. In 2000, an estimate of 120.000 genes was still proposed (Liang et al, 2000). In 2001, however, the *International Human Genome Sequencing Consortium* (IHGSC) reduced the official estimate to ~ 31.000 genes. And in 2004, in the landmark paper that presented the finished version, covering 99% of the human genome, a more or less final estimate was given of ~22.500 genes (IHGSC 2004, cf. van Ommen 2005, 931). This was something of a disappointment indeed, or even a narcissistic offence (Freud 1917/1947; cf. Zwart 2007)), not only in comparison to previous estimates, but also in comparison to the number of genes on the genomes of other model organisms such as Drosophila melanogaster (~14.000 genes), Caenorhabditis elegans (~19.000 genes) and Arabidopsis thaliana (~25.000 genes). Indeed, it raised the question "what *does* set us apart from flies and worms" (Van Ommen 2005).

Dialectically speaking, the realisation of the HGP entailed an important experience, namely that the human existence proves decidedly more complex than was initially expected. The HGP undermined (or "negated" in dialectical terms) rather than confirmed a genetic reductionist understanding of the genome as our "blueprint". In order to know ourselves, a more comprehensive portrayal is required, which not only encompasses molecular genomics, but also envisions how we come to terms with our socio-cultural environment. We are not only the product of our genes, but abut forged by culture as well. Human existence results from a dialectical interplay between two types of texts: by the "language" of the genome, but also by the "symbolic order" (Lacan): i.e. the multiple (political, scientific, religious and moral) forms of discursivity that constitute human civilisation or *Sittlichkeit* (Hegel 1970).

The surprisingly small number of genes raised the philosophical question how humans are able to create a highly complex, artificial environment, a technological world or "technotope", equipped with a genome that contains such a small number of genes? While we are exploring and unraveling the structure of the universe and reshaping our environment at an unprecedented scale and pace, the genetic basis for our unique talents and creativity remains unclear. On the level of our genome, we do not seem that different at all. Our uniqueness and otherness is hardly reflected by our genes. One conclusion may be that, apparently, we are not that unique and different as a species after all. As Venter phrased it during his White House speech: "We [...] have many genes in common with every species on Earth [...] we're not so different from one another" (National Human Genome Research Institute, 2000).

But this is not the only conclusion we might draw. Another possibility is that we should look for the source of our uniqueness and complexity as human beings elsewhere. As the Gospel of Saint John phrases it: in the beginning was the word (λόγος), but DNA is not the only text which shapes human existence. Our intelligence and creativity also depends on our techno-cultural environment, on our susceptibility and exposure to culture, to words, to discursivity: to the symbolic order, the discourse of the Other, as Jacques Lacan phrased it (1966, 524).

Thus, human existence evolves at the intersection of two types of texts. On the one hand the molecular textuality disclosed by genomics and post-genomics research, commencing from the genome (a portmanteau of *gene* and *chromosome*), but closely interacting with the metabolome, the transcriptome and various other –omics layers (up to the exposome and the environome: Cheng & Cooper 2001; Miller & Jones 2014) that are studied by genomics, metabolomics, transcriptomics and many other –omics fields. The understanding of the genome as a text or code was initiated by Erwin Schrödinger in his classic *What is life* (1944/1967), but reinforced by the discovery of DNA by Watson and Crick (two researchers who were explicitly inspired by Schrödinger's book: Zwart 2013) and its alphabet of nucleotide letters (A, C, G and T). More recently, the understanding of life in terms of textuality was reconfirmed by the introduction of the CRISPR/Cas9 technique, allowing gene editing "by the letter", "letter by letter", as Doudna and Sternberg (2017, p. 93) phrase it, enabling the correction of "single-letter mistakes" (p. 100) in DNA with "single-letter accuracy" (p. 212). The genome is the primordial layer from where multiple circuits and complicated networks of molecular messages pervade living organisms.

But this bio-molecular textuality is complemented by a second type of textuality: the discursivity of the socio-cultural ambiance, again a multi-layered and stratified phenomenon. And also with regard to the textuality of our socio-cultural environment, a primordial layer can perhaps be discerned, consisting of primordial or initiating texts, the equivalent of the genome in the field of the humanities, and likewise referred to as the language of God, namely key textual sources that served as grounding documents for whole cultures, for national languages, ranging from the Bhagavad Gita via the Hebrew Bible up to the New Testament and the Quran. Such documents reflect seminal efforts by human authors to respond to an experience calling: the experience of being addressed by a voice from elsewhere, by a speaking Other.[3]

Human genomics and post-genomics evolve in a world populated by 5.8 billion people (84 percent of the world population) who report themselves as religiously affiliated. If anything, this astonishing figure reflects a tenacious human predisposition or receptiveness for spirituality.[4] It has been argued, by Hamer (2005) and others, that this worldwide susceptibility of responsiveness of human beings to the spiritual dimension of existence, resulting in multiple forms of religiosity, is based on the presence of a so-called god gene (VMAT2), hardwiring an inclination towards spirituality into human DNA. But regardless of whether a genetic susceptibility for religious experiences does exist, a number of decisive examples have been recorded where exposure to a Divine λόγος gave rise to experiences of awakening and conversion (or delusion, if you like), inaugurated by susceptible voices such as Jeremiah (626 B.C.), Jesus (30 A.D.) and Mohammed (610 A.D.). Such events (and the various documents resulting from them) seem to point to a basic human susceptibility to be addressed: by textuality in general, but especially by texts of a specific spiritual nature, urging us to question *who* we are and *where* we come from: the very questions that spurred scientists like Collins (2006) into sequencing the human genome in the first place. In other words, not the genome as such, but rather the *quest* for the genome, the *desire* to know and read the human sequence may tell us something about who we are.

But how should the interaction between these two forms of textuality (between bio-molecular codes and socio-cultural discursivity) be envisioned? As Pierre Teilhard de Chardin (1955) once phrased it, scientific portrayals of humankind (genetic, anatomical, physiological, neurological, genetic, etc.) consistently seem to fall short. They seem to lack a key dimension, namely self-consciousness or world-openness. Humans are animals, but they also represent a leap, a discontinuity, a metamorphosis, a crisis, an awakening. And it is precisely here that the basic human responsiveness to λόγος, to words, to *the word* even (the calling by the Other), seems to play a decisive role. Due to this exposure to λόγος, for those susceptible to it, existence becomes part of a cultural journey, adding an additional existential layer over and above the biological and socio-economic dimensions of human existence. Moreover, Teilhard believed that, at some point in the imminent future, the current collision between science and religion will be sublated into convergence (an event which

3 "Koranic revelation is considered to be a reception of the hyperoriginary text of the Other [which] had already been written" (Benslama 2009, 13)

4 http://www.pewforum.org/2012/12/18/global-religious-landscape-exec/

he referred to as the Omega point). From a Teilhardian point of view, the HGP, focussed on the biological textuality of existence, was certainly a decisive milestone on the pathway leading up to this convergence. And yet, at present, we rather seem faced with an ever widening split or gap between techno-scientific and spiritual understandings of human existence.

But maybe genomics can help us to articulate this experience of collision or convergence at with a higher level of resolution as it were. The phrase "In the beginning was the genome" resonates with the conviction that the starting point of life is DNA: the giant molecule which orchestrates the functioning and development of life on the cellular level. The nucleus of every cell, of every fertilised ovum, contains a molecular text composed of sequences of four letters (A, C, G and T). For religious scientists such as Collins, it was via these letters that God breathed life into the mayhem of abiotic matter: the genome as our plan or programme. But the experience of the HGP confirmed that, in order to really come to terms with human existence, we must continue to pay attention to other texts as well, to other instances of λόγος; the texts of culture. In other words, we are products of processes of co-creation, of a dialogue involving various types of text: on the one hand the language of the genome, on the other hand the languages of civilisation. It is from this perspective that I will reread Aristotle's *De Anima*.

Rehabilitating Aristotle

In modern scientific circles, Aristotle (the proverbial giant on whose discursive shoulders both Christian and Islamic medieval thinkers stood) no longer enjoys a good reputation.[5] Notably his views on physics tend to be regarded as blatantly misguided. It has been claimed that the modern scientific revolution was only possible because researchers during the early modern period dared to step beyond the Aristotelian worldview. As Rovelli puts it, Aristotle's science is either not considered as science at all or as a failure. [6] Psychoanalytically speaking, the emergence of modern scientificity required an act of intellectual patricide (Rovelli 2013).

But in the era of DNA and genomics, this verdict is under reconsideration (Mauron 2011). Max Delbrück, one of the founding fathers of molecular biology,

5 For a more extended analysis of the importance of Aristotle for Islamic and Christina understandings of the body and the tension with the modern scientific view see Zwart & Hoffer (1998)

6 https://www.academia.edu/5739248/Aristotles_physics

already argued that, in retrospect, Aristotle should be credited for discovering "the principle implied in DNA" (1973, p. 55) and that molecular biology echoes Aristotelian conceptions, as is suggested by the title of his paper (*Aristotle-totle-totle*). Aristotle, Delbrück argues, discovered DNA because he discerned that living beings are composite creatures, composed of form and matter. Or rather: they are matter shaped by (and brought to life by) form. According to Aristotle, the soul is the form and principle of life. But whereas Delbrück predominantly refers to Aristotle's biological writings, the ancient Greek philosopher-biologist from Stagira developed his 'hylemorphic' understanding of life even more poignantly in *De Anima* (Περὶ Ψυχῆς), translated as *On the Soul* (Aristotle 1986). Therefore, I will briefly recapitulate this text, one of the key documents of oriental and occidental metaphysics, intensely studied by both Islamic and Christian scholars, such as Ibn Sīnā and Ibn Rushd (known in the West as Avicenna and Averroes) and Thomas Aquinas (1922, Prima Pars Q76).

According to Aristotle, the soul (ψυχή) is the principle (ἀρχή) of life (Aristotle 1986, 402a, 415b). It is the *form* (εἶδος) or *formula* (λόγος) of living beings. All organisms are composite entities: fusions of form (ψυχή) and matter (ὕλη), resulting in the realisation or actualisation (ἐντελέχεια, 412a) of the living being's formula or plan (λόγος, 412b, 415b).[7] The body is regarded as the instrument (ὄργανον) of the soul (415b). While plants grow and reproduce (as realisations of their "vegetable" soul), animals also perceive and move (as realisations of the sensitive part of their soul). But it is only in humans that Aristotle discerns the presence of a thinking soul (νοῦς). A scholarly paper, for instance, is a realisation of the thinking soul, a uniquely human dimension, realising itself in thinking (*νοεῖν*), which can be both passive (receptive) and active (self-directed).

At this point, however, a basic ambivalence seems at work in Aristotle's text. On the one hand, he regards thinking as a continuation of visual perception in the sense that, whereas via eyesight we perceive the things themselves (as compounds of matter and form), the human mind assesses their form (εἶδος) stripped of matter, so that thinking is a more abstract version of sense perception. In other words, whereas perception focusses on external things (πράγματα), the soul reflects on their inner images (φαντάσματα). But Aristotle also suggests that the thinking soul focusses, not on the visual shape or form, but rather on the formula (λόγος): the plan of things. Seen from this perspective, Aristotle argues, thinking is more similar to considering letters (γραμματείον) before they are actually written down on tablets (430a). In other

7 "The soul is the first principle (ἀρχή), the realisation (ἐντελέχεια) of that which exists potentially: its essential formula (λόγος)" (415b: 14-15).

words, thinking (in the sense of: mentally considering formula) is comparable to writing a text that has not yet been written: a writing that is not yet realised as actual writing (on a tablet).

The tension between these two versions of thinking, namely thinking as working with mental images (φαντάσματα) versus thinking as working with mental characters (γράμματα), corresponds with a similar ambiguity already described above concerning the concept of form, which may either be interpreted as form in the more *visual*, morphological sense (εἶδος), or as form in the sense of *formula* (λόγος): the plan that is realised in the actual living entity. This tension or difference is not clearly spelled out by Aristotle, but it is important to emphasise this in view of later developments, because in contemporary philosophy the distinction between the *imaginary* (focussed on images or φαντάσματα) and the symbolic (focussed on symbols or γράμματα) has become quite decisive, while the textuality of life discussed above clearly builds on the latter rather than on the former. Aristotle notices the difference, for instance when he explains that, when we see a beacon, we initially recognise it as fire, until it begins to move, for then we realise that it actually is a signal which signifies something (for instance: the approach of the enemy). This distinction between fire as a (natural) shape (or image) and fire as a (conventional) signal (or symbol, i.e. an element in an alphabet of signals) is not further pursued by Aristotle, but it became increasingly important, not only in contemporary debate, but also in Western culture as such. For whereas ancient Greek culture was still predominantly a visual culture (even in the textual domain oriented on visual, imaginative genres such as epic poetry), one could argue that Christian and Islamic scholars (notably during the medieval era) represented a much more scriptural or textual approach, so that the focus shifted from thinking-as-processing-φαντάσματα to thinking-as-processing-γράμματα. Thus, the basic tension between images and words, between imaginative and discursive thinking continues to run as a basic epistemological thread through the history of culture as such,[8] and has been reinforced by decidedly scriptural (textual) cultures, such as Christianity and Islam.

8 Carl Gustav Jung (1911 / 2001) introduced a distinction between two modes of thinking: namely imaginative and discursive thinking. Whereas the latter evolves on the basis of logic and the causality principle, the former relies on association. Historically speaking, Jung argues, discursive thinking is a fairly recent phenomenon. It was introduced by critical minds such as Socrates (the founding father of logic as a philosophical discipline) and further elaborated by Aristotle and scholasticism. Without this intellectual trend (the gradual conversion of the Western mind to discursive thinking), the emergence of modern science would have been unthinkable, Jung argues.

A final important distinction in *De Anima* is the one between passive (receptive) and active (self-directed) thinking, notably because Aristotle at a certain point suggests that, whereas the passive soul is perishable like the body (being connected to it as its form), the supra-individual and truly *actively* thinking soul is independent from perishable living and thinking individuals) and therefore imperishable and everlasting (430a). This seems to suggest that, although no living being can sustain itself without a soul, the thinking soul as such may operate in the absence of a body. To what extent does Aristotle's conception of the soul allow us to deepen our understanding of the human genome in the context of the textuality of human existence?

To begin with, I endorse Delbrück's view Aristotle's hylemorphic conception of life can be regarded as a remarkably lucid anticipation of the principles of genomics. From an Aristotelean perspective, the genome can be considered as the formula, the program or plan (λόγος) which guides the development of living beings from their embryonic state up to their full realisation (ἐντελέχεια) as flourishing, self-sustaining and reproducing adults who have fully actualised their potential form (εἶδος). So, yes, from an Aristotelian viewpoint, the genome can meaningfully be regarded as the text of life, producing living beings from the chemical mayhem of their abiotic surroundings (i.e. inorganic matter). And prominent genomics researchers such as Craig Venter (2013) even argue that, whereas no living organism can exist without its DNA, DNA can be isolated from living beings as pure information, the pure formula of life, everlasting and immortal, processed in computer systems, or even used to reassemble replicas of living organism elsewhere for instance: on other planets, so that microbes in principle can be beamed to Mars, in order to produce an aerobic atmosphere and terraform the planet.

The view that DNA (as carrier of the genome) is the text of life notably applies to the vegetative and sensitive dimensions of bodily existence, however: to metabolism, first and foremost, albeit in continuous interaction with the ecosystem (life as a continuous dialectical dialogue between nature and nurture). But when it comes to understanding the noetic dimension of the soul (the thinking soul or νοῦς), the explanatory power of the genome becomes less obvious. As indicated, a genetic basis for our creativity and intelligence cannot be detected in our genome as such. Although in has been claimed that the human genome contains certain genes that may explain textual-cultural behaviour, dubbed the language gene (FOXP2) and the god gene (VMAT2) for instance, the presence of such genes can only account for a basic susceptibility or responsiveness to textuality. Our noetic or discursive existence as such cannot be explained on the basis of genomics ('nature', natural textuality) alone, but must be prompted or activated (realised) by exposure to other kinds of text as well. Our basic ability to be addressed requires something else besides the genome and the FOXP2 protein (depicted on the left), encoded by FOXP2 gene. It presupposes the existence of a world of language, a socio-cultural ambiance, enabling and facilitating language use, providing a scaffold for the development of intelligent and responsive textual behaviour. This world of language provides a textual infrastructure, a discursive scaffold allowing cultural and moral existence to unfold, in response to the language of the Other, which is already there as a cultural ecosystem. And whereas the ability to be addressed corresponds with Aristotle's concept of passive thinking, our active responsive contribution to and participation in this socio-cultural world of language concurs with what Aristotle refers to as active thinking (discursivity as such), *realising* or actualising itself via us, but as a symbolic order which is already operating and will continue to function when we as individuals leave the scene (Lacan 1974/2005).

Thus, rereading Aristotle's *De Anima* likewise prompts us to recognise the basic bi-textuality of human existence. Human self-consciousness emerges at the interface between two types of texts, namely the natural textuality of the body (as studied by molecular biology and genomics) and various social-cultural forms of textuality (analysed by the humanities, from linguistics up to religious studies). In order to understand human existence, physiology must be complemented by philology. Ideally, a state of harmony or at least compatibility and mutual adaptation between both dimensions can be achieved (Lacan 1959-1960/1986, p. 107; Lacan 1956-1957/1994, p. 25), but from the history of culture it is clear that the tension between these two types of text is not that easily to solve, and rather gives rise to chronic experiences of frustration, malaise and failure. But perhaps a reframing in terms of bi-textuality can help us to elucidate the basic split or incongruence that runs through the human condition and was articulated by Sigmund Freud (1930/1948) as discontent in culture. The question is: how to combine or reconcile our openness to the languages of culture (our cultural or even spiritual ecosystems) with the bio-molecular languages of the genome (our biological program)?

Civilisation and its Discontents Reframed

The phrase "In the beginning was the genome" conveys the idea of DNA as the commencement of life: the biopolymer which generates and orchestrates molecular messages. The nucleus of every cell contains a molecular text composed of sequences of four letters (A, C, G and T) and via them life is breathed into the mayhem of entropic abiotic matter: the genome as our programme.

But the experience of the HGP confirmed that, in order to come to terms with human existence, we must pay attention to other texts, other instances of λόγος, as well namely oral, scriptural and digital discourses of culture so that we are products of processes of co-creation, and interaction between various different types of text: on the one hand the languages of molecular biology, on the other hand those of civilisation.

Human beings are driven by two types of texts. They are continuously fuelled by myriads of biochemical messages produced by molecular circuits informed by DNA, but also relentlessly besieged by voices coming from society and culture. And this entails an existential challenge. Various filters and defence systems have evolved to allow us to cope with this unsettling over-abundance, and most of the interactive processing occurs unconsciously, allowing us to focus our attention on tiny samples of (internal or external) signals (Freud 1920, /1940, 27). Various forms of tension, contradiction and confusion may nonetheless result from our susceptibility to these incommensurable types of text. In the course of human history, a split or gap has evolved between our (slowly evolving) Palaeolithic genome and the contrasting demands of our global civilisation (evolving at a tremendous and accelerating pace), a gap which seems too fundamental for genome editing technologies to bridge (Stammers 2017). The pastoral paradise (in which genome and culture once were compatable) seems irretrievably lost (if it ever existed), so that we are facing a chronic incompatibility between the molecular information circuits of embodied nature and the textual cacophony emerging from our socio-cultural ambiance, giving rise to discontent in civilisation, as a persistent and collective human symptom (Freud 1930/1948). Whatever the circumstances, humans always seem to be looking for something *more* and something *else* than that which is provided by the immediate material environment.

In this dynamical relationship, a dialectical triad can be discerned. Initially, some level of coherence between genome and cultural Umwelt may have existed (first moment: M1), but at a certain point this pre-established harmony was disrupted: the birth trauma of human culture (M2). According to Jacques Lacan (1966, 1974/2005), language played a crucial role in this, introducing a new and perhaps uniquely human dimension: the desire for things we may conceive or imagine rather than see, smell or grasp. Over the past millennia, this has given rise to a neo-environment: a techno-sphere or socio-sphere, over and above the atmosphere, geosphere and biosphere. But this has failed to appease the tension or gap between what we seek (desire) and what we find (the entities, either natural or artificial, that actually surround us). And this explains the turbulence of human existence in a polarised force-field between two incommensurable types of text: the molecular code of the genome and the textual codes of culture. But perhaps, by increasing our literacy on both sides of the equation, (that is, by simultaneously strengthening our fluency in molecular life science research as well as our erudition in the cultural realm) these two textual poles of human existence may eventually become reconciled again (M3)?

As indicated, most of the interaction between these two types of texts takes place unconsciously. As embodied biological beings we are under the sway of the biological unconscious, orchestrated by the genome: a dimension we share with plants, animals and microbes taking care of the metabolism that continuously takes place, within cells as well as within the body as a whole, in close interaction with the environment. The *extimate* microbiome (both *intimate* and *external*) plays an important part in this, as an organ composed of bacteria functioning as a "collective unconscious" (Dinan et al 2015). But over and above the biological unconscious, which we share with other living beings, humans are also prompted by a textual unconscious, which is psychic unconscious, is not a fluid reservoir of bodily or animalistic drives (which would make it biological again), but rather textual and highly organised (1975, 79). The psychic unconscious is structured like a language, as Lacan (1981 and elsewhere) phrases it. It is not the seat of primordial instincts, but rather consists of chains of signifiers (1966, 501 ff.). This unconscious is a discourse-producing machine speaking to us, albeit in an oblique or indirect manner. For whereas others (children, spouses, colleagues, civil servants, etc.) address us more or less directly, via words and gestures, there is another 'Other', addressing us through dreams, neurotic symptoms and mistakes (slips of the pen), or via inner voices (Socrates' δαιμόνιον for instance), often articulating seemingly irrational desires or incomprehensible concerns (the voice of conscience which, in the case of neurotic patients, may become a paralysing, over-compelling and over-demanding super-ego). In the case of psychotic patients, the unconscious may really surface as a strange, enigmatic language: as an audible, uncanny voice. But non-pathological individuals may likewise be overwhelmed by the language of the Other: by sudden artistic inspirations (the muses), or by scholarly brainwaves (εὕρηκα-experiences) or by religious revelations (Moses on Mount Sinai, Jesus in Gethsemane, Mohammed in the Hira cave).

For Jacques Lacan, the unconscious is a text-processing, "typographical" realm (1998, 147), but different from biological (genomic) textuality (Zwart 2013). Whereas molecular messages coming from the convey informational signals, cultural languages (conveying truth and meaning) seem a uniquely human phenomenon. Whereas the needs, growth patterns and functions of the body are to a certain extern governed by biological messages and codes, human desire is under the sway of the language of civilisation. For Lacan, we are "speaking animals", liberated from nature to some extent, but burdened by language, or even sick with language (1974/2005, 90, 93; cf. 1961-1962, 42). And whereas in animals the genome and the environment seem fairly adapted to one another (as in expressed by notions such as fitness), in humans we basically see a *failure* to adapt, because of our exposure to conflicting messages coming from elsewhere, expelling us from biological forms of existence in which we were once embedded (the biosphere) and opening up a cultural and spiritual realm of truth and meaning, – the symbolic order or noosphere, to use the term coined by Teilhard (1955).

Lacan's analysis of the story of the Sacrifice of Isaac (Genesis 22: 1-13) may serve as an example here (1974/2005). Abraham is not spurred on to climb the mountain in order to sacrifice his son by animalistic or metabolic drives (say, hunger). Rather he responds to a sense of calling: a Divine command, the word of "the Other", not someone who is physically or tangibly present (a spouse, a neighbour, etc.), but a voice coming from "elsewhere". The various physiological phenomena (arousal, metabolism, etc.) that allow him to comply with this command are biological (and genome-based) no doubt, but the crucial invocation that spurs him into action is a phenomenon beyond biology. When he is about to sacrifice his child, however, a voice (a messenger) once again intervenes, so that Abraham once again responds, this time by revoking his original intention. Whereas predators will go for their prey without further ado, humans may deliberate about their sacrifices, and reconsider their choices, due to their openness to reason and language, to their ability to be addressed, by words, by λόγος. Language allows humans to transcend the biological parameters of their existence, so that the biological Umwelt is transformed into a literate human world, replete with language.

Basically speaking, the language of the genome (the bio-molecular messages spurring us to develop certain responses to environmental cues: M1) is negated, by the language of conscience and culture (M2), and this gives rise to various tensions and conflicts. In order to arrive at a viable situation, however, this negation of nature by culture must be negated or ('sublated') again (the negation of the negation, as Hegel phrases it: M3). By developing a profound understanding of our biological nature (starting with the human genome) as well as of the dynamics of human cultural existence (the spiritual dimension), both moments (nature and culture; nature and spirituality) may become reconciled again, on a higher level over complexity, so that desire may become *sublimated* into culture and negativity becomes *sublated* into responsibility (M3).

Genomics and Iconoclasm

In the current era, as global mass media attention predominantly focusses on tensions between religion and science, between spirituality and secularisation, it seems relevant to highlight something which may easily be overlooked, namely a basic affinity between monotheism and the scientific world-view, captured by the term iconoclasm. In a religious context, iconoclasm refers to the tendency to discard idolatrous or iconic manifestations of religiosity (practices of producing and worshipping icons, statues, idols, etc.) in favour of scriptural, symbolic sources. Iconoclasm was inaugurated by Pharaoh Akhenaten (ca. 1353–1336 B.C.), and subsequently transferred as a quintessential feature to Judaism, Islam and Christianity (notably Protestantism).

Psychoanalyst of science Gaston Bachelard pointed out, however, that iconoclasm is also a distinctive feature of modern science (1947, 77; 1953, 122), in the sense that science not only challenges narcissistic self-images, but also disrupts established (imaginary) world-views. Dialectically speaking, the objective of science is to understand nature or natural entities (M1), but instead of letting nature be (as happens in the case of artistic meditation or poetic exaltation), Bachelard explains how science actively transforms natural entities into something noumenal and abstract (bio-chemical molecules, captured in formula, symbols, equations, etc.) with the help of laboratory equipment. In dialectical terms, the concreteness and immediacy of natural entities becomes (M1) negated or abolished (M2) by scientific knowledge production. Although research begins with self-constraint (letting things be, observing rather than consuming them), they are eventually transformed into something than can be technically manipulated. In other words, research entails negativity. The initial object (the natural phenomenon or *Gestalt*) becomes obliterated through measurements and quantification, so that a noumenal (physical, chemical, molecular) essence is revealed. Thus, the visible Gestalt (a tree, for instance) gives way to chemical letters and symbols (CO_2, H_2O, $C_6H_{10}O_5$, etc.), and the living organism becomes "obliterated" (Zwart 2016). The living thing is broken down into basic components that can be represented with the help of letters: the symbolic alphabet of chemical compounds (H_2O, CO_2, etc.), genes (FOXP2, VMAT2, etc.), nucleic acids (A, C, G and T), amino acids (Ala, Arg, Asn, Asp, Leu, Lis, Met, etc.) and so on. Due to this symbolisation or literation of nature (or even obliteration) of nature, natural entities as living entities disappear from view, thus exemplifying the iconoclastic tendency of science. Only via iconoclastic symbolisation, the logic (λόγος) of scientific reason is able to reveal the basic textuality (λόγος) of molecular systems.

This example why the output of human genomics is not a recognisable portrait of a human being, but rather a stream of letters, something essentially textual: a formula, a sequence, a code. If a human genome sequence printed on canvas, it resembles a modernistic artwork, as we have seen. This iconoclastic tendency, at work in modern science, but also in modernistic art, reverberates with monotheistic precursors. From a historical perspective, however, iconoclasm is the exception rather than the rule. Cultures tend to speak to their adherents not only through words, but also via images. Ancient Greek culture, as we have seen, was primarily a visual culture and ancient sculptures of Greek and Roman culture, erected in public spaces, conveyed a moral message: become an athlete, transform yourself into a work of art, so that such statues actually functioned as exemplary idols. Such artworks entailed a form of moral propaganda (Lacan 1959-1960/1986). In contemporary commercials (displaying superbly healthy men and women for instance) such messages are still abundantly present (inciting us to become beautiful, healthy, fit, athletic, etc. by following a certain diets or fitness programs for instance). Monotheistic religions such as Judaism, Islam and Protestantism, however, are iconoclastic, relying on the Word (the Quran and the Bible) rather than on exemplary images. Monotheistic religions address their adepts via commandments and other normative, apodictic formula, while Catholicism can be seen as an intermediate form, a compromise between the imaginary and the symbolic. According to Hegel, Islam represents the most radical effort to abolish the imaginary and realise symbolic sublimity (Hegel 1970).

Modern science, however, is now reframing the normative dimension, by relying on high-tech forms of symbolisation, with the help of personality tests, IQ tests, BMI indicators, blood sample readings and so on. The idea that, in the near future, health science will increasingly address individuals in terms of personalised and digitalised data. Health gadgets will inform us whether our personal physiological performance (our body language, as recorded by smart wearable gadgets) concurs with societal expectations of normalcy: the molecularised version of the super-ego (Zwart 2016). As a result, human populations will increasingly be governed in an algorithmic manner, with the help of health data: algorithmic governance as the final stage of biopower (Rouvroy & Stiegler 2016). Although human existence continues to be bi-textual, this bi-textuality is being radically reframed, namely as the tension between molecular messages coming from the body (transmitted by iPhones, smart watches and so on) and the standards of normalcy of the terabyte age, based on big scientific data collected by millions of citizens (Zwart 2016). Thus, the language of the Other (which gave rise to moral and spiritual experiences in the past) gives way to a secularised susceptibility to a different type of text: the super-ego of secularised culture urging us to become entrepreneurs of our health data and responsible managers of our personal health.

Conclusion

Human existence, we may conclude, is a dialectical interplay between two types of texts, biomolecular and socio-cultural ones. To deepen our understanding of this dialectics, a consistent dialogue between contemporary science (genomics and post-genomics) and the humanities (including religious studies) is indicated. This requires a shift of focus from ethical issues in the applied sense of the term towards the broader cultural ambiance of the science-society debate, for instance by reflecting on the impact of genomics on human self-understanding. If such a dialogue would focus solely on applied ethics deliberations, we may easily fall into the trap of seeing science as liberating and progressive, while metaphysical and religious world-views are framed as conservative and restrictive. A focus on applied ethics, moreover, may entail a plea for strategies of avoidance and compartmentalisation, delisting metaphysical, spiritual and religious issues from the agenda of the debate. I would rather advocate a strategy of retrieval: zooming out somewhat from frontstage bioethical quandaries towards the more fundamental backdrop issues. This allows us to discern how the bi-textuality of human existence is currently undergoing a transition, now that not only the physiological, but also the normative dimension is being reframed in biomolecular and terabyte terms.

References

Ahmed, Arzoo, and Mehrunisha Suleman. 2019. "Islamic Perspectives on the Genome and the Human Person: Why the Soul Matters". Included in this volume.

Aquinas, Thomas. 1922. *Summa Theologica*. Taurini: Marietti.

Aristoteles. 1986. *De Anima / On the Soul* (Loeb: Aristotle 8). Cambridge: Harvard University Press. London: Heinemann.

Bachelard, Gaston. 1947. *La formation de l'esprit scientifique: Contribution à une psychanalyse de la connaissance objective*. Paris: Vrin.

Bachelard, Gaston. 1953. *Le matérialisme rationnel*. Paris : Presses Universitaires de France.

Benslama, Fethi. 2009. *Psychoanalysis and the challenge of Islam*. Minneapolis: University of Minnesota Press.

Cheng, Andrew and Brian Cooper. 2001. "Introduction". *The British Journal of Psychia-*

try. 178(40): s1-s2

Collins, Francis. 1999. "Medical and societal consequences of the Human Genome Project". *New England Journal of Medicine*. 341:28-37.

Collins, Francis. 2006. *The language of God. A scientist presents evidence for belief.* New York: Free Press (Simon & Schuster).

Collins, Francis, 2011. *The Language of Life: DNA and the revolution in personalised medicine.* New York etc: Harper.

Davies, Kevin. 2001/2002. *Cracking the genome: Inside the race to unlock human DNA*. Baltimore and London: John Hopkins University Press.

Delbrück, Max. 1971. "Aristotle-totle-totle". In *Of Microbes and Life*, edited by Jacques Monod, Ernest Borek, 50-55. New York: Columbia University Press.

Dinan, Timothy, Roman Stilling, Catherine Stanton, John Cryan. 2015. "Collective unconscious: How gut microbes shape human behaviour". *Journal of Psychiatric Research* 63: 1-9.

Freud, Sigmund. 1917/1947. "Eine Schwierigkeit der Psychoanalyse". *Gesammelte Werke XII*, 3–12. London: Imago.

Doudna, Jennifer and Samuel Sternberg. 2017. *A crack in creation: gene editing and the unthinkable power to control evolution*. Boston/New York: Houghton Mifflin Harcourt.

Freud, Sigmund. 1917/1947. "Eine Schwierigkeit der Psychoanalyse". *Gesammelte Werke XII*. 3-12. London: Imago

Freud, Sigmund. 1920/1940. "Jenseits des Lustprinzips". *Gesammelte Werke XIII*, 1-70. Freud, Freud, Sigmund. 1930/1948. "Das Unbehagen in der Kultur." *Gesammelte Werke* XIV, 419-513. London: Imago.

Gilbert, Walter. 1992. A vision of the grail. In *The Code of Codes. Scientific and Social Issues in the Human Genome Project*, edited by Daniel Keyles and Leroy Hood, 83-97. Cambridge: Harvard University Press.

Hamer, Dean. 2005. *The God Gene: How Faith Is Hardwired Into Our Genes.* Anchor Books

Hegel, Georg, Friedrich Wilhelm. 1970. *Vorlesungen über die Philosophie der Geschichte.* Werke XII. Frankfurt am Main: Suhrkamp.

IHGSC. 2001. "Initial sequencing and analysis of the human genome". *Nature* 405: 860–921.

IHGSC. 2004. "Finishing the euchromatic sequence of the human genome". Nature 431: 931–45.

Jinek Martin, Krzysztof Chylinski, Ines Fonfara, Michael Hauer, Jennifer Doudna, Emmanuelle Charpentier. 2012. "A Programmable Dual-RNA-Guided DNA Endonuclease in Adaptive Bacterial Immunity". *Science* 337(6096): 816-821

Jung, Carl Gustav. 1911/2001. *Symbole der Wandlung*. Düsseldorf: Walter Verlag

Lacan, Jacques. 1953-1954/1975. *Le séminaire I: Les Écrits Techniques de Freud.* Paris: Éditions du Seuil.

Lacan, Jacques. 1955-1956/1981. *Le Séminaire III: Les psychoses.* Paris: Éditions du Seuil.

Lacan, Jacques. 1957-1958/1998. *Le séminaire V: Les formations de l'inconscient.* Paris: Éditions du Seuil.

Lacan, Jacques. 1961-1962. *Le Séminaire IX: L'identification*. Unpublished: http://www.valas.fr/.

Lacan, Jacques. 1966. "L'instance de la lettre dans l'inconscient ou la raison depuis Freud". In Écrits, 493-528. Paris: Éditions du Seuil.

Lacan, Jacques. 1974/2005. *Le triomphe de la religion. Précédé de discours aux catholiques.* Paris: Éditions du Seuil.

Lacan, Jacques. 1975-1976/2005. *Le séminaire XXIII: Le sinthome.* Paris: Éditions du Seuil

Mauron, Alex. 2001. "Is the Genome the Secular Equivalent of the Soul?". *Science* 291 5505: 831-832. DOI: 10.1126/science.1058768

Miller, Gary and Dean Jones. 2014. "The nature of nurture: refining the definition of the exposome." *Toxicological Sciences* 137 (1): 1–2. DOI:10.1093/toxsci/kft251

National Human Genome Research Institute. 2000. "June 2000 White House Event." Last modified August 29 2012. https://www.genome.gov/10001356/

Ommen, Gert-Jan van. 2005. "The human genome, revisited". *European Journal of Human Genetics* 13: 265–70.

Rouvroy, Antoinette and Bernard Stiegler. 2016. "The digital regime of truth: from the algorithmic governmentality to a new rule of law". *La Deleuziana: Online Journal of Philosophy* 3: 6-29.

Rovelli, Carlo. 2013. "Aristotle's Physics" Aix-Marseille University. https://www.academia.edu/5739248/Aristotles_physics

Schrödinger, Erwin. 1944/1967. *What is life? The physical aspect of the living cell / Mind and matter*. London: Cambridge University Press.

Song, Robert. 2003. "The Human Genome Project as soteriological project". In *Brave New World? Theology, Ethics and the Human Genome*, edited by Celia Deane-Drummond, 164-184. London: Clark.

Stammers, Trevor. 2017. "Genome Editing - Creation, Kinds and Destiny: A Christian View. Paper presented at the conference". *Islamic Ethics and the Genome Question*, Research Centre for Islamic Legislation and Ethics. Doha, Qatar, April 4, 2017

Teilhard de Chardin, Pierre. 1955. *The Phenomenon of Man.* New York: Harper.

Teilhard de Chardin, Pierre. 1955. *Le Phénomène humain*. Œuvres 1. Paris: Editions du Seuil [translated as: *The Phenomenon of Man* (Bernard Wall). New York: Harper 1959; *The Human Phenomenon* (Sarah Appleton-Weber). Eastbourne: Sussex Academic Press, 2003.]

Venter, J. Craig. 2013. *Life at the speed of light: From the double helix to the dawn of digital life*. New York: Viking.

Watson, James. 2002. "A personal view of the project". In *The Code of Codes. Scientific and Social Issues in the Human Genome Project*, edited by Daneil J. Kevles and Leroy Hood, 164-173. Cambridge: Harvard University Press).

Zwart, Hub. 1993. *Ethische consensus in een pluralistische samenleving: de gezondheidsethiek als casus*. Amsterdam: Thesis Publishers.

Zwart, Hub and Cor Hoffer. 1998. *Orgaandonatie en lichamelijke integriteit: een analyse van christelijke, liberale en islamitische interpretaties*. Best: Damon. ISBN: 90 5573 083 1

Zwart, Hub. 2007. "Genomics and self-knowledge. Implications for societal research and debate". *New Genetics and Society*, 26 (2): 181-202.

Zwart, Hub. 2009. "Genomics and identity: the bioinformatisation of human life". *Medicine, Health Care and Philosophy: a European Journal* 12: 125-136.

Zwart, Hub. 2010. "The adoration of a map: Reflections on a genome metaphor". *Genomics, Society & Policy* 5(3): 29-43.

Zwart, Hub. 2013. "The genome as the biological unconscious – and the unconscious as the psychic 'genome': a psychoanalytical rereading of molecular genetics". *Cosmos and History: the Journal of Natural and Social Philosophy* 9 (2): 198-222.

Zwart, Hub. 2015. "Human Genome Project: history and assessment". In *International Encyclopaedia of Social & Behavioural Sciences*, (2) 311–317. Oxford: Elsevier.

Zwart, Hub. 2016. "The obliteration of life: depersonalisation and disembodiment in the terabyte age". *New Genetics and Society* 35 (1) 69-89. DOI: 10.1080/14636778.2016.1143770

Zwart, Hub. 2017. "From the Nadir of Negativity towards the Cusp of Reconciliation: A Dialectical (Hegelian-Teilhardian) Assessment of the Anthropocenic Challenge". *Techné: Research in Philosophy and Technology* 21:2–3. DOI: 10.5840/techne20176565

CHAPTER 8

Creation, Kinds and Destiny: A Christian View of Genome Editing

Trevor Stammers[1]

> Men ought not to play God before they learn to be men,
> and after they have learned to be men they will not play God
> (Paul Ramsey, *Fabricated Man: The Ethics of Genetic Control*, 138)

The discovery of the double helix structure of DNA by Crick, Franklin and Watson in 1953 caused a paradigm shift in our understanding of the nature of both humankind and creation as a whole. Subsequently in 2003, the mapping of the human genome by Frances Collins and his colleagues and the ensuing development of techniques to alter it, raise fundamental questions about our destiny – whether we ourselves can and should shape it in a way previously outside our ability and known only to God.

The Christian understanding of the Fall – the movement of humanity from an initial state of perfection or at least of being 'very good' (Genesis IV 31) in God's sight, to a state of obvious imperfection - has always raised questions of normalcy in relation to our current 'fallen' state compared to what was originally intended by the Creator. This paper explores Christian visions of the ethical possibilities of genome editing using Bonhoeffer's understanding of the nature of the material world and the effects of the Fall upon it as a model and continues with an exploration of wider implications of other elements of the creation account.

It will be argued that, far from supporting the popular understanding of genetic determinism with its implications for the concepts of both free will and human responsibility, our greater knowledge of genomics weakens such a determinist view. The paper concludes with a consideration of the *telos* of humanity in relation to gene editing and an examination of the concept of the

1 Reader in bioethics at St. Mary's University, Twickenham, London and Director of its Centre for Bioethics and Emerging Technologies, trevor.stammers@stmarys.ac.uk

DOI:10.1163/9789004392137_010

genome as a 'secular soul' and the religious elements of the genomic editing quest.

Origins: Creation and Fall

Christianity has its roots inextricably embedded in the Old Testament. Jesus either quotes from or refers to it, dozens of times in the Gospel accounts. On one such occasion, when responding to a question about divorce (Matthew 19:4-6, Mark 10:6-8), Christ quotes from the Genesis creation account to answer his critics. The creation narrative plays a key role in the New Testament and in Christian theology as a whole so I begin here with an overview of the 20th century theologian, Dietrich Bonhoeffer's understanding of *Creation and Fall* in his 1937 work of that title, as an example of Christian understanding of the nature of the world which emerges from the Genesis accounts.

Bonhoeffer on Creation

Bonhoeffer, writing of course prior to the discovery of DNA but post-Darwin, highlights several important elements of Creation according to Genesis. Firstly, God is distinct from his creation; the creation is not a fragment of God. He does not give birth to the universe but speaks it into being. He creates by his word alone. "God is never in the world in any way except in his absolute transcendence of it" (Bonhoeffer 1937, 19).

So God speaks creation into being and sees that every element of it is 'good' (Genesis 1v25) or 'very good' (Genesis 1v31). Now Bonhoeffer immediately stresses that this does not mean, "that the world is the best of all conceivable worlds. It means that the world lives completely in the presence of God, that it begins and ends in him and that he is Lord" (1937, 22). Furthermore that "which is created by the Word out of nothing, that which is called forth into being, remains sustained by the sight of God" (1937, 23). God does not wind up the universe like a clock and leave it to tick on of its own accord; rather as the New Testament has it, "he holds all creation together" (Colossians 1v17) and "he sustains all things by his powerful word" (Hebrews 1v3).

God also speaks life into being. – plants and vegetation, sea life, birds and land animals, all 'according to their kind' (Genesis 1v14). Even so, Bonhoeffer reminds us "It is not the Creator's own nature which he here instils in the living and life-creating. The living and creative is not divine: it is and remains the creaturely" (1937, 34). However when it comes to the creation of humankind, there is another element involved. "Only in something that is itself free can the One who is free, the Creator, see himself" (1937, 34) comments Bonhoeffer.

"If the Creator wills to create in his own image, he must create it in freedom; and only this image in freedom would fully praise him and fully proclaim the honour of its Creator" (1937, 34).

So we read that God does indeed create humankind in his own image, male and female, from the dust of the earth. The human body is fashioned out of earth just as the earth gave rise to other animals but God breathes his life uniquely into this creature and man becomes 'a living soul' (Genesis 2v7). Humans alone are created in the 'image of God' - the *imago dei.*

There are many contemporary theologians who attempt to play down the importance of the *imago dei.* "The actual meaning of 'image of God' has varied so much during Christian history that no clear, single reference emerges and it seems to mean what people want it to mean." (Page 2003, 71) Much the same could be said however of the breadth of meaning of many other terms such as 'human dignity' or 'autonomy' but it may well be this relates to the richness of meaning of these concepts rather than implying they have no meaning at all.

Bonhoeffer for example, singles out two prime elements of what it means to be 'in the image of God'; firstly that it means to be free and in particular, free to worship the Creator and secondly that it is mankind who has the delegated authority of God to rule over creation in responsible way. "I belong to this world completely. It bears me, nourishes, and holds me. But my freedom from it consists in the fact that world to which I am bound ...is subjected to me and that I am to rule over [it]" (Bonhoeffer 1937, 78).

How does this very brief synopsis of Bonhoeffer's view of the creation narratives tie in with our contemporary knowledge of genomics? Surprisingly well in my view. The account emphasizes firstly, that all living things, including human beings, are created out of the earth. The fact then that the Human Genome Project (HGP) has shown us that there is similarity between the DNA of all species is no challenge to belief in a Creator. For some people, however, genomic similarity appears problematic in this regard. "Perhaps this is an unwelcome challenge to our opinion of ourselves. As humans we have long regarded ourselves as the pinnacle of creation" (Seller 2003, 340). Perhaps so, but the Bible certainly does not encourage us to have too high an opinion of ourselves (Romans 12v3) and in any case it is our ability to commune with our Creator that makes us special and not our genome *per se.* It should therefore not be a concern to us that as a species we share 50% of our DNA with a banana and over 98% with a chimpanzee (O'Connell 2009). We came from the same clay after all.

Secondly, our physical embodied form is affirmed along with the rest of creation as being very good. It is not a mistake that we have bodies just as other animals do but rather this is God's intention. Therefore we are not to regard our bodies as a prison from which to escape but as a 'temple of God' (I Corinthians 3v16, 6v19), through which we are to live for his worship and praise. Our bodily well being is therefore important and we have good reason to use wisely our knowledge of what makes it healthy, including the new possibilities arising from genomics.

Thirdly, however, the creation account gives clear indications that despite our material similarities with the rest of living things, we are different. Christians, along with those of other faiths which believe humanity has a special relationship to God being made in his image in a way like no other creature, have no option but to be 'guilty' of speciesism. Not because we believe other species should be treated in any way we like – there are many scriptural warnings against inhumane treatment of animals (e.g. Deut 25v4; Proverbs 12v10; I Timothy 5v18) – but because we alone have the freedom to rule over and care for the rest of creation by virtue of being made in God's image and receiving his delegated authority to do so (Genesis 1v26; 2v15). How this might be rightly exercised in regard to genomic editing can only however be considered after taking into account the reality of the Fall.

Bonhoeffer on the Fall

Though as we have seen, for Bonhoeffer a key element of being made in the image of God is the reality of our human free will, we are not entirely free to do as we please. God also sets a limit on that freedom with a prohibition that Adam and Eve were to adhere to, in the form of a tree from which they were not to eat (Genesis 2v17). Adam "who is addressed as one who is free, is shown this limit, that is to say his creatureliness, and by this prohibition is his being confirmed in its kind" (Bonhoeffer 1937, 51). That is to say that Adam, though in the image of God, is not God; temptation comes to him in due course in the voice of the serpent instilling first doubt - "Has God said", then denial - "you shall not die" and finally defiance- "God knows when you eat it your eyes will be opened and you will be as God" (Genesis 3v4).

Bonhoeffer sees the Fall as a rejection of contentment with the *imago dei* – being in the image of God- resulting in an attempt to be *as or like* God – *sicut deus.* Prof. Neil Messer puts it like this "The attempt to be 'like God'...springs from a forgetfulness or denial of our creaturely limits, an assumption we can and may do anything we choose. The problem with projects done in this spirit is not so much the prospect of failure... as the price of success." (Messer 2011, 38-39). The price of success for Adam is the ultimate one, as Bonhoeffer explains: "It is true that man becomes *sicut deus* through the fall but this very *sicut deus* can live no longer; he is dead" (1937, 70-71). Alienated from God by disobedience, Adam reaps the bitter consequences of human shame (Genesis

3v7) and banishment from the presence of God (Genesis 3v23). Not only does mankind undergo spiritual death – separation from God, the earth too from which humanity was fashioned is also cursed (Genesis 3v17). "All other creatures rise up against *sicut deus* man, the creature that tries to live out of his own self ....since they are subject to man, they fall with the Fall of man. Nature is without a lord and therefore it is itself rebellious and desperate" (Bonhoeffer 1937, 87).

In the light of Bonhoeffer's analysis, one of the ways in which we might attempt to discern between the ethically permissible and impermissible in bioengineering projects, including genome editing, is to look at whether they are appropriate for us to undertake as creatures made in God's image or whether they extend our attempts to usurp God's place and to be like him. Making this distinction however is rarely easy but may be helped by exploring a number of other key debates in bioethics related to genome editing to which I now turn.

Identity, Healing and Enhancement

"It is a profound misunderstanding of the human condition to think we can optimise ourselves in such a way that all human suffering is abolished", insists Prof Maureen Junker-Kenny (2003, 127). Most reflective healthcare professionals intuitively recognise this to be true. As a physician for over thirty years, I would reckon that around a half of the suffering I encountered in my patients in general practice was existential, rather than stemming from disease. The grief of parents whose son was murdered, the heartache of a mother whose son had not spoken to here for decades are just two specific examples of the two main general areas of suffering highlighted by Junker-Kenny (2003, 126), - firstly the consciousness of human finitude, the awareness that we will die and secondly the heartache of unrequited recognition by others which can never be eased by ourselves but only in the response of others to us, which we cannot control no matter how 'perfect' our genome 'is. It is not good to be alone' (Genesis 2v18) is the first thing in the creation account that God declared was not good. Our relationships with others remain a fundamental human need in spite of all our technological advances.

The Medical Model of Health

Though questions about how we define health, disease and normalcy have long been a source of debate, the advent of genome editing has undoubtedly given them a new urgency. The medical model, championed amongst others by Boorse, seeks to confine such definitions within supposed objective scientific parameters. Disease is a state that "interferes with the performance of some natural function…characteristic of the organism's age" and becomes an "illness only if it is serious enough to be incapacitating" (Boorse 1976, 61).

This model of understanding of health is the one within which gene editing has already achieved early success which is likely to accelerate. However Prof. Richard Hare has raised several problems posed by the medical model. Firstly he suggests that our intuitive understanding of disease and illness is that they are bad for us to have. Boorse therefore has to rely on a "the rather wobbly notion of natural function" (Hare 1986, 178) in order to avoid the intrinsic evaluative element of 'badness' we all have.

Once Hare's evaluative element is allowed however, two other important variables arise:

1) who is making the evaluation? An adult with Down's syndrome for example, may view Down's in a very different light from the clinical geneticist advising a couple at high risk of the condition in their children
2) on what grounds are they making it? An autistic adult may find the condition *per se* distressing or they may not consider autism 'bad' for them but rather the social difficulties associated with it.

These questions are already important in the light of genetic screening but they will become even more so should genomic editing advance as predicted. Eliminating 'abnormality' is already a reality (which will accelerate further with increasingly efficient means of detection of Down's) but it is debatable as to whether this has made or will make for a healthier society.

Views of Creation: Augustine and Wyatt

If the medical model of health provides too narrow an understanding of disease, the well-known World Health Organisation definition of "Health is a state of complete physical, mental and social well-being and not merely the absence of disease or infirmity" (World Health Organisation 1948) arguably offers far too wide a concept of health.

Eliding health and well being in this way leads to the expectation that social disorder can also be eliminated through medical means. As Messer has pointed out in relation to genome editing, "this understanding would eliminate the distinction between genetic therapy and enhancement and would encourage us to use genetic manipulation…to address any kind of social ill" (2003, 102). Furthermore this would necessitate labelling social dissidents as "sick" and coercion rather than compassion might become the response to sickness as a whole – a 'tyranny of health' to use Callahan's striking phrase (1973, 77)

Even so the distinction between therapy and enhancement is not easy to draw. Varying Christian attempts to do so originate from different interpretations of the creation and fall narratives throughout church history. Augustine of Hippo (AD 354-430) in his *City of God* explicates the fall as entailing the ruin of all humanity (as the offspring of Adam) from a state of perfection by Adam's sin of disobedience. Augustine draws not only on the Genesis account but also St Paul's exposition of it in his letter to the Romans (Chap 5v12-20). This Augustinian schema underpins Prof John Wyatt's analogies of the restored masterpiece and the Lego kit, to attempt to differentiate medical therapy from enhancement.

According to Wyatt, "Our bodies do not come to us value free. They are instead wonderful, original artistic masterpieces which reflect the meticulous design and order imposed by a Creator's will and purpose" (2009, 98). This original masterpiece has however become defaced and flawed by the effects of the fall and Wyatt contends that the task of medicine from a biblical anthropological perspective is to renew the body back to the Creator's original intentions, just as an art restorer does in her work on a painting. This is what therapy entails. It does not preclude the use of innovative technology but the purpose is always to restore to the original.

Wyatt contrasts to this what he dubs the Lego kit view of humanity, which is very different. "There is no right or wrong way to put the pieces together. There is no masterplan from the designer. There is no ethical basis of Lego construction. You can do what you like. In fact, as the advert says 'The only limit is your imagination'" (2009, 35). Furthermore, since there is no natural order within a random, mechanistic view of humanity, the difference between natural and enhanced becomes obliterated completely.

Views of Creation: Irenaeus and Cole-Turner

The flawed masterpiece half of Wyatt's analogy depends on the Augustinian view of the state of original *perfection*. A different view from Augustine's however was taken by an earlier Christian theologian, Irenaeus (130-202 AD). Both Augustine and Ireneaus considered that mankind fell and is hence in need of redemption. However the Irenaean view on creation is that it is still a work in progress. The first stage of creation – that of being 'in the image of God' is complete. However in this stage humanity is not mature. For Irenaeus, the command to 'be fruitful and multiply' (Genesis 1v28) implies future growth and development, as he explains in *Against Heresies* (4.11.1). He thus understands the description of Adam and Eve as 'naked and unashamed' (Genesis 2v25) to refer not to complete unawareness of shame in a sinless state but rather to their prepubescence. Hence according to Ireneus (*Against Heresies* 4.11.1 and

4.38), God made Adam good but immature so the second stage of creation requires us to grow into the likeness of God by exercising our free will, which includes the possibility of choosing evil.

Thus for Irenaeus, God's declaration of his creation as 'very good' did not mean the world was free from pain and suffering but that it was perfectly suited to God's purpose of developing us into his likeness. Ironically the very thing that constitutes the essence of sin in Bonhoeffer's view – mankind seeking to be like God – becomes the very purpose of God for mankind in Irenaean thought. For Irenaeus, "Adam and Eve could not have been morally and spiritually mature because it is in the very nature of such maturity that it cannot happen apart from over the course of a lifetime of moral choices and experiences" (Schneider 2012: 165).

It is this model of the Irenaean Adam that has proven very attractive to many contemporary theologians as a path to reconciling the Genesis accounts not only with Darwinian evolution but also with more modern evolutionary theories derived from the mapping of the human genome such as that, for example of a group of scientists who, from comparing mitochondrial DNA of many races, conjectured that all humans are descended from one female living in Africa about 240,000 years ago and appropriately called "Eve" (Cann *et al.* 1987). But aside from its possible implications concerning human origins, the Irenaean account also leads to a very different moral viewpoint from that of Wyatt on the scope of genetic engineering. Such a view is exemplified in the work of Prof Ronald Cole-Turner.

Cole-Turner sees gene editing and synthetic biology having a naturally legitimate role for mankind as partners with God in co-creating our own development. "At the very least, the question of the human creature as creator (or 'co-creator' as some have suggested) who contributes to the divine work of creation through new technology remains an open question, more urgent than ever. Some people of religious conviction see science as a new source of theology and technology as a new avenue of service in the grand work of creation" (Cole-Turner 2009, 198). That Cole-Turner himself may well be one of those 'people' he refers to, appears likely from his stating elsewhere that "we humans do play something of a cooperative role in the creative process and… we should intend to do so", though he does immediately add "Before we can dignify this role with the label of 'co-creation', however, we need to have a clear idea of what the creator intends for the creation" (1987, 345).

He goes on to quote approvingly Prof Arthur Peacocke, a scientist and Christian apologist who suggests that we are "co-explorers" with God. Cole-Turner then suggests, "We might be that part of creation through whom God works to explore new possibilities as yet *unknown to God*" (1987, 348). In his enthusiasm to give an apparently unshackled theological mandate for scientific advance, he does not even seem to consider the possibility that a God who doesn't know something that his creatures know, might not be God at all.

To be fair to Cole-Turner however, he does clearly acknowledge in a more pastorally-oriented reflection, that the hubris of unbridled confidence in scientific progress is not without its dangers. His point is not that we merely "exaggerate these technologies' powers or the speed of their development, but that by exaggerating them we distort the limited but legitimate value of the technologies. They become a dangerous obsession under which our anxieties flourish. Thinking we are on track for a technology that can control life, we run the risk of losing whatever small capacity we have to live at peace with our uncontrolled imperfections, with illness and disability" (Cole-Turner 2002, 44).

It is indeed a difficult path to tread between on the one hand realising the hope of healing of disease resulting from the ethical application of new genome editing technologies and on the other hand, falling into the danger of thinking that we can become masters of our own destiny entirely without God

Genes Are Us?

Christian theology in common with many other world faiths, contends that we are more than the sum of our constituent parts, including our DNA base pairs. However it is not perhaps surprising that by contrast, scientists have often given the impression that we are determined by our genes in a very mechanistic way. Francis Crick and James Watson, the discoverers of DNA structure, themselves subscribe to this view. Crick dubbed it *The Astonishing Hypothesis* – "that "You", your joys and your sorrows, your memories and your ambitions, your sense of identity and *free will*, are in fact no more than the behaviour of a vast assembly of nerve cells and their associated molecules" (Crick 1994,3 italics mine).

As we have seen, thought differing in significant ways, both the Augustinian and Irenaean views of the Fall agree it involves the exercise of free will. The advent of genetic determinism is not the first challenge to the reality of free will that science has presented however and the type of 'knight's move' involved in every case is exemplified in Crick's use in the quote above of the phrase 'no more than'. Certainly the 'nerve cells and their associated molecules' including those in DNA are necessary, but are they *sufficient* for all that Crick attributes to them, such as our ambitions?

Cybernetic Determinism

The 'smoke and mirrors' intrinsic to Crick's approach is magnificently exposed in another context by Hans Jonas' in his 1966 essay on '*Cybernetics and human purpose*'. Norbert Weiner first defined cybernetics in its modern sense in 1948 as "the science of communication and control in the animal and the machine" (Weiner 1948, 6); he proposed, "Society can only be understood through a study of the messages and communication facilities which belong to it" (Weiner 1950). Jonas seeks to demonstrate that the claim of cyberneticists that purpose and teleology can be evolved from mechanical premises alone is "spurious and mainly verbal" (1966, 111). He does so by arguing that the cyberneticists mistake 'carrying out a purpose' for 'having a purpose'. Machines may (and do) carry out purposes but the purposes they carry out are human purposes.

At the conclusion of his detailed argument, Jonas turns to biology. He suggests that though at a superficial level, the sensory-motor pattern of behaviour in animals does resemble a feedback loop in a machine, it is in fact entirely different because living beings are creatures of need. "This basic self-concern of all life, in which necessity and will are bound together, manifests itself on the level of animality as appetite, fear and all the rest of the emotions. The pang of hunger, the passion of the chase, the fury of combat, the anguish of flight, the lure of love – these and not the data transmitted by the receptors…make behaviour purposive" (Jonas 1966, 126).

To have purpose, a goal, a *telos* requires more than mere input and output. Jonas' conclusion is still strikingly relevant today in an age of genomic reductionism – "According to cybernetics, society is a communication network for transmitting, exchanging and pooling of information and it is this that holds it together. No emptier notion of society has ever been propounded. Nothing is said on what the information is about and why it should be relevant to have it" (Jonas 1966: 126).

Genetic Determinism

A similar problem occurs in relation to the reductionist view of the transmitting, exchanging and pooling of genetic information. It is fascinating to read in literature published before the mapping of human genome, how the concept of the gene defined as 'a unit of heredity containing the information for one protein' became known as the central dogma of molecular biology (Suzuki and Knudtson 1989, 52). At a CIBA symposium in 1989, one senior scientist working on genome mapping stated that "Genetics investigates the plan of the organism. The plan is embodied in the collection of genes that is handed down in the germ line to specify the construction of the organism ….The manifesto – if not the programme - of molecular genetics must remain the computation of organisms from their DNA sequences" (Cited in Jockemsen 1997, 77).

With this kind of outlook, it is easy to see how genetic determinism as expressed by Crick above can take root. But even prior to human genome mapping, there was evidence that the relationship of the structure of a particular protein and the structure of a particular organ, let alone an entire organism, is not a direct one (Tauber and Sarkar 1992). Jockemsen points out two other epistemological problems of the genetic determinist view of a direct causal relation between gene and trait. Firstly this view elides two very different types of knowledge – the molecular biological knowledge with the knowledge involved in recognition and judgement of clinical diagnosis. He makes the important point that observation of the correlation between gene defects and clinical diagnostic features does not prove genes are causally related to traits. "It only proves that gene *defects* can disturb the development of the organism. The gene or gene defect acquires meaning only in the context of the existence and functioning of the entire organism. In other words, the observation that concrete genetic information is a necessary precondition does not make that information a sufficient precondition for a 'normal' development." (Jockemsen 1997, 79). Secondly, if the DNA sequence contains a message, this presupposes a meaning in the message which cannot be generated by the mechanism which translates it. Furthermore the DNA has not generated the translation mechanism since in order to be expressed it needs that mechanism. The genetic message itself "needs an explanation – both a final and causal one" (Jockemsen 1997:79) and for people of the Abrahamic faiths, that final cause is God.

If this were not enough in itself, new developments have given rise for even more reasons to be cautious about genetic determinism. Before the HGP, it was thought that there were around 80 000 coding genes for proteins. When the actual number turned out to be around 25 000, the rest of the DNA was initially written off as redundant and labelled as 'junk' (Parrington 2016, 72). However, the publication in 2012 of the Encyclopaedia of DNA Elements (ENCODE) project soon changed that. ENCODE (2012) demonstrated that 80% of the DNA has newly recognised biochemical functions (Parrington 2016, 91), many of which of are carried out non-coding RNAs (ncRNAs) which are involved in regulation of protein coding genes by either facilitating or down-regulating their expression. There is also evidence that these ncRNAs and their effects are influenced by environmental factors including smoking (Hou *et al* 2011). So with both a) the vast majority of the DNA not coding for proteins and b) environmental factors influencing the ncRNAs' control of protein- coding, the central dogma of molecular biology looks less and less central than it did.

This is even without taking into account epigenetic effects. The term epigenetics has had various definitions. Perhaps the simplest to understand is "the study of mitotically and/or meiotically heritable changes in gene function that *cannot* be explained by changes in DNA sequence" (Riggs and Porter 1996, 29). However this definition has recently been refined in both more positive and comprehensive terms defining epigenetic events as "the structural adaptation of chromosomal regions so as to register, signal or perpetuate altered activity states" (Bird 2007, 398).

The key point to grasp is that changes to the DNA *other* than the widely known mutations of DNA sequencing, can influence changes in the organism, some of which are inheritable. These changes may entail small molecules being added to or removed from both the DNA itself and/or the histone complexes, around which the nucleic acid sequences are spiraled. Environmental factors can also affect these epigenetic changes.

Finally even for those genes that do code for a single protein, recent findings have shown that the protein is not all there is to the expression of that gene. The gene for enzyme monoamine oxidase A (MAOIA) provides perhaps the best-known example. This enzyme inactivates neurotransmitters such as serotonin, which therefore accumulate when the gene for MAOA is defective. A Dutch study (Brunner *et al* 1993) of a family in which male members exhibited aggressive behaviour showed that using DNA mapping that the gene for MAOA had mutated and the resulting enzyme was ineffective so neurotransmitter levels were much higher than normal in these men. Women were unaffected because the gene is sex linked. However another study (Caspi *et al* 2002) examined another gene variant known as MOAO-L that produces a low functioning rather than non-functioning enzyme variant. Caspi's team followed up abused children over many years and found that presence of MOAO-L alone was *not* associated with high levels of aggression over 25 years, unless the children had been subject to maltreatment themselves. In contrast, those children with a normal MOAO gene appeared more resistant to developing aggressive behaviour even if they were subject to childhood abuse. In short, MOAO research to date suggests that the idea of mutation leading inexorably to phenotypic change is rarely, if ever, all there is to human behaviour.

Where does this then leave us theologically in relation to our human responsibility before God? It surely confirms that though our genes do influence everything about us, they do not determine everything we do. Our environment and our human wills also have a large part to play in what we do with our lives. "Our whole being is influenced by our genes. But not everything about us is explained by our genes. Environment and personal responsibility play a role. Theology has a stake in maintaining that the role played by personal responsibility is genuine and significant. It is not epiphenomenal or illusory" (Cole-Turner 1992, 170)

The Secular Soul and Two Types of *Telos*

What is illusory is to place all our hopes in our genes "fueling the expectation that the last word about the human genome will be last word about human nature" (Mauron 2001). In an article critiquing the popular rise of the concept of the genome as the 'secular soul', Prof Alex Mauron writes, "the notion that our genome is synonymous with our humanness is gaining strength. This view is a kind of "genomic metaphysics": the genome is viewed as the core of our nature, determining both our individuality and our species identity. According to this view, the genome is seen as the true essence of human nature, with external influences considered as accidental event" (Mauron 2001).

His article surely raises the question of whether the concept of the soul as a metaphor for the genome might also indicate the religious element of the quest to map and explore it. Demonstrating the inadequacies of genetic reductionism does not go far enough in a critique of it in that it does not consider the reasons why this reductionism has gained such a hold. Its historical roots may go back to the seventeenth century when earlier ideas of human knowledge as largely passive and received by illumination or rumination, began to wane. What is now termed a constructionist epistemology then began to emerge where knowledge is constructed by measurement of and experimentation with the world. In his seminal work on this paradigm shift, Funkenstein comments, "applying knowledge-through-construction to the whole world was as inevitable as it was dangerous. It was dangerous because it makes mankind to be 'like God knowing good and evil' " (1986, 290), - echoing the language of ancient Augustinian Fall.

Prof Robert Song powerfully points out that the Human Genome Project is a game-changer in the search for eliminating human suffering. At the same time as it provides knowledge of how the body is constructed, it also shows how it might be reconstructed. He suggests the HGP has become " a surrogate form of salvation, ... It develops, for example, a doctrine of creation, which conceives nature as a raw material available for technological manipulation, while its anthropology defines human beings in terms of self-defining freedom above the contingencies of bodily life. It espouses eschatological hope, which lies in the dream of escape from finitude and locates the means of salvation to that end in the application of technical reason..." (Song 2003, 178-79).

Divergent Paths of Purpose

Whilst of course there are some overlaps of this alternative secular creation narrative with a Christian understanding of human stewardship and authority over the rest of nature, it is the means of reaching its *telos* or end purpose which makes the secular vision very different from the Christian vision revealed in scripture and indeed I would suggest sets the two completely at odds with other.

In an illuminating paper on how human potential relates to genetics and Christian theology, Lysaught makes reference to the seventeenth century paradigm shift considered above, in pointing out that that post-Enlightenment humanistic progress and Christian theology are both teleological and the intended goal in both is towards a future of wholeness and completion. However "in the narrative of genetic potential, the future utopia is achieved by the temporal eradication of human imperfection. In the scriptural vision—instantiated in Eden, the promised land, occasionally in Jerusalem, the kingdom of God, the heavenly banquet, and the new creation—communal human flourishing is achieved simply when individuals and communities choose to dwell with God. They might still be fat, not very smart, short, slow runners, or manifest myriad other flaws, but when persons in Scripture choose to acknowledge God's power and to live as God's people, perfection, wholeness, goodness (holiness, righteousness, and justice) come to be. The definition of human perfection in Scripture is not perfection of mind or body but rather: being in relationship with God" (Lysaught 2011, 235).

Moreover, contrary to popular and persistent myths about Christianity, it is not the perfect whom Christ calls to be his people but rather those who acknowledge their sickness (Matthew 9v12; Mark 2v17; Luke 5v31). "God chose the foolish things of the world to shame the wise; God chose the weak things of the world to shame the strong. God chose the lowly things of this world and the despised things—and the things that are not—to nullify the things that are" (I Corinthians 1 v27-28).

Conclusions

In deciding if various aspects of genome editing are ethical from a Christian perspective, there are many factors to consider. Does it increase the ability to view life as a gift to be received with thanks or a commodity to be grasped as a right? Does it acknowledge human nature as something to be restored to wholeness or something to be transcended by enhancement? Does it recognise the limitations of its likelihood of success in changing phenotype predictably as it discovers the increasing importance of environmental factors and the complexity of epigenetic mechanisms? What is the *telos* of the genome editing project and perhaps most crucially of all does it reflect wise stewardship and care for the world as creatures made in the image of God or does it seek to enable us to become as God in recreating ourselves into something else?

References

Brunner H G, Nelen M, Breakefield X O, Ropers H, van Oost B A. 1993. "Abnormal behavior associated with a point mutation in the structural gene for monoamine oxidase A". *Science* 262: 578-80

Bonhoeffer, Dietrich. 1937. *Creation and Fall* (English translation 1959). London: SCM Press

Boorse, Christopher. 1976. "What a theory of Mental Health should be". *Journal for the Theory of Social Behaviour* 6: 61-84

Callahan, Daniel. 1973. "The WHO definition of Health". *Hastings Centre Studies* 1.3: 77-87

Cann, Rebecca L., Mark Stoneking, and Allan C Wilson. 1987. "Mitochondrial DNA and Human Evolution". *Nature* 325:31–36.

Caspi, Avshalom. Joseph McClay, Terrie E. Moffitt, Jonathan Mill, Judy Martin, Ian W. Craig, Alan Taylor, and Richie Poulton. 2002. "Role of genotype in the cycle of violence in maltreated children". *Science* 297: 851-854

Cole-Turner, Ronald S. 2009. "Synthetic biology; Theological Questions in Biological Engineering". In *Without Nature?: A New Condition for Theology*, edited by Albertson D, King C, 136-152. Fordham University Press

Cole-Turner, Ronald S. 2002. "Biotechnology: A Pastoral Reflection". *Theology Today* 59: 41-54

Cole Turner, Ronald S. 1992. "Religion and Human Genome". *Journal of Religion and Health* 31: 161-173

Cole-Turner, Ronald S. 1987. "Is Genetic Engineering Co-Creation?". *Theology Today* 44: 338-349

Crick, Francis. 1994. "*The Astonishing Hypothesis*". New York: Simon and Schuster

ENCODE Project Consortium. 2012. "An integrated encyclopedia of DNA elements in the human genome". *Nature* 489: 57-74

Funkenstein, Amos. 1986. *Theology and the Scientific Imagination from the Middle Ages to the Seventeenth Century*. New Jersey: Princeton University Press

Hare, Richard M. 1986. "Health". *Journal of Medical Ethics* 12: 174-181

Hou, Lifang. Dong Wang, and Andrea Baccarelli. 2011. "Environmental chemical and microRNAs". *Mutation Research* 714: 105-112

Irenaeus, (c180) *Against Heresies* http://www.newadvent.org/fathers/0103.htm

Jockemsen, Henk. 1997. "Reducing People to Genetics". In *Genetic Ethics: Do the Ends Justify the Genes?*, edited by John Fredric Kilner, Rebecca Davis Pentz, Frank E Young. 75-83. Milton Keynes: Paternoster Press

Jonas, Hans. 1966. "Cybernetics and Purpose: A Critique". In *The Phenomenon of Life*, edited by Vogel, L and H Jonas (2000), 108-134. Westport Connecticut: Greenwood Press

Junker-Kenny, Margaret. 2003. "Genes and the Self: Anthropological questions in the Human Genome Project" In *Brave New World: theology, ethics and the human genome* edited by Celia Deane-Drummond, 116-140. London: T and T Clark

Lysaught, M Therese. 2011. "The Last Shall Be First: Human Potential in Genetic and Theological Perspectives". *Theology and Science*, 9: 223-240

Messer, Neal. 2011. *Respecting Life: Theology and Bioethics.* London: SCM Press

Messer, Neal. 2003. "HGP, Health and the 'Tyranny of Normality'". In *Brave New World: theology, ethics and the human genome,* edited by Celia Deane-Drummond, 91-115. London: T and T Clark

Mauron, Alex. 2001. "Is the genome the secular equivalent of the soul?" *Science* 291: 831-832 DOI: 10.1126/science.1058768

New Advent. 2017. "Against Heresies (Book IV, Chapter 11)". Accessed from http://www.newadvent.org/fathers/0103411.htm

New Advent. 2017. "Against Heresies (Book IV, Chapter 38)". Accessed from http://www.newadvent.org/fathers/0103438.htm

O'Connell S. 2009. "Are human beings impossible to ape?" *Daily Telegraph.* Accessed 30.6.2009 http://www.telegraph.co.uk/news/science/evolution/5695045/Are-human-beings-impossible-to-ape.html

Page Ruth. 2003. "The Human Genome and the Image of God". In *Brave New World: theology, ethics and the human genome,* edited by Celia Deane-Drummond, 68-85. London: T and T Clark

Parrington, John. 2016. *The Deeper Genome: why there is more to the human genome than meets the eye.* Oxford: Oxford University Press

Ramsey, Paul. 1970. *Fabricated Man: The Ethics of Genetic Control.* New Haven: Yale University Press

Riggs, Arthur D. and Thomas N. Porter T. 1996. "Overview of epigenetic mechanisms". In *Epigenetic mechanisms of gene regulation,* edited by Russo VEA, Martienssen R, Riggs AD, 29–45. Cold Spring Harbor, NY: Cold Spring Harbor Laboratory Press

Russo, Vincenzo E. A., Robert A. Martienssen, and Arthur D. Riggs, (eds). 1996. *Epigenetic Mechanisms of Gene Regulation.* Woodbury: Cold Spring Harbor Laboratory Press

Schneider, John. 2012. "The Fall of Augustinian Adam: Original Fragility and Supralapsarian Purpose". *Zygon* 47: 949-969

Seller, Mary J. 2003. "Genes, Genetics and the Human Genome". In *Brave New World:*

theology, ethics and the human genome. edited by Celia Deane-Drummond, 27-44 . London: T and T Clark

Song, Robert. 2003. "HGP as Soteriological Project". In *Brave New World: theology, ethics and the human genome,* edited by Celia Deane-Drummond, 164-184. London: T and T Clark

Suzuki D, Knudtson P. 1989. "*Genetics*". Cambridge, Massachusetts: Harvard University Press

Tauber, Alfred I., Sahtora Sarkar. 1992. "The Human Genome Project: Has Blind Reductionism gone too far?". *Perspec Biol Med* 35: 220-35

Weiner, Norbert. 1950. *The Human Use of Human Beings.* Cambridge, Massachusetts: Da Capo Press

Weiner, Norbert. 1948. *Cybernetics: Or Control and Communication in the Animal and the Machine.* Cambridge, Massachusetts: MIT Press

World Health Organisation. 1948. http://www.who.int/about/definition/en/print.html

Wyatt, John. 2009. *Matters of Life and Death; Human Dilemmas in the light of the Christian Faith.* London: Leicester IVP

CHAPTER 9

Living with the Genome, by Angus Clark and Flo Ticehurst, within the Muslim Context

Ayman Shabana[1]

Modern genetics has ushered a new phase in human history and revolutionized human understanding of how living organisms are constituted and how they function. More particularly, it has revealed the processes associated with the transmission of inheritable features and characteristics to subsequent generations. The new knowledge that it generates brings promises of unprecedented preventive as well as therapeutic possibilities, especially as far as inherited diseases are concerned. These unprecedented possibilities are by no means limited to the fields of health care and life sciences as they touch many other aspects of our lives. However, as much as this genetic revolution has given rise to new exciting possibilities, it has also raised important ethical questions pertaining to the production and application of genetic knowledge. It is within this context of the double-edged nature of modern genetics that *Living with the Genome: Ethical and Social Aspects of Human Genetics* has to be placed. Although published in 2006, the book still provides a useful introduction to the range of ethical, legal, and social implications of modern genetic research and technology. It comprises 42 articles on a wide range of topics, which are drawn from the *Encyclopedia of the Human Genome* (one of the co-editors of the book, Angus Clarke, was also the editor of the "Ethics and Society" section of this reference work). The book is intended to enhance the readership of these topics by making these articles available to a wider audience beyond specialists in human genetics.

In terms of its basic subject matter, genetics aims to study how living organisms both change and maintain their basic characteristics over time. This

1 Associate Research Professor at Georgetown University's School of Foreign Service in Qatar (SFS-Q), as2432@georgetown.edu. This publication was made possible by NPRP grant # NPRP8-1478-6-053 from the Qatar National Research Fund (a member of Qatar Foundation). The statements made herein are solely the responsibility of the author.

study is undertaken at three distinct levels: life of the cell, life of an individual organism, and history of the population of a particular species. Genome stands for the entire set of genes (of an individual organism or of an entire population) and genomics refers to the study of genes in this collective sense. Modern genetics traces its roots to important discoveries during the 19th century (Mendel's attribution of inheritance to certain particles, discovery of chromosomes, development of statistics and its application to hereditary processes, and Darwin's theory of evolution), which inspired further developments during the 20th century (identification of DNA as the chemical basis of heredity and development of molecular biology). The past few decades have witnessed increased interest in the deployment of genetics in medical research and practice with the hope of identifying genes associated with particular diseases and developing effective ways not only to treat but also to prevent the occurrence of such diseases at the individual and collective (population) levels. Genetics is also used in the development of new drugs as well as their administration for certain diseases or even tailoring them for individual patients. Use of genetic examination in health care and medical practice is already replacing traditional physiological and biochemical methods and is expected to increase even further in the future. Important examples include confirmation of the diagnosis of certain diseases in individual patients and screening for certain disorders within particular groups or populations. Yet, as noted above, despite these remarkable therapeutic potentials of modern genetics, it raises a host of ethical, legal, and social concerns pertaining to proper use as well as implications of genetic information. The articles selected for inclusion in this volume explore these concerns with varying degrees of length and depth. They are divided into six main parts (prefaced by general editorial introductions): The Human Genome Project: Genetic Research and Commercialization; Genetic Disease: Implications for Individuals, families and Populations; Disability, Genetics and Eugenics; Genetics and Society: Information, Interpretation and Representation; Genetic Explanations: Understanding Origins and Outcomes; and Reproduction, Cloning and the Future.

The richly diverse collection of contributions that *Living with the Genome* contains offers only a glimpse of the range of concerns that biomedical technology has engendered. With the extremely fast pace of technical advances in this field, new discoveries or inventions spark new questions and launch new debates to examine their (bio)ethical implications. Still, this book remains a useful starting point as it captures some of the most important issues that remain as relevant now as they were when the book was published more than a decade ago. The main limitation of the book, however, is its limited scope of coverage as it focuses almost exclusively on the Western geographic as well as intellectual context, with very few exceptions (e.g. chapter on the Maternal and Infant Health Care Law in China, 147; brief references to discriminatory sex selection practices in India, 293; and Muslim ban on gamete donation, 276). This is somehow understandable because these discussions coincided with the rise of the early waves of genetic and reproductive technology in the West, particularly in the second half of the 20th century. But, considering the increasingly globalized nature of our world and the global influence of Western medicine, experts and practitioners worldwide are also joining these bioethical discussions. At the practical level, the availability of the latest applications of biomedical technology, especially to those who can afford it regardless where they happen to live, has also stirred similar debates over important bioethical concerns. The book, therefore, calls for comparative analysis of the various ethical, legal, and social issues that it highlights in order to reflect the diversity and richness of particular societies, cultures, and religious traditions. Such comparative analysis would also highlight important parallels and similarities in various social and cultural contexts. One interesting example that the book discusses is the case of the deCODE project in Iceland (56-63), which can provide important lessons for countries, especially those with small populations. The case highlights the critical role that regulators should play in ensuring compliance to ethical standards and proper conduct of genetic research. Another example is the successful use of genetic screening (premarital and prenatal) in Cyprus for inherited hemoglobin diseases (114-121).

Within the Muslim context, bioethical debates can be traced to two main factors: globalization of the medical curriculum (including bioethics), and the arrival of various applications of biomedical technology. Bioethical deliberations in the Muslim world, however, have drawn heavily on the Islamic normative tradition, particularly on the Islamic legal tradition. Researchers often point out that bioethical discourses in the Muslim world are dominated by the Islamic legal discourse, which is evident in the increasing volume of legal opinions (*fatwas*) on almost each of the issues addressed in this book. One of the main problems with these disparate *fatwas* is lack of a consist methodology for the examination of bioethical issues. The past few decades have witnessed serious efforts on the part of jurists and medical experts to provide systematic examination of bioethical issues with the goal of developing guidelines that should inform professional practice as well as national policies and legislation. These collaborative efforts have been facilitated by a number of national and transnational institutions as well as a number of academic and research centers. Although, for the most part, the development of a comprehensive Islamic bioethical framework remains work in progress and the general state of bioethics differs from one national context to another, empirical research shows that in order for any treatment of bioethical issues to be taken seriously, it has to engage this evolving body of Islamic normative literature. The bulk of this literature revolves, for the most part, around some key documents in the form of resolutions, decisions, or institutional *fatwas*, particularly ones that are issued by prominent national as well as transnational institutions. Below I give a summary of the main parts of this book together with brief comments in light of this Islamic normative literature.

The first part comprises eight articles offering a historical account of the Human Genome Project (HGP) with a particular focus on the debate over the commercialization of genetic information, which was one of the driving forces behind this project until its successful completion in 2003 (two and half years prior to the scheduled deadline in 2005). This debate was fueled by two competing visions for the project. The first was championed by the private sector, as represented by Celera Corporation, and the second was represented by an international consortium consisting of major public and state-sponsored research entities. While the first advocated patenting and monopolizing genome sequencing data as new inventions, the second insisted that genomic data should remain freely available as a non-commercial shared human resource for further research and development. Other contributions in this section provide various perspectives on the rationale, objectives, and implications of the HGP, which overall are not quite sanguine. Contrary to the usual hype emphasizing the miraculous achievements of genetic technology these contributions identify and highlight significant issues that tend to be glossed over due to the usual and unquestionable embracement of technology and its equation with progress. Chief concerns that run through these contributions include: commodification of genetic data, ownership and subsequent use of genetic material, distinction between a novel invention and mere discovery of nature; implications of the control of the human gene pool for future generations; obtaining informed consent in genetic research; and exaggerated expectations of/for gene therapy. At the global level these concerns reflect larger tensions between multinational capitalist interests on the one hand and rights of indigenous people and their claim over native resources on the other. In general, these concerns are also echoed in normative Islamic pronouncements. While these pronouncements praise the remarkable potentials of genomic research, they urge careful evaluation of any procedure involving intervention in, or manipulation of, the human genome. They also warn against any commercial exploitation or monopolization of genetic materials. These reserved sentiments are reiterated in several documents such as: the recommendations of a seminar organized by the Kuwait-based Islamic Organization for Medical Sciences (IOMS) in 1998 under the theme of "Genetics, Genetic Engineering, Human

Genome, and Genetic Therapy: An Islamic Perspective," the resolution of the Islamic Fiqh Academy (IFA), affiliated with the Muslim World League in its 16th session that was held in Mecca in 2002, and the resolution of the International Islamic Fiqh Academy (IIFA), affiliated with the Organization of Islamic Cooperation in its 21st session that was held in Riyadh in 2013.

The second part of the book includes seven articles covering a number of ethical concerns associated with genetic counseling. The process aims to enhance understanding of a genetic condition and to explore possibilities to avoid or cope with such a condition. It is often pursued for either health risk or reproductive purposes. Although it focuses primarily on scientific explanation of a particular genetic condition, it may also address religious and metaphysical questions that patients and their families feel they must address. This is particularly important in multi-religious or multi-cultural contexts reflecting various meanings for universal experiences of illness and suffering. Considering the large scope of Islamic ethical-legal regulations, which cover various aspects of a person's life, including matters of health and illness, genetic counseling acquires added significance within the Muslim context. Practitioners need to develop familiarity with distinctive religious and cultural features that may create potential tension with mainstream (Western) bioethical standards. For example, genetic counseling is guided by three main ethical principles: autonomy of the individual or couple, right to full information, and utmost level of confidentiality (p. 115). Several studies, however, point out significant difficulties in the implementation of autonomy in non-Western contexts, including Muslim ones. For example, these studies show that Western emphasis on individual freedom might not be compatible with certain religious and cultural norms placing more emphasis on communitarian ethics. Moreover, such emphasis on autonomy may clash with particular attitudes concerning controversial issues such as abortion, euthanasia, or cremation.

One of the unique consequences of genetic testing has been increasing public awareness of genetic risk for, or susceptibility to, particular health conditions not only for the individual undertaking the test but also for close family members. This, in turn, raises important questions on whether/how information concerning risk for others could be communicated. Research shows that the perception of susceptibility to genetic diseases is dynamic and varies according to several factors such as family history, gender, age, or economic standards. Some studies even question the utility of genetic testing for certain conditions, especially when positive results may lead to a fatalistic attitude of resignation rather than proactive behavioral changes (p. 105). Medical practitioners, therefore, should be sensitive to the religious conceptualization of illness as a means for spiritual refinement. In Islam, for example, illness is seen as an opportunity for personal growth and also a way to gain reward in the hereafter. Delivery of effective genetic counseling would, therefore, require careful attention to contextual factors such as cultural background and religious attitudes towards certain medical conditions or procedures.

The third part consists of five articles discussing the relationship between disability and genetics. Each of these contributions addresses the issue from a particular perspective ranging from a historical investigation of problematic precedents, lingering traces of these precedents in contemporary practices, distinction between Western and non-Western outlooks on this issue, perception and implications from a human rights perspective, and examination of particular disability groups. Historical accounts of the contentious relationship between disability and genetics often start with the indelible eugenic practices during the 20th century in Europe and the United States ranging from positive eugenics to enforced sterilization and even euthanasia. Some of the important traces of eugenics in contemporary research revolve around exploration of genetic explanation for criminal behavior as well as genetic screening especially in regions witnessing higher rates of genetic diseases due to inbreeding or lower rates of migration, where genes causing diseases tend to cluster over time. Within the Muslim context, the issue of consanguineous marriage has stirred extensive debates. Scriptural sources delineate certain prohibited degrees within close family relationships, which define the boundaries of incest. No categorical prohibition, however, is indicated beyond these prohibited degrees. Although it is generally discouraged, actual practice has always varied from one region to another depending on dominant cultural norms. The past few decades witnessed extensive efforts throughout Muslim-majority countries, with the help of international organizations, to raise public awareness about the significant genetic risks that consanguineous marriage involves, especially in places where it is commonly practiced. On the other hand, traces of positive eugenics could be found in the creation of sperm and ova banks, particularly ones obtained from individuals possessing desirable physical and cognitive traits (p. 144-145). Several Muslim countries have already established biobanks not only to encourage organ donation but also to facilitate the procurement of tissue samples for scientific research. A significant portion of Islamic bioethical discourses is dedicated to the elucidation of proper guidelines that should govern organ transplantation. However, in light of Islamic regulations pertaining to lineage and family relationships, these guidelines often include specific reservations and restrictions concerning transplantation of reproductive organs, gamete (sperm and ova) donation, and also milk banking.

The fourth part focuses on several social issues related to the employment of genetics in the creation of medical profiles. It consists of seven contributions dealing with topics including emergence of the gene as a significant cultural icon, perception of the interplay between genetic and environmental factors, confidentiality of genetic information, implications of genetic data for insurance purposes, and role of genetic factors in the confirmation of racial and cognitive stereotypes. Heightened media attention to the role of genetics in the definition of one's identity has transformed the gene into a powerful cultural symbol. The gene has increasingly been used as the locus of personhood and has almost acquired the sacred status attributed to the soul or other similar entities in different cultures. The concepts of the soul (*nafs*) and spirit (*rūḥ*) feature prominently in Islamic bioethical discourses, particularly on issues associated with beginning and end of human life. This is mainly due to the fact that Islamic scriptural sources define inception as well as end of human life in terms of the infusion or extraction of this metaphysical entity. Consecration of the role of genetics, therefore, raises questions concerning the continued relevance of the classical religious conceptualization of the soul or spirit as the primary factor to settle questions related to issues such as abortion or the new definition of death on the basis of brain (stem) function. On the other hand, some of the most important ethical concerns that modern genetics gave rise to are associated with boundaries of individual privacy and confidentiality of personal (genetic) information. The main ethical challenge in this regard remains how to reconcile concerns for individual privacy with others' right to have access to shared genetic information. These privacy and confidentiality concerns may have significant social and even economic implications such as one's ability to obtain affordable health insurance. This, in turn, raises questions of social justice as well as equitable distribution of social goods and services among members of the society.

The fifth part focuses on the general theme of genetic explanations particularly of behavioral traits. It consists of seven contributions dealing with the manner in which genes are used for explanatory purposes within the context of the HGP, evolutionary accounts of natural selection, and emergence of counter evolutionary accounts such as creationism and intelligent design, genetic reductionism and determinism, and reinforcement of racial and ethnic characteristics. Most of these contributions point out the limitation of exclusive reliance on genes for explanatory purposes and call instead for a more nuanced account for the interplay of genetic as well as environmental factors. From a religious perspective these discussions may also inspire new reflections on classical theological debates on divine destiny and human freedom. While notions of genetic determinism can be read as a modern extension to theological determinism, man's role in the creation and manipulation of environmental factors can be seen as a reflection of human agency and freedom.

The final part, comprising eight contributions, is devoted to the interaction between genetics and modern reproductive technologies in terms of regulation, range of choices, feminist perspectives, particular procedures, distributive justice, and impact on the future. The reproductive revolution that modern biomedical technology has unleashed forces a reexamination of the regulatory aspects of parenthood, particularly within cultural contexts where genetic and social definitions of parenthood are not coextensive (p. 184-187). Countries vary widely with regard to the legislative model they adopt but any regulatory model that a country ends up choosing would depend on a number of contextual considerations and "nuances in tradition, religion, culture, economics, and wealth" (272). In general, while a liberal approach would issue from a permissibility presumption on the basis of fundamental rights and freedoms, a restrictive approach would be driven by concerns such as disrupting natural order and playing God. Ultimately, the range of possibilities that modern genetic and reproductive technologies generate would raise moral questions on fair distribution of benefits and burdens in society as well as on potential implications for future generations. In addition to these questions, ethical debates surrounding assisted reproductive technology (ART) within the Muslim context also address questions such as inter gender interaction in the clinical setting and also involvement of a third party in the procreative process. Islamic bioethical discourses often resort to classical ethical-legal concepts such as need (*ḥājah*), necessity (*ḍarūrah*), and utility (*maṣlaḥah*) to evaluate particular cases and scenarios. In light of these concepts a distinction is often made between medical (therapeutic) and non-medical uses of these technologies. In the case of cloning, for example, a distinction is made between therapeutic cloning, which is perceived as potentially useful, and reproductive cloning, which is depicted as dangerous and harmful. A similar distinction is made also with regard to fetal sex selection, which is unanimously allowed for therapeutic purposes while being permitted for family balancing only in limited situations and on an individual basis. Anthropologists also point out different views throughout Muslim-majority countries, which are sometimes developed along sectarian lines. The most famous example is the Sunnī-Shīʿī divide on gamete donation, which is opposed by most Sunnī scholars while being allowed by some Shīʿī scholars. This distinction is rooted in the conceptualization of gamete donation and its analogy with adultery. While critics equate the process with adultery, supporters limit the definition of adultery to actual physical contact.

In conclusion, as its title indicates, this book introduces some of the important challenges that modern genetics create, which are here to stay. People need to learn how to live and cope with them. A fuller exploration of the ethical and social aspects of human genetics, however, requires further comparative studies reflecting additional complexity and nuance associated with particular religious or cultural contexts.

References

Clark, Angus and Flo Ticehurst (eds.). 2006. *Living with the Genome: Ethical and Social Aspects of Human Genetics.* 327. New York: Palgrave Macmillan

PART 4

Contributions in Arabic

الفصل 10

الجينوم والطبيعة البشرية: مقاربة تحليلية في ضوء الفلسفة والعلم التجريبي والأخلاق الإسلامية

سعدية بن دنيا [1]

مقدمة:

إن العودة إلى التفسير الديني الإسلامي لمسألة أصل خلق الإنسان هي عودة ضرورية لأن عملية الخلق هذه لا تنتمي إلى هذا العالم ومادته ممّا يعني أن مصدر إلهي بالبديهة وقد جاء في الحديث النبوي، قوله صلى الله عليه وسلم «إن الله خلق آدم على صورته، طوله ستون ذراعاً» (البخاري/6227)، لكن بعد عملية الهبوط إلى الأرض أصبحت هذه الطبيعة البشرية داخل مجرى التاريخ وآثاره أي داخل محيط من القوانين الطبيعية والتحوّلات التي تخضع لها الكائنات الحية أثناء محاولة تكيّفها مع تلك القوانين، بدليل ما تبقّى من (الستين ذراعاً) التي خُلق عليها آدم أول مرة.

لقد أصبح سؤال ماذا بقي من الطبيعة البشرية التي خُلق عليها الإنسان أوّل مرة؟ سؤال العلم وأحد أهم رهاناته وتحديّاته، وبهذا المعنى برزت النظرية الداروينية في التطور كواحدة من نظريات تفسير الطبيعة البشرية وفهم النوع الإنساني في حدوده البيولوجية، وخارج السجال والجدال الذي حدث بين الإسلام المعاصر ومضمون هذه النظرية، فإن ما هو مهمّ فيها أنها فتحت أسئلة معرفية جوهرية حول الحياة البيولوجية للإنسان وتركيبتها والتحوّلات التي يمكن أن تطرأ عليها بالتناسب مع الزمن وبالتفاعل مع البيئة.

---

1 أستاذة التعليم العالي في الفلسفة، جامعة عبد الحميد بن باديس، مستغانم، الجزائر، saadia.bendenia@univ-mosta.dz

يعمل علم الجينوم بالخصوص على وضع قراءة دقيقة للرموز البيولوجية في جسم الإنسان والتي من شأنها أن تقدّم وعوداً جادة بحلّ مشكلة المرض باستباق حدوثه جينياً أو إصلاح الخلل الذي يمثّل طفرة بيولوجية غير متوقعة أو تحسين البنية الخارجية والداخلية للجسد (عمليات التجميل، زرع الأعضاء..)، غير أن هذا التطور العلمي يواجه تحدّيا أخلاقياً، سواء على المستوى البيوإيتيقي فيما يتعلق بأخلاقيات الطب وممارسته والآثار غير الإنسانية على الإنسان أو على مستوى الأخلاق الإسلامية فيما يتعلق بتغيير الخَلق الطبيعي للطبيعة البشرية المصمّمة إلهيا، وهنا تقدم المعرفة الفلسفية فهماً مختلفا ومتجددا يجعلنا نعيد التفكير في الطبيعة البشرية كظاهرة وليس كمعطى ثابت.

تهدف هذه الدراسة إلى فهم العلاقة بين الجينوم والطبيعة البشرية في ضوء القراءات الفلسفية والعلمية، وفي ضوء الأخلاق الإسلامية ومحاولة جعلها فضاء خصبا لتناول الجينوم كظاهرة متعددة الأبعاد والآفاق، إذ الغرض ليس دراسة مدى تطابق علم الجينوم مع الأخلاق الإسلامية بل إيجاد مجال معرفي يجعل هذه الأخلاق نفسها تستثمر في الجينوم وجعله أفقا علميا لتطور أخلاق عملية ذات أبعاد إسلامية حضارية، حيث يمكننا أن نتوقع، في ضوء التطورات العلمية، جينا مسؤولا عن «العنف» وآخر عن «الكراهية» وغيرها.

أولا: نظرة الفلسفة إلى الطبيعة البشرية:

لقد نظر الفلاسفة منذ القدم إلى الطبيعة البشرية (Human Nature) من وجهات نظر عديدة لما تكتسيه من أهمية بالغة، وكذا لتعقيدها والغموض الكبير الذي يكتنف جوانبها المختلفة، وكان نتاج ذلك أن كانت آراؤهم وأبحاثهم متباينة، ولقد برزت في هذا الإطار مدارس كثيرة ومختلفة يسودها التمايز في التخصص والتوجه الفلسفي، تروم كلها الوصول إلى فهم دقيق للطبيعة البشرية وسبر أغوارها.

إن السؤال الكبير الذي تطرحه هذه التصورات الفلسفية يتمحور حول تحديد ماهية الطبيعة البشرية وهل هي جوهر ثابت أم متغير؟ ويتصل ببحث النفس البشرية وخلودها ووحدتها وصلتها بالجسم، كما يتعلق في صميمه بما يجب أن يوجه إليه السلوك الإنساني، ويتوجه نحو فهم أفضل لطبيعة البشر.

من بين أوائل التفسيرات الفلسفية الجادة والرصينة للطبيعة البشرية ما أسس له سقراط في فلسفته التصورية والذي عنى عناية بالغة بمسألة البحث في ماهية الإنسان، بحيث يرى أن «المعرفة الحقة هي معرفة النفس فقط، والمهمة الأساسية للمعرفة - معرفة الذات- إعرف نفسك» (أوسموس 1979، 29)، فالبحث عن تفسير للعالم الطبيعي لم يكن الهم الفكري لسقراط بخلاف البحث عن النفس الذي استحوذ على اهتمامه وعُدّ نقطة ارتكاز لجل مسائله وطروحاته الفلسفية.

جادل سقراط، «في جمهورية أفلاطون، بوجود ثلاثة أقسام متميزة للروح، جزء يختص بالرغبة، وجزء عقلاني، وثالث أطلق عليه «تيموس» (Thymos)، وهي كلمة يونانية عادة ما تترجم إلى الحيوية، والتيموس هو الجانب المتكبر من شخصية الإنسان، أي الجزء الذي يطالب أن يعترف الآخرون بقيمة المرء» (فوكوياما 2006، 62)، وذلك مؤداه أن العقل هو الجزء الإلهي في الإنسان بل هو دايمون (روح خفي) (Démon) يشكل كل إنسان وبما أنه إلهي فمن صفاته الخلود (أفلاطون 1973، 288).

ولئن استطاع سقراط توجيه دفة البحث الفلسفي من المسائل المادية للعالم الطبيعي إلى إدراك النفس والبحث في طبيعتها، فإن مسار البحث الفلسفي في هذا الشأن لم يرتق إلى مستوى المذهب الفلسفي الشامل إلا على يدي تلميذه أفلاطون الذي أولى البحث في مشكلة النفس أهمية قصوى، على اعتبار أنها «الأساس الجدير لإحياء المادة بمعنى أن يعطيها الحياة» (Durozoi 1987, 15)، فلقد كانت النفس تشكل عند اليونان بعامة مبدأ الحياة، بل تكون دراستها في نظرهم دراسة الحياة وظواهرها، وهكذا الحال عند أفلاطون الذي يميز كأستاذه سقراط بين النفس والجسد، فالنفس في نظره جوهر متميز عن الجسم، «إنها جوهر عقلي متحرك من ذاته على عدد ذي تأليف» (التكريتي 1982، 44).

على خطى أفلاطون، مضى أرسطو إلى القول بوجود طبيعة بشرية في مقابل الطبيعة المادية وأولى مسألة النفس اهتماما كبيرا ووضع دراستها في المرتبة الأولى بالنسبة لسائر المعارف (أبوريان 1976، 118)، على أنه لم يجعل النفس مفارقة بطبيعتها كأفلاطون، بل هي عنده صورة للجسم متحدة به اتحادا جوهريا.

في سياق الحديث عن العدالة الاجتماعية، «جادل أرسطو بأن المعتقدات البشرية بخصوص الصواب والخطأ ـ وهي ما نسميه بحقوق الإنسان ـ كانت ترتكز في نهاية الأمر على الطبيعة البشرية ذاتها، بمعنى أنه بدون فهم الكيفية التي نتواءم بها الرغبات والأهداف والسمات والسلوكيات

الطبيعية في كل بشري مكتمل، لن يمكننا فهم الغايات البشرية، والحكم على الصواب أو الخطأ، أو الطيب والخبيث» (فوكوياما 2006، 62)، وهذه النظرة سيكون لها أيضا تأثيرات كبيرة على المفكرين الاجتماعيين المعاصرين.

وعموما فإن أرسطو قد أنتج، «ومن قبله سقراط وأفلاطون، حوارا حول الطبيعة البشرية استمر في التعاليم الفلسفية الغربية حتى أوائل العصر الحديث عندما ولدت الديمقراطية الليبيرالية» (فوكوياما 25،2006)، أين استمد أفكاره العديد من الفلاسفة المحدثين وعلى رأسهم الفيلسوف ديكارت الذي تمثّل الدرس الفلسفي الأرسطي لكشف خبايا الطبيعة البشرية. إن المهمة التي يضطلع بها ديكارت هي «تغيير نظام الفكر، وهو يدعونا إلى تناول كلمة الفكر بالمعنى الواسع والذي كرسه في كتبه تأملات أو في المبادئ « (Ricœur 1967, 42)، لقد نادى ديكارت «بالذهن كنقطة البدء في طريق المعرفة» (الدر 1983، 20)، مستمدا في ذلك أفكار أرسطو ليعلي من شأن العقل وينظر إلى الإنسان على أنه فكر.

لقد قام ديكارت بنقلة فكرية باقتراض العقل كجوهر للتفكير مستقل عن الجسد (تشومسكي، فوكو 2015، 29، 31)، كما أن قوله «بآلية الجسم كان فتحاً عظيماً إذ أقبل العلماء يدرسون عمليات الجسم الفيسيولوجية ليعرفوا كيف تعمل آلة الجسم، ولقد استفاد الإنسان من هذه الدراسات في طبعه وعلمه» (الدر 1983، 21)، وهكذا أفاد ديكارت الفلاسفة والعلماء على حد سواء في تفسير الطبيعة البشرية، وكان له من رصانة الطرح ما جعل أفكاره ونظرياته تهيمن طويلا.

كان الفيلسوف جون لوك الذي وجد في نفس الفترة الزمنية، من أشد معارضي ديكارت، إذ سرعان ما قاد فتحا عظيما في ميدان التجريب، وأثبت نقدا متاخما للكوجيتو الديكارتي، فإذا كان ديكارت قد ركز على الفكر معتبرا العقل أساسا للمعرفة والذات الإنسانية (أنا أفكر، إذن أنا موجود)، فإن لوك يعتبر التجربة ينبوعا أول للمعرفة بقوله: (علينا أن نختبر ونجرب لنعرف).

يرى جون لوك أن الطفل يولد صفحة بيضاء، والبيئة هي التي تشكل عقله وما يمكن أن يتشكل لديه من آراء، يقول لوك في هذا الصدد: «إن عقل الطفل يكون خاليا تماما من الأفكار، قبل أن يستقبل أي إحساسات من حيث إنها نتيجة مثير لأعضائه الجسمية، فهو أشبه بخزانة فارغة من الأدراج أو صفحة بيضاء لم يطبع عليها شيء» (زكريا 1962، 77)، وهذا يعني أن طبيعتنا البشرية لوح فارغ تخط عليه البيئة ما تشاء، وهو بذلك يختلف عن ديكارت الذي عرف الطبيعة البشرية من خلال فعلي التفكير والتأمل.

يذهب الفيلسوف الإنجليزي دافيد هيوم مذهب جون لوك في التوجه التجريبي، ولقد بحث الطبيعة البشرية انطلاقاً من هذا المنهج، حيث يرى أن «الخبرة الحسية هي المصدر الوحيد لما نعلم» (هيوم 2008، 11)، وهو في ذلك يقول: «إن الملاحظة والانهمام بالسؤال ما المعرفة؟ وكيف تعرف وتنهي النظر إلى الذات بوصفه جوهرا مفردا وتستبدله بالنظر إليه بوصفه مجرد، صبدجيكت (Subject) للمعرفة» (هيوم 2008، 13)، ومعنى ذلك أنه لا يوجد شيء حقيقي غير التجربة الحسية التي يتشكل من خلالها ما يملك من معارف وأفكار، وهو بذلك يرفض اعتماد أي تصورات قبلية مسبقة عن الطبيعة البشرية، ومن ثمة يطرح أفكارا فلسفية جريئة في مقابل المذهب العقلاني القائل بإرجاع مصادر معارفنا عن العالم إلى الأفكار الفطرية ومبادئ العقل القبلية.

إن المسألة الأساسية التي سيعالجها دافيد هيوم هي كالتالي: كيف يصبح الفكر طبيعة بشرية؟ يرى هيوم أن الطبيعة «البشرية هي علم الإنسان الوحيد» (دولوز1999، 17)، كما يبين في كتابه بحث في الطبيعة البشرية: «أن الشكلين الاثنين اللذين يتأثر بهما الفكر، هما قبل كل شيء الانفعالي والاجتماعي، وهذان الشكلان يتضمن أحدهما الآخر» (دولوز1999، 5)، وعلى هذا فقد اعتمد في تعريفه للطبيعة البشرية على المبادئ الأساسية التي توجه النفس مثل: الانفعالات والأحاسيس.

حسب هيوم «يمكن تناول علم الطبيعة البشرية، بمنحيين مختلفين، لكل منهما أهليته الخاصة: في المنحى الأول ينظر إلى الإنسان بوصفه مولودا للفعل أصلا، «(هيوم 2008، 21)، وذلك مؤداه أن الفلاسفة في هذا المنحى إنما ينظرون إلى الإنسان على أنه كائن أخلاقي ويبذلون الجهد الأكبر لتقييم ردات أفعاله وحثه على انتهاج طريق الفضيلة والابتعاد عن الرذيلة، وعلى هذا فإن الهدف الأسمى المرتجى في هذا المستوى هو ضبط وتقويم الأفعال الإنسانية.

أما المنحى الثاني، فينظر الفلاسفة من خلاله إلى «الإنسان بوصفه كائنا عاقلا، ويجتهدون لأن يصلحوا فاهمته أكثر مما يهذبوا عاداته «(هيوم 2008، 22)، وتكمن أصالة نظرية هيوم في كونها تجمع بين الجانبين: الأخلاقي والعقلاني وتؤكد على أهميتهما في فهم وتحليل الطبيعة البشرية. بيد أن هيوم لا يكتفي بهذين الجانبين ويضيف جوانب أخرى، وذلك لأنه يعتقد أن التأثر الأخلاقي والعقلاني هو فقط جزء من الطبيعة البشرية، وقد حاول رصدها وتحليلها من زوايا مختلفة اجتماعية ونفسية وكذا سياسية، وهو ما تفرضه خصيصة الطبيعة البشرية نفسها.

مع أواخر القرن التاسع عشر أخذت الدراسات حول الطبيعة البشرية منعرجا جديدا كان العالم النفسي سيغموند فرويد أحد أعلامه البارزين، بحث فرويد الطبيعة البشرية وفسرها من خلال اللاشعور الذي يعني به «مكمن الرغبات المكبوتة والخبرات الماضية» (كردي 2015، 26)، والواقع أن فرويد قد تابع في طرحه هذا «داروين في مقولته أن غرائز الإنسان هي الامتداد الطبيعي لغرائز الحيوانات السابقة له في الصعود» (العجمي 1983، 18) غير أن فرويد يختلف عن داروين في نظرته حول العقل الباطن، وعموما لاقت فرضيته تطورا كبيرا وشملتها إضافات وتحويرات عديدة وأخذت معاني كثيرة وواسعة، أضفى من خلالها الباحثون على الطبيعة البشرية في علم النفس كما في الفلسفة أبعادا كثيرة كل حسب مذهبه وتوجهه.

وإذا كان فرويد وأنصاره قد ذهبوا إلى القول بأن ما يميز الطبيعة البشرية هو اللاشعور، فإن الماركسيين يقولون إن العمل هو صميم الماهية البشرية، أي أن الكائن البشري يحقق إنسانيته بالعمل. وهذا ما عبر عنه كارل ماركس نفسه في كتابه «الاقتصاد السياسي والفلسفة»، حيث يقول «إن كل التاريخ المزعوم للعالم ليس إلا عملية خلق للإنسان بواسطة العمل البشري» (إبراهيم 1971، 259)، وهو ينفي في هذا السياق - ومن منطلق ماديته التاريخية والجدلية - أي ثبات للطبيعة البشرية، ويرى أنها في تغير مستمر(روز 1990، 91)؛ «بيد أن المادية الجدلية، وإن كانت تأبى أن تنسب إلى الإنسان ماهية ثابتة، إلا أنها لا ترى مانعا من القول بأن الإنسان صنيعة الطبيعة، وأنه إلى - حد ما - أثر من آثار البيئة» (إبراهيم 1971، 256)، ومعنى هذا أن الموجود البشري جزء لا يتجزأ من الصيرورة الكونية، وهو يخضع لقانون التطور الذي ينطبق على سائر الكائنات الحية، وبهذا فإن مفهوم الطبيعة البشرية على نحو ما تصوره الماركسيون هو مفهوم نسبي، ذلك أن البشرية في حد ذاتها ظاهرة متحولة متغيرة تقبل الترقي والتطور.

هذا ويذهب الكثير من المفكرين المعاصرين «للاعتقاد بأن البشر يتمتعون بمرونة غير محدودة تقريبا، أي يمكن لبيئتهم الاجتماعية تشكيلهم بحيث يكون سلوكهم قابلا لكل شيء، وهذا يعني أن الطبيعة البشرية ليست ثابتة، وإنما هي في تغير مستمر لأنها حصيلة لمجموع الخبرات التي يتلقاها الفرد، والتأثيرات التي تحيط به في البيئة الخارجية. وهنا يبدأ «التحيز المعاصر ضد مفهوم الطبيعة البشرية، وكثير من المؤمنين بالتفسير الاجتماعي للسلوك البشري لديهم دوافع خفية قوية، فهم يأملون استخدام الهندسة الاجتماعية لخلق مجتمعات عادلة ومنصفة وفقا لمبدأ أيديولوجي مجرد»

(فوكوياما 2006، 25-26)، ومن ثمة فإن التبريرات التي يسوقونها لتفسير السلوك البشري إنما تهدف إلى تغطية مصالح سياسية خاصة، وهي نتاج لغايات اجتماعية صرفة.

لكن لقد فات هؤلاء أنه بالرغم من «أن السلوك البشري مرن ومتباين، فهو ليس كذلك على نحو لا نهائي» (فوكوياما 2006، 26)، فعند مستوى معين تقوم الغرائز الطبيعية والعناصر المتأصلة في الطبيعة البشرية بإعادة إثبات ذاتها لتؤكد أن جوهر الطبيعة البشرية إنما هو السجية الداخلية الثابتة، ومن ثمة فإن التفاعل الخارجي والاستعداد الطبيعي الأصيل هما معا من أسس الطبيعة البشرية.

وهكذا نجد أن محاولة معرفة الطبيعة البشرية قاسم مشترك بين مختلف المدارس الفلسفية على اختلاف توجهاتها الفكرية، فجميعها درست الطبيعة البشرية، وتعرضت لهذا المفهوم بالتحليل والتفسير، بغية إدراك كنه الطبيعة الداخلية للإنسان. حاولت المدارس الفلسفية الحديثة والمعاصرة تقديم تصور جديد للطبيعة البشرية على نحو أدق يتجاوز القضايا الفلسفية الكلاسيكية كمسائل الوجود والميتافيزيقا، فأغلبها يتساءل عن عناصر تكوين هذه الطبيعة، وأيها الذي يمثل تأثيرا أقوى، وتأتي الإجابات وفق الخلفية الفلسفية التي تتبناها هذه المدرسة الفلسفية أو تلك، فهي عند البعض عناصر فطرية لا تتغير، وعند البعض الآخر مفهوم يحيل إلى التطور أو مفهوم نسبي متغير.. إلخ، ومن ثمة لم تنجح في تقديم تفاسير حاسمة للسلوك الإنساني.

ومنه فإن مفهوم الطبيعة البشرية ذو طبيعة إشكالية (concept problématique)، يتضمن جوانب عديدة نأمل في الإمساك بها، لكن الغموض والتعقيد يشوبها، ومن ثمة تنفتح على عدة قراءات وتأويلات، ذلك أن الفلسفة إنما تنظر إلى الطبيعة البشرية على أنها ظاهرة وليس كمعطى ثابت وهذا يدعم أن الطبيعة البشرية في حد ذاتها متغيرة، وبصفة عامة فإن الفلسفة تساعدنا على إعادة النظر في معاني الإنسان والحياة والموت والمصير والجسد والكينونة.

ثانيا: المنظور البيولوجي للطبيعة البشرية:

لم ينته الجدل الكبير حول الطبيعة البشرية عند حدود المدرسة الفلسفية السائدة في الأروقة الفكرية، وإنما تعداها إلى فضاءات العلم والبيولوجيا التي تشهد تطورا مذهلا لعلوم الفيزيولوجيا والأبحاث البيوطبية، ، وفي هذا السياق نجد تضاربا واسعا بين المفكرين والعلماء حول المضامين

المكتنفة لمفهوم الطبيعة البشرية، وهو جدل أكاديمي من نوع خاص يجمع بين النظرة العلمية الثاقبة والمنهج التجريبي الحيوي، ليؤسس لطرح علمي يسعى إلى فهم العلاقة القائمة بين الفرد البشري والطبيعة وتفسير السلوك البشري، فكيف نظر علماء البيولوجيا إلى مفهوم الطبيعة البشرية؟ وما هي حدود المقاربة العلمية البيولوجية لهذا المفهوم في ظل إنجازات علوم الوراثة والتطورات الحاصلة في علوم الحياة؟ ثم ما هي أهم الإشكاليات التي يطرحها العلماء في هذا الشأن؟

لقد شهد القرن السابع عشر مبعث علم البيولوجيا الذي وجه البحث العلمي نحو دراسة الكائنات الحية بعد أن كانت تقتصر على دراسة المادة الجامدة، وكانت الجهود الأولى المبذولة تهدف إلى القيام بالتجريب على الكائن الحي وفق أسس علمية وتقنية مضبوطة، غير أن أطماع العلماء سرعان ما توسعت نحو البحث عن إمكانية تغيير فيزيولوجيا الطبيعة البشرية.

لقد بقيت النتائج العلمية المحصلة في هذا الإطار محتشمة إلى أن استطاع داروين بنظريته في التطور أن يعمق دراسة الكائنات الحية على أساس مبدأ التنوع، بحيث يرى أن «الصفات الوراثية تنتقل إلى الأبناء ليس بفعل العوامل البيئية فحسب و إنما عن طريق الوراثة أيضا» (حسنين 2004، 110)، وذلك راجع في رأيه إلى «أن كل أشكال الحياة مرتبطة ببعضها البعض» (Watson, Andrew 2003, 14)، وهذا يعني أن الصفات الوراثية الموجودة في الآباء تمتد إلى أبنائهم، ومن ثمة فإن العوامل البيئية ليست العامل الحاسم في التركيبة البشرية كما كان يعتقد سابقا، بل ثمة عناصر وراثية تقف معها جنبا إلى جنب في تحديد سلوكات الإنسان.

إن داروين بطرحه لفكرة الارتقاء يحدث تحولا كبيرا في فهم الطبيعة البشرية، إذ يقصي تمايز الإنسان ورقيه عن سائر الكائنات الأخرى، ويهوي به إلى الدرك الأسفل، عندما يساوي بينه وبين الحيوان، ولقد أدت افتراضاته إلى طرح قضية مهمة دقيقة وخطيرة جدا فيما يتعلق ب «حيونة» الإنسان، من خلال طرح السؤال التالي: «هل الإنسان ينحدر من القرد أوعلى الأقل من أحد الحيوانات القديمة القريبة له؟» (بوكاي، 113)، الأمر الذي أفضى إلى إثارة تصادم كبير بين العلماء ورجال الدين حتى عدت هذه القضية من أهم المسائل الخلافية بين العلم والدين لما لها من علاقة بقضية الخلق.

لقد مهدت أفكار داروين عن أصل الإنسان الطريق أمام علم الوراثة، إلا أن افتراضاته لم تتأكد إلا عندما « قام راهب يسمى جريجور مندل سنة 1867 بسلسلة من التجارب على النباتات والتي فتحت الباب لعصر الوراثة الحديث (Moore 2015, 28)، ولقد أفاد هذا الاكتشاف

الباحثين في فهم عملية انتقال الصفات الوراثية والتعبير عنها، ومن ثمة إيجاد عناصر أساسية لتفسيرات وراثة الكائن الحي.

لم تتأكد أبحاث مندل - على الصعيد العلمي - إلا سنة 1900 عندما أعاد عدد من علماء النبات اكتشاف القوانين العامة ذاتها في ظروف أخرى، والتي بدأ معها علم الوراثة، ولقد كان الدانماركي ولهلم جوهانسن هو من أطلق في 1905 اسم الجينات على عوامل مندل (أوفراي 2012، 15-16)، وبذلك تم معرفة أن الجينات كائنات غير مرئية تتكون منها التركيبة الوراثية، كانت البحوث الأولى تقتصر على الحيوان والنبات « أما الإنسان - الذي يتصف تكاثره بالبطء والاستقلالية والخصوصية - فلم يكن المادة الطيبة للبحث العلمي» (كيفلس، هود 1997، 14)، لكن بمجرد إعادة اكتشاف القوانين المندلية بدأت الجهود تتجه صوب « الطبيعة البشرية، أي تلك الخصائص النمطية للنوع والمشتركة بين جميع البشر بوصفهم بشرا، وهذا هو، في النهاية، ما يتعرض للخطر في خضم ثورة التقنية الحيوية» (فوكوياما 2006، 131).

بدأت التجارب على الجين تتقدم شيئا فشيئا، ومنذ سنة 1944 ظهرت مؤشرات قاطعة تظهر أن الدنا هو الجهاز الجيني الذي يحمل مخطط الحياة العضوية كاملة، لكن بنية الدنا بقيت مع ذلك مبهمة وأثارت العديد من التساؤلات، لم يكن للعلماء معطيات كافية لتفسير كيف يمكن لهذه المعلومة كلها أن تكون مخزنة في سلسلة وحدات كيميائية مرتبطة ببعضها البعض، كما لم تكن هناك إجابات واضحة تبين كيفية انتقال المعلومات الوراثية إلى الأبناء عندما تنقسم الخلايا (Watson 2003, 15-16) ، ومن ثمة بقي السؤال ما هي طبيعة الشفرة الوراثية غامضا ومستعصيا على الإجابة.

لكن الحدث البارز هو بالطبع اكتشاف الأمريكي جيمس واطسن والبريطاني فرانسيس كريك عام 1953 لبنية الدنا (Moore 2015 ,28)، إذ توصلا إلى أن الجينات عبارة عن لولب مزدوج من جديلتين من الحمض النووي الديوسكي ريبوزي (الدنا) الذي يحتوي على الصفات الوراثية، ثم تواصلت جهود العلماء نحو البحث عن تحديد أكثر دقة لهوية الدنا وطبيعة الجين، وهو ما تحقق بالفعل عندما قام دانيال كوهن وعالمه الجينومي المفضل إيليا شيمكوف بتحديد تموضع الجينات ونشر خريطة فيزيائية كاملة للكروموزوم 21، في مجلة الطبيعة (Rabinow 2000, 77)، ومن ثمة تأكد أن «الدنا تحمل المعلومة الجينية لدى كل الكائنات كما أن الشفرة الوراثية كونية»

(Jordan 2007, 31)، أي أن المعلومات الوراثية التي يحملها الدنا تحكم في وراثة كل الكائنات الحية، كما أن الجينات مخزن المعلومات الوراثية.

في خضم هذا الكشف العلمي الهائل نتج جدل[2] علمي كبير حول دور الجينات في تحديد السلوك البشري، تتمحور إشكاليته المركزية حول نصيب كل من الموروث والمكتسب في الخصائص التي تحملها الطبيعة البشرية، وهو يرتكز أساسا في السؤال عن سبب كون البشر على ما هم عليه من صفات وسلوكات؟ أيرجع إلى محددات وراثية بيولوجية أم إلى عوامل بيئية ثقافية؟ وهو جدل لا يزال قائما حتى اليوم بين أنصار الحتمية البيولوجية و الحتمية الاجتماعية.

يرى الحتميون البيولوجيون أن الجينات هي العامل الحاسم في تشكيل السلوك البشري، حيث أن «حيوات البشر وأفعالهم هي نتائج محتومة للخصائص البيوكيمياوية للخلايا التي تكون الفرد، وهذه الخصائص تحددها بدورها على نحو منفرد مكونات الجينات التي يحملها كل فرد وفي النهاية، فإن السلوك البشري ـ وبالتالي المجتمع البشري ـ محكوم بسلسلة من العوامل المحددة تجري من الجينات إلى الفرد حتى مجموع تصرفات كل الأفراد» (روز 1990، 18)، وذلك مؤداه أن الوراثة هي العامل الحاسم في تكوين شخصية الإنسان، مما يعني من منظور بيولوجي محض أن الإنسان يملك طبيعة بيولوجية فطرية ثابتة لا تتغير.

وفي مقابل هذا الطرح «ثمة دعوة نقيضة كثيرا ما تقوم في مواجهة الحتمية البيولوجية، وهي أن البيولوجيا تتوقف عند الميلاد لتحل الثقافة محلها بعد ذلك» (روز 1990، 23)، ويمثل هذا الموقف أنصار الحتمية الاجتماعية الذين يعتقدون أن خصائص الشخصية لا تورث، كما أنها تكون غير موجودة البتة عند الولادة (آدلر 2005، 163)، فهي ليست تعبيرا عن قوى أو ميول وراثية، وإنما يتم اكتسابها بفعل التربية والبيئة الاجتماعية التي ينشأ فيها الفرد.

فالحتميون البيولوجيون يرون إذن أن «الطبيعة البشرية مثبتة بجيناتنا» (روز 1990، 18)، وكل سلوكاتنا قد وضعت شفرتها سلفا في التحديد الجيني، وعلى العكس من ذلك تذهب الحتمية الاجتماعية «إلى النظر إلى الطبيعة البشرية وكأنها تكاد تقبل التشكل إلى حد ما لا نهائي» (آدلر

2 تعود الجذور الأولى لهذا الجدل العلمي إلى بدايات علم الوراثة أين انتقد العديد من علماء البيولوجيا والأحياء ـ ومنهم عالم الأحياء الأمريكي توماس مورغان ـ قوانين مندل، ورفضوا في البدء التسليم بأن انتقال الخصائص الوراثية يتم عبر عامل داخلي جيني، معتقدين أن انتقال الخصائص عبر الأجيال إنما يتم بفعل أثر البيئة المحيطة أنظر: شارل أوفراي،، ص17 وما بعدها.

2005، 23)، من منطلق أن الوراثة لا تلعب أي دور مهم في التأثير على نفسية الفرد وشخصيته، فسلوكاته لا تورث أبدا، وكل ما في الأمر أنها تبدأ في التشكل في مرحلة مبكرة، ولهذا تبدو كما لو كانت مورثة.

وفي السياق نفسه يذهب أصحاب الحتمية الاجتماعية إلى « إنكار البيولوجيا، والاعتراف بالتكوين الاجتماعي فحسب، وهكذا تحول عجز الطفولة ووهن الشيخوخة...إلى مجرد تسميات تعكس التفاوت في القوة» (روز 1990، 23)، بحجة أن ثمة أوجها من السلوك البشري يمكن تفسيرها بمعزل عن الجينات.

والواقع أن كلا الفريقين قد وقع في أخطاء فادحة نتيجة الاحتكام إلى أحد العاملين الوراثي أو الاجتماعي والابتعاد عن الأسباب العلمية، فقد أثبتت اكتشافات البيولوجيا في النصف الأول من القرن العشرين « أن الحياة نتيجة لامتزاج دينامي بين البيئة المحيطة ومكونات الأجسام الحية، ومن بينها الحمض النووي والجينات التي يحملها، وهي تضطلع بدور مهم، ولكن ليس في عزلة عن غيرها من العوامل والمكونات، بل في تفاعل وتعاون معها» (أوفراي 2012، 65)، وهو جزء من سوء فهم عام عن الجينات، وهذا بالضبط هو السبب في وجود مثل هذه الصعوبات العميقة في مفهوم «الطبيعة البشرية.. فهي طبيعة يتم بناؤها بيولوجيا واجتماعيا في الوقت نفسه» (روز 1990، 26)، ذلك أن الإنسان نتاج الوراثة والبيئة معا، ويتضمن ذلك كلا من جيناته والبيئة الفيزيائية والاجتماعية التي يوجد فيها.

ومع التطورات الكبيرة الحاصلة في العلم، وبالضبط «علوم الحياة التي حققت عددا من الاكتشافات المهمة بخصوص الطبيعة البشرية» (فوكوياما 2006، 130)، دلت التجارب أن الجينات، وفضلا عن تحديدها للاستعدادات الوراثية، تلعب أيضا دورا مهما في تحديد الكثير من الصفات النفسية والاجتماعية، وعليه توصل علماء الوراثة إلى حقيقة علمية مؤكدة مفادها «أن الجينات تحمل برامج العمل الفعلية للحياة البشرية» (فوكوياما 2006، 98)، حيث أنها «تحمل مخطط الحياة العضوية كاملة» (Watson 2003, 15)، وذلك يعني أن الجينات هي مفتاح الطبيعة البشرية، وهي مكمن الهوية الحقيقية للإنسان بالمعنى البيولوجي.

توالت الفتوحات العلمية في المجالات التطبيقية الخاصة بعلم الوراثة ، و«أخذ علم الأحياء الحديث في نهاية الأمر يعطي محتوى تجريبيا ذا معنى لمفهوم الطبيعة البشرية» (فوكوياما 2006، 25)، إذ اتجهت الاهتمامات البحثية نحو البحث عن معرفة أدق للطبيعة البشرية، بالاستناد إلى

أسس تجريبية كيميائية حيوية، «وهذا يعني أن مهمة العلماء لا تقتصر على فهم العالم والإنسان، وإنما يهتمون أيضا بتغييره وتطويره» (صبحي 1993، 148)، من خلال فهم عمل جيناته وتفصيلات تراكيبه الوراثية بالأساس، بغية الاستفادة منها عمليا في معرفة الأبعاد العميقة للعضوية البشرية، ودراسة إمكانيات تغييرها وتعديلها نحو الأحسن.

لقد ظهر إذن أن «الموضوع الحقيقي للعلم هو الطبيعة البشرية" (دولوز1999، 27)، وأن مفهوم الطبيعة البشرية عينه مفهوم ذو نظرة معقدة، مستعصية، ومتدرجة قد تطور من المفهوم التقليدي إلى المفهوم البيولوجي أي الخصائص العضوية، وإن كنا نرى أن تحديد هذا المفهوم أمر بالغ الصعوبة، إذ يوجد توتر جدلي كبير بين الفهم التصوري والفهم التجريبي.

وأن تطور العلوم قد أعطى الطبيعة البشرية معنى ذا طبيعة تجريبية، وأنه كلما ازداد ما يخبرنا به العلم عن الطبيعة البشرية، كانت معرفتنا بالبناء العضوي البشري علمية، دقيقة، لكن غير محدودة بدرجة مذهلة. غير أن السؤال الذي يطرح نفسه وبإلحاح بين ثنايا هذا التحليل هو: هل يمكن تعديل الطبيعة البشرية أو تغييرها فعليا بإقامة التجارب على البشر في المخابر أم أن ذلك لا يعدو أن يكون مجرد فرضية علمية؟

ثالثا: سؤال الجينوم والطبيعة البشرية:

غير بعيد عن السجال النظري الذي ساد الأوساط الفلسفية والعلمية حول مفهوم الطبيعة البشرية، أدت التقنية الحيوية المعاصرة إلى الاقتراب كثيرا، وبطريقة جد مبتكرة، من بيولوجيا الإنسان، مما أتاح للعلماء الكشف عن العديد من أسرار الطبيعة البيولوجية البشرية، غير أنها طرحت تساؤلات مهمة وخطيرة حول تداعيات التجريب على البشر، وإمكانية تعديل الجينات الموجودة في الدنا البشري، ومن ثمة القدرة على تغيير الطبيعة البشرية ذاتها، وفي هذا السياق جادل البعض بأننا «لا نمتلك الآن القدرة على تعديل الطبيعة البشرية بأي طريقة ذات شأن، وقد نكتشف أن الجنس البشري لن يتوصل إلى هذه القدرة على الإطلاق» (فوكوياما 2006، 109)، في المقابل أبانت اكتشافات الهندسة الوراثية البشرية والتجارب العلمية الخاصة بمشروع الجينوم البشري عن تطور كبير وغير مسبوق في امتلاك تقنية جد متطورة قادرة على إحداث تحويرات عميقة في صلب الطبيعة البشرية، وبأن إمكانيات تعديل البشر وراثيا باتت أمرا واقعا،

فهل سيحطم العلم الحياة ويفسد الطبيعة؟ هل يمكن لتطبيقات الجينوم أن تطمس هوية الإنسان وتخل بطبيعته البشرية؟ أم ثمة إمكانيات لأن نستثمر في مشروع الجينوم البشري ونزيد من فرص تعديل وتحسين الطبيعة البشرية بصورة واعية بدلا من أغراض الهدم وآثار الترويع؟

أ. أبحاث الجينوم وهندسة الطبيعة البشرية:

بعد اكتشاف جيمس واطسن وفرانسيس كريك لشفرة الحياة الدنا، «صارت الجينات رسميا كائنات مادية تضطلع بمهمة نقل المعلومات الوراثية من جيل إلى جيل» (أوفراي 2012، 26)، أعقب ذلك تجارب واكتشافات عديدة لكن ظلت الشفرة نفسها، أي اللغة التي يعبر بها الجين عن نفسه تحتفظ في عناد بغموضها. كان السؤال الأساسي الذي يراود العلماء في هذا الشأن هو «كيف تسهم الجينات في اشتغال الأجسام الحية عبر التدخل في ظهور الخصائص التي ترتبط في الغالب بالبروتينات من قبيل الإنزيمات والهرمونات؟» (أوفراي 2012، 27)، بقي الغموض يكتنف هذه الجزئية المهمة من البحث إلى أن توصل كريك بمساعدة مجموعة من الباحثين سنة 1960 إلى أنه لا بد من «وجود شفرة تقابل بين الترتيب في سلسلة من ثلاثة النكليوتيد في الحمض النووي، وبين كل من الأحماض الأمينية العشرين التي تكون البروتينات» (أوفراي 2012، 27)، هذه الشفرة الجينية هي جزيء المواءمة RNAأو ما يسمى أيضا رنا الناقل (RNA transfer) (ريدلي 2001، 63) الذي يربط بين عالمي الدنا والبروتينات في نقل المعلومات داخل المادة الحية.

ومع تطور العلوم الطبية والبيولوجيا الجزيئية أصبح الاستكشاف في وراثة الإنسان يقدر من أجل ذاته، فلقد أدرك العلماء مدى الحاجة إلى تطوير علم راسخ يوفر عناصر أساسية للتغيرات التي تخص الكائن البشري على أسس علمية دقيقة، ومن ثمة أنشأت معامل تتعلق بوراثة الإنسان، كانت الكشوف الوراثية تتوالى بسرعة وكانت «تعبر عن قدرة الإنسان الكبيرة على السيطرة على الطبيعة، والطبيعة الإنسانية بوجه خاص» (صبحي 1993، 148)، وهكذا أدت التجربة الوراثية إلى تغير مفاهيم الجين، ليطرح السؤال الجوهري: هل في مقدورنا أن نفهم كل الجينات التي تكون التركيبة البشرية؟

لقد أدت هذه التطورات إلى ثورة، ليس في الهندسة الوراثية فحسب بل ثورة في علم الأحياء، ومع زيادة النظم «التجريبية للكائنات والتقنيات التي يتبعها الوراثيون تغير- خلال ثلاثة أرباع

القرن مفهوم الجين وتعمق، ومعه بالتكامل الخرائط الجينية والتتابعات» (كيفلس، هود 1997، 50) وبذلك طرحت فكرة إنجاز خريطة وراثية للإنسان من خلال مشروع علمي ضخم وكبير جدا هو مشروع الجينوم البشري (Human Genome project).

يعد مشروع المجين البشري جهدا علميا هائلا مولته الولايات المتحدة الأمريكية وحكومات أخرى لفك شفرة المتوالية الكاملة للدنا البشري، وكذا فك شفرة متوالية دنا كائنات أدنى (فوكوياما 2006، 98)، وهو في الأصل يجمع تطور علم الوراثة منذ تحول القرن العشرين بل هو نتيجة هذا التطور، بدأ العمل فيه 1990 للاقتراب من الطبيعة البشرية بشكل دقيق (مصباح 1999، 67)، فهو تطبيق تكنولوجيا علمية للوصول إلى هدف معين هو المحتوى المعلوماتي للجينوم، بالبحث عن الجينات ومعرفة ما تفعله بالكائن الحي بعامة وفي البشر بخاصة، ومن ثمة التعرف على المعلومات الكاملة عن شفرة الجينات البشرية.

لقد فتح هذا الإنجاز العلمي آفاقا واسعا لمعرفة الطبيعة البشرية عن كثب، ذلك أنه سيوفر بالفعل حصيلة من التطبيقات نافعة للغاية من خلال المعالجة الجينية الجسدية وهندسة الخط الجيني، كما طرح إمكانيات تحسين الصفات البشرية في إطار ما يسمي إجينيا((eugénisme (Pastermak 2003, 13)) ، ومكّن العلماء من معرفة دور بنية الجينات في تطور الأمراض الوراثية وكذلك في ظهورها، وكيف تسهم الجينات في تسبب الأمراض النادرة، ومن ثمة سيرفع من نوع الحياة بالتقليل من تفشي العديد من هذه الأمراض المؤلمة للعائلات والمكلفة للمجتمع (كيفلس، هود 1997، 26؛40).

لكن على الرغم من هذه الإيجابيات الكثيرة وغيرها، طرحت أبحاث الجينوم البشري الكثير من العقبات البيولوجية والمشاكل الأخلاقية، كون الجينوم البشري، وفي غضون تطور التقنية الحيوية الجارف، سيهدف إلى التغيير الوراثي للطبيعة البشرية (فوكوياما 2006، 18)، وكون الهندسة الوراثية البشرية الثاوية تحت إطاره ستؤدي إلى عواقب وخيمة من خلال فكرة «تعديل البشري وراثيا» (درويش، العلي 2008، 248)، وفي هذا السياق وسمها العديد من العلماء ورجال الدين بأنها أبحاث منافية عن الكرامة الإنسانية، ولأن «البشر لا يولدوا إلا بشرا (...) حسب القاعدة الشيء يولد الشيء نفسه» (Pastermak 2003, 05)، طرح جدل كبير حول إذا ما كانت أبحاث الجينوم ستغير من طبيعة الكائن البشري جوهريا، وكذا حول إن كانت هناك دواعي دفينة يحملها هذا المشروع للسيطرة على الطبيعة البشرية بسبب الطموح، وما ستؤول إليه

هذه التقنيات؟، وسواء كانت العواقب مقصودة «أو غير مقصودة، والتكاليف غير منظورة، فإن عمق المخاوف التي تغري الناس بخصوص التقنية ليس خوفا نفعيا على الإطلاق، لكنه الخوف من أن تتسبب التقنية الحيوية، في النهاية، في أن نفقد بشريتنا بصورة ما» (فوكوياما 2006، 130-131)، ونفقد معها مقومات وجودنا الإنساني الأخلاقية والدينية. وبذلك كان الهاجس الرئيس الذي يصاحب مشروع الجينوم البشري هو الخوف من أن تناط أبحاثه وتطبيقاته بتغيير الطبيعة البشرية، وأن تستخدم في غير مصلحة الإنسان وحياته الوجودية والقيمية، وهو ما يعني إخراج سؤال المجين البشري من الدائرة الضيقة للعلم إلى الدائرة الأوسع للفلسفة وفلسفة الأخلاق التطبيقية بالخصوص.

فما هي الاحتمالات التي سيحملها هذا المشروع لتعديل السلوك البشري أو التحكم فيه على المستوى الكلي؟ هل ستؤدي تقنياته إلى نتائج دراماتيكية أم ثمة احتمالية لأن تستثمر بصورة فاعلة؟

ب. تطبيقات[3] الجينوم البشري البحثية والطبية وعلاقتها بالطبيعة البشرية:

يوجينيا (تحسين) الطبيعة البشرية:

إن فكرة سلسلة الجينوم البشري هي بأبسط معنى محاولة لتحديد الجينات التي تجعل منا بشرا، وفي هذا الإطار ركزت الجهود العلمية الأولى لمشروع الجينوم على تحسين النسل من خلال البحث جينيا عن كل ما يساهم في تحسين النوع الإنساني، وترجع هذه الفكرة في أصولها إلى الفيلسوف اليوناني أفلاطون(كيفلس، هود 93،14،2007)، أما المفهوم المطور لهذا العلم فهو اليوجينيا (eugenics) وقد صاغه فرانسيس جالتون، ويرتكز على فكرة الاستيلاد المتعمد للناس من أجل صفات وراثية منتقاة.

لقد طرحت اليوجينيا إمكانات رفع الجينات الإيجابية والبحث في الصفات الجيدة وخفض معدل الصفات السلبية لاستيلاد سلالات بشرية أفضل، لكن «الإنسان هو أكثر من جيناته.، بل أكثر كثيرا من أن يكون مجرد شفرة جينية» (ريدلي 2001، 10)، أو عينة تجربة في معامل الوراثة التحسينية، ولهذا أثارت بحوث اليوجينيا «قضية إن كانت الهندسة الوراثية المحسنة للسلالة

3 للجينوم البشري عدة تطبيقات من أهمها تطبيقات الطب الشرعي كالبصمة الوراثية، علم مجين الدواء، تطبيقات الأمراض الوراثية...إلخ، غير أننا سنقتصر في هذا الشق من البحث على التطبيقات التي لها علاقة جد مباشرة بالطبيعة البشرية.

أو المخلة بها قد تصبح يوما ما واسعة الانتشار لدرجة أن تؤثر في الطبيعة البشرية ذاتها» (فوكوياما 2006، 106)، كما وتسببت في قدر «هائل من الخلاف لأنها تحدى مفاهيم راسخة عن المساواة بين البشر وعن القدرة على الاختيار الأخلاقي... وستغير فهمنا للشخصية والهوية البشرية» (فوكوياما 2006، 109)، وأكثر المخاوف شيوعا في هذا الصدد أن تؤدي فكرة تحسين الصفات الجسمية والفكرية للأجيال اللاحقة إلى تصادم اجتماعي وأخلاقي - ديني بين طروحات العلماء الرامية إلى تحسين البشر وراثيا وبين العقائد الدينية المسيحية والإسلامية التي تعتقد أن الله خلق الإنسان فقط على صورته (الحفار 1984، 29)، وهو موضوع جدلي كبير أفرز اعتراضات كبيرة على اليوجينيا من عدة نواحي دينية وفلسفية.

العلاج الجيني للطبيعة البشرية:

يسعى العلماء من خلال علم الجينوم إلى معرفة الجينات التي تسبب الأمراض الوراثية وتحديد هوية زمرة كاملة من الجينات التي تؤثر في نمو الجسم أو عجزه عن أداء وظائفه، ومن ثمة إصلاح الخلل في الجينات أو تطويرها أو استئصالها تحت إطار العلاج الجيني القائم على استخدام الجينات التي يتكون منها الجينوم البشري للإنسان (الخادمي 2013، 286) لمعالجة العديد من الأمراض الشائعة، وكذا الوقاية من الأمراض المحتملة والمتوقعة.

يعتمد هذا الأسلوب الجديد في العلاج والتداوي على «عملية إدخال مورثات سليمة إلى الخلايا لتصحيح عمل المورثات غير الفعالة بغية علاج المرض» (باشا، 67)، وإزالة الاخلال الوراثية الناجمة عنه. وعلى الرغم من أن العلاج الجيني لا يزال في مراحله المبكرة، إلا أنه قدم للبشرية منافع كبيرة نذكر منها الاكتشاف المبكر للأمراض الوراثية وبالتالي تفادي وقوعها من الأصل أو الإسراع في علاجها، ومعرفة التركيب الوراثي بما فيه القابلية لحدوث أمراض معينة كضغط الدم والنوبات القلبية ونحوها (القره داغي 2006، 314)، لكن على الرغم من كل هذه المنافع وغيرها تثير هذه التقنية عقبات بيولوجية وأخلاقية، حيث أنها تعرض حياة الإنسان للمخاطر في بعض الحالات خاصة في تحضير الناقلات الفيروسية (باشا، 73)، ومن المحتمل كذلك أن تؤدي إلى تعريضه إلى نكسات نفسية واجتماعية من خلال قراءة جينومه الخاص والكشف عن بعض الأمراض الوراثية، مما قد يؤثر فعلا على عمله الوظيفي وعلى زواجه وعلى كثير من أموره الخاصة (القره داغي 2006، 315).

وهكذا إذن تثير هذه السلبيات تساؤلات واضحة حول إمكانات اعتماد العلاج الجيني دون آثار جانبية، ومدى مشروعيته بمعزل عن الأضرار، لا سيما أن الإنسان لا يريد أن يحافظ على صحته فقط، بل يطالب بصحة أفضل (super santé)، ولهذا لم نتأكد بعد نجاعة هذا النوع من العلاج الذي لا يزال في أطواره الأولى مثل الكثير من المحاولات العيادية البشرية التي لا تزال في مرحلتها التجريبية.

التعديل الجيني للطبيعة البشرية:

بعد الألق الكبير الذي حققته تطبيقات الجينوم البشري في مجال العلاج الجيني، لم يعد شفاء المرض هو الذي يستأثر بتفكير الإنسان وخياله فحسب، بل أيضا التحكم في شكله ومميزاته، ومن ثمة «ستكون ملامح الإنسان وأوضاعه الذهنية قابلة للتعديل والتغيير، وسيتأثر الذكاء والصفات بالوسائل الكيميائية» (الحفار، 262)، وهو التحدي العلمي الذي فرضه أسلوب التعديل الجيني البشري، والذي ينشأ عنه «تعديل الخلايا في الجنين المبكر، ما قبل الزرع، أو عن طريق إدخال مورث جديد في بويضة الأنثى أو الحيوان المنوي للذكر» (درويش، العلي 2008، 247)، وهو لا يختلف عن العلاج الجيني للخلايا الجسدية إلا من حيث كونه علاجا جينيا لنوع آخر من الخلايا الجنينية والجنسية، وكلاهما يتم بالمستويات التقنية نفسها من إصلاح أو إضافة أو استئصال أو استبدال (أبو جزر 2008، 66).

غير أن التعديل الجيني للخلايا الجنسية لا يقتصر تأثيره على الخلية المعالجة، وإنما يؤثر على الذرية بعد ذلك (أبو جزر 2008، 66)، «بحيث يصبح هذا التغيير المورثي مسجلا في الخلايا الجنسية وهو ما من شأنه أن ينتقل إلى الأجيال اللاحقة.. وهو في هذه الجزئية أكثر فعالية من العلاج بالخلايا الجسدية، لأن تأثيراته لا تقف عند حدود الشخص الذي تم علاجه، بل سيخلص المريض وسائر نسله من هذا العيب الوراثي» (درويش، العلي 2008، 247-248)، الأمر الذي سيؤدي إلى التخلص من الكثير من حالات تشوه المواليد.

لكن على الرغم من كل هذه المزايا أثار التعديل الجيني جدلا كبيرا لجملة من الاعتبارات أهمها أنه يمكن التلاعب جنينيا بالخلايا الجنينية البشرية لإدخال الجينات العلاجية، «لقد تعودنا أن نفكر في الجينات على أنها وصفات، تنتظر سلبية استنساخها حسب هوى الاحتياجات الجماعية للكائن ككل، أي أن الجينات خدم للجسد، أما هنا فنحن نلاقي حقيقة مختلفة، فالجسم

هو الضحية والألعوبة وميدان المعركة ووسيلة النقل لطموحات الجينات» (ريدلي 2001، 129)، وأخطر تهديد تمثله هذه التقنية في هذا الصدد «هو احتمالية تغيير الطبيعة البشرية، وبالتالي تنقلنا إلى مرحلة «ما بعد البشري» من التاريخ» (فوكوياما 2006، 18)، مع ما يترتب على ذلك من انعكاسات أخلاقية واجتماعية، وانزلاقات دينية وقيمية.

التغيير الجيني للطبيعة البشرية:

في نطاق تحسين النسل في إطار هندسة الجينات مكنت أبحاث الجينوم الباحثين من استنساخ فصائل جديدة من النبات ومن «استنساخ حيوانات ثديية من الخلية الجسمية، وإنتاج نسخ طبق الأصل من هذه الحيوانات» (الحفار 1984، 262)، لتتوجه الأطماع العلمية بعد ذلك إلى البحث في إمكانيات معالجة الجينات الموروثة لخلق نسخ جديدة معدلة من الذوات البشرية من أجل تكاثر نوع مختلف وأفضل من البشر.

لقد أتاح الاستنساخ في مجال الحيوان إنتاج العديد من الأدوية ومن أهم ما تم في هذا المجال استنساخ الجين المسؤول عن الأنسولين في جسم الإنسان، كما فتح الاستنساخ في مجال النبات المجال أمام برامج عديدة لزيادة تكاثر أنواع من النبات المعرضة للانقراض (مصطفى 2012، 288، 290)، وعلى ضوء التجارب الناجحة في استنساخ أجنة الحيوان، فإن استنساخ الإنسان لم يعد مستحيلا (القره داغي 2006، 377)، إذ يأمل الباحثون في علم الهندسة الجينية تنشئة أجيال جديدة من البشر، أجناس متفوقة ومختلفة على حد سواء، وعلى الرغم من أن ذلك لم يتم إلى الآن، إلا أن عملية الاستنساخ البشري قد قطعت نطاقا واسعا فيما يمكن أن نسميه تكنولوجيا النسل والجهود تتجه إلى تغيير صفات الكائن البشري تجريبيا من خلال تغييرات فيسيولوجية ووراثية وباستعاضة بعض أجزائه بالآلات (الحفار 1984، 118)، مما يعني نهاية الإنسان كإنسان، وخلق فصيلة جديدة من قبله هو بالذات، إن الإشكاليات التي يثيرها استنساخ البشر عويصة والنزاع على أشده فعلا بين علماء الأحياء وعلماء الدين، لأنه مسخ للجينات وطمس وتغيير نهائي للطبيعة البشرية.

لقد أفضت تطبيقات الجينوم البشري والتطورات التي أفرزها التطور البيولوجي - الطبي في العموم إلى طرح معضلات أخلاقية كثيرة وإشكاليات جديدة تتعلق بالتلقيح الاصطناعي، زرع الأعضاء البشرية، تغيير الجنس، الإجهاض، الموت الرحيم (الأوتانازيا) (ديران 2015، 32)،

كما طرحت مفاهيم وتقنيات جديدة مثل إطالة الحياة، النسالة البشرية، الأطقم الوراثية، أطفال الأنابيب، الأمهات الحاضنات، كراء الأجنة والجراحة التجميلية، وكل هذه المسائل التي هي ليست محل إجماع، عجلت بالمطالبة الأخلاقية لعلوم الحياة والطب بشكل عام ولأبحاث الجينوم بوجه خاص من أجل ترشيد النتائج العلمية وتوجيهها نحو مصلحة الإنسان، واستنهاض مجال حيوي إنساني يضمن سلامة الحياة البشرية.

ج- المساءلة الفلسفية والأخلاقية لأبحاث الجينوم:

لقد قدمت البيوإتيقا مساءلة أخلاقية وسياسية للنتائج التي أفرزها التقدم العلمي في حقل البيولوجيا، فهي «تشير إجمالا إلى التفكير المهيمن منذ عشرين سنة، على مختلف الحقول الفرعية، حول المسائل المطروحة من قبل التقدم البيوطبي» (ديران 2015، 35)، وهي حاليا من أهم الفروع المعول عليها في نقد وتقييم النتائج العلمية والتقنية الحاصلة في علم الوراثة الجينية وعلم الجينوم وانعكاساتها على كرامة الذات الإنسانية والواقع البشري الجديد.

من جهة أخرى قدمت الفلسفة المعاصرة تساؤلات إيتيقية مهمة حول التقدم العلمي المذهل لتطبيقات الجينوم البشري، حاول الفلاسفة من خلالها تهذيب الأبحاث التقنية الجينومية والمساهمة في إيجاد حلول للمشكلات العويصة الطارئة التي أصبحت تهدد حياة الإنسان وتخدش قيمه وكرامته. ومن أبرز الإشكاليات المطروحة في هذا السياق البحث عن أنسب تعريف للطبيعة البشرية بما يتواءم مع مصلحة وقيمة الفرد البشري، وكذا ترشيد وأخلقة هذا الواقع في ظل جموح المتخصصين في علم الجينوم إلى إجراء المزيد من التجارب على الإنسان وتطبيق أحدث التقنيات العلمية.

يعد الفيلسوف الألماني يورغن هابرماس (1929) من أبرز الفلاسفة المعاصرين الذين قدموا جهودا رصينة في هذا المضمار من أجل الوصول إلى حل للمشكلات الأخلاقية الجديدة التي أمست تؤرق الحال البشري، تطرق هابرماس وبالضبط في مؤلفه الشهير مستقبل الطبيعة الإنسانية: نحو نسالة ليبرالية إلى نتائج وإفرازات التطور التقني لثورة الجينوم في مسائل الوراثة الحيوية ونتائجه الوخيمة.

جادل هابرماس بشأن التدخل النسالي ورأى أنه مسألة معقدة ستفضي إلى ابتداع إنسان مشوه هجين (هابرماس 2006، 20)، وهذا ما يجعل البحث الجينومي يمد بتأثيراته إلى عمق

كينونتنا الذاتية -على حد قول هيدجر[4]، ويؤدي بدوره إلى إنزلاقات خطيرة على المستوى القيمي والإتيقي، كما تطرق أيضا إلى مسألة تطبيق التكنولوجيا الجنينية، وإلى الجدل الكبير حول استخدامات الجينوم البشري والمشاكل الأخلاقية التي خلفها لاسيما في مجال الخرطنة الجينية و البرمجة الوراثية (هابرماس 2006، 22)، وهو في ذلك يشدد على ضرورة إعادة النظر في هذا الفضاء العلمي والتقني لاسيما بعد الإخفاق الكبير الذي حازه الدين والمجتمع وتراجعهما الشديد في ترشيد الحياة الإنسانية.

فما هي نظرة الأخلاق الإسلامية للجينوم البشري وتطبيقاته؟ وما هي التحديات المعرفية التي ستطرحها -بالموازاة مع الأخلاقية الفلسفية- في عصر ما بعد الجينوم؟

رابعا: الجينوم والطبيعة البشرية (رؤية إسلامية):

إن التصور الإسلامي المعاصر للطبيعة البشرية «موروث» عن النصوص الشرعية في الكتاب والسنة، وعن المصادر التاريخية للفقه وأصوله، أي تصورات الأخلاق الإسلامية النابعة من رؤى فقهية وأصولية، وعن الأدبيات التراثية التي تربط الروح و الجسد كقاعدة أساسية لكل فهم للإنسان في طبيعته البيولوجية والنفسية أو جعل الروح جوهرا في كل ما يحرك الكائن البشري، بيد أن الغرض ليس دراسة مدى تطابق علم الجينوم مع الأخلاق الإسلامية، بل إيجاد مجال معرفي يجعل هذه الأخلاق نفسها تستثمر في الجينوم وجعله أفقا علميا لتطور أخلاق عملية ذات أبعاد إسلامية حضارية.

ومن ثمة فإن السؤال المطروح في هذا العمق هو: هل سيصمد هذا التصور أمام ثورة العلم وعلم الجينوم بالخصوص، هل سيغير علم الجينوم نظرتنا إلى الطبيعة البشرية فيصبح ما نسميه «تعديلا جينيا» هو من صلب تلك الطبيعة نفسها؟

4 لقد تساءل هيدجر عن ما الشيء الذي أصاب عمق كينونتنا الذاتية، كي يصبح العلم موضوع شغفنا، ليستطرد بأن النشاط العلمي الكاسح في العالم المعاصر ليس إلا طريقة خاصة في التعامل مع الكائن البشري والبحث في مكوناته الجوهرية.

أ. نظرة الإسلام إلى الطبيعة البشرية:

لقد اختلف الفلاسفة والعلماء في تحديد ماهية الطبيعة البشرية لشدة تعقيدها، ولاقتصار كل واحد منهم على جانب من جوانبها، أما الإسلام فقد قدم لنا تصورا شاملا عن الحقيقة الطبيعية للإنسان، يقوم على أساس من التكامل بين كل جوانبها، ونتيجة لذلك يتفرد الكائن البشري بخصائص وصفات تجعله يتمايز عن غيره من الكائنات.

إن الإسلام يرفع بعمق من شأن الطبيعة البشرية، فلقد خلق الله تعالى الإنسان وفضله على سائر المخلوقات بأن نفخ فيه روحا شرفها بأن نسبها إلى نفسه، «وإذ قال ربك للملائكة إني خالق بشرا من صلصال من حمإ مسنون، فإذا سويته ونفخت فيه من روحي فقعوا له ساجدين» (سورة الحجر/ الآية 28، 29)، كما حباه بالعقل وهو البعد الجوهري في طبيعته البشرية، به اختص بأمانة التكليف واستحق الاستخلاف ووهب قابلية التمييز بين الخير والشر، وهو الجوهر الذي يتفرد به عن سائر الكائنات من جماد وحيوان.

يتحدد جوهر الإنسان أيضا بالتكامل بين خصائصه المادية والروحية، فأما الجانب المادي فيقصد به التكوين العضوي «ويتمثل في التراب والماء أو ما يتركب منهما، وهو الطين وقد نتج عن ذلك التكوين البيولوجي للإنسان المشتمل على أجهزته وحواسه وأعضائه وحاجاته الأساسية» (يوسف 2013، 13)، قال تعالى «ومن آياته أن خلقكم من تراب ثم إذا أنتم بشرا تنتشرون» (سورة الروم /الآية 20)، وقال أيضا: «وهو الذي خلق من الماء بشرا فجعله نسبا وصهرا وكان ربك قديرا» (سورة الفرقان/ الآية 54)، في حين أن الجانب الروحي «هو الذي تتحقق وتتسم به وظيفة الإنسان الوجودية، والصحة لا تعتبر كاملة إذا فقدت بعدها الروحي» (يوسف 2013، 13)، فقد شاء الله تعالى أن تكون الروح -التي لا يؤمن بها الماديون- إحدى وسائل ارتفاع الإنسان بالحياة (العجمي 1983، 63).

إن الكرامة عنصر رئيس أيضا في تركيبة الطبيعة البشرية، قال تعالى: «ولقد كرمنا بني آدم وحملناهم في البر والبحر» (سورة الإسراء/الآية 70)، وإنما تشريف وتكريم الله للبشر في خلقهم على أحسن هيئة وأكملها (ابن كثير 2002: 5)، قال تعالى: «يا أيها الإنسان ما غرك بربك الكريم الذي خلقك فسواك فعدلك، في أي صورة ما شاء ركبك» (سورة الانفطار/الآية 08).

وإذا كانت الاتجاهات العلمية والفلسفية قد شهدت نشاطا هائلا من الجدل والمناقشة حول تأثير البيئة والوراثة في الطبيعة البشرية، وأيهما أقوى في تحديد السلوك البشري، فإن الإسلام يقر بأهمية وتأثير كلا العاملين كمحددات أساسية لسلوك الإنسان وتشكيل شخصيته، وفي هذا الإطار

يمكن إيراد الآية الكريمة «هو الذي خلق من الماء بشرا فجعله صهرا ونسبا وكان ربك قديرا» (سورة الفرقان/ الآية 54) وقد جاء في تفسير القرطبي بأن المقصود بالماء في هذه الآية الكريمة هو النطف، وقد أثبت العلم الحديث أن النطفة تحمل الخصائص الوراثية التي يتأثر بها الإنسان في مختلف مراحل عمره، كما جاء في التفسير المقصود بالنسب جهة الأب، والمقصود بالمصاهرة جهة الأم، ومن ثمة فإن الآية الكريمة تجمع بين الإشارة إلى الجانبين البيولوجي (الوراثي) والاجتماعي (المصاهرة).

فإلى جانب التأثير المهم للوراثة، فإن الإسلام يولي أهمية كبيرة لتأثير البيئة في تشكيل الطبيعة البشرية، وفي هذا الصدد يقول الرسول (ص): «ما من مولود إلا يولد على الفطرة، فأبواه يهودانه، أو ينصرانه أو يمجسانه»، بيد أن هذا لا يعني بأي حال من الأحوال أن الإسلام يميل إلى الحتمية الاجتماعية، حيث أن للوراثة دوراً فاعلاً لا ينكره الإسلام، كما للبيئة والاكتساب، دورهما المهم أيضا، فللفطرة دورها المؤثر، غير أنها قابلة للتعديل والتغيير بالإيمان والعمل، يقول الله تعالى: «إن الله لا يغير ما بقوم حتى يغيروا ما بأنفسهم» (سورة الرعد/ الآية11).

ولما كانت الطبيعة البشرية حاملة لقيم الكرامة فطرة وجبلة، ومخلوقة على أحسن تقويم وأعدله، أمرنا الله بصونها وحفظها والإبقاء عليها عل نحو ما هي عليه من دون تغيير أو تحويل عما فطر الله «لا تبديل لخلق الله» (سورة الروم/الآية 30)، فالأولى إذن أن يرتفع الإنسان ويسمو عن كل ما يعيب طبيعته أو يفسد خلقته بالمحافظة على حياته وسلامة جسمه، خصوصا في ظل التطور العلمي الحديث.

لكن ما مدى إطلاقية هذا التصور في خضم ثورة الجينوم؟

ب. المنظور الفقهي والأخلاقي للطبيعة البشرية في ضوء تطبيقات الجينوم البشري:

إن التطور الهائل الذي يشهده علم الجينوم اليوم يطرح تحديات خطيرة حول تغيير الجينوم البشري، ومن ثمة إمكانية «تغير وتحول الطبيعة البشرية»، وهي إشكاليات مركزية تتجاوز الأبعاد العلمية إلى البعد الأخلاقي والأنطولوجي، فمجال تطبيقات الجينوم قد أفرز نتائج خطيرة وهي تتناقض مع الأغراض الإنسانية والقيمية المعلن عنها في البداية، وتأسيسا على ذلك سنتطرق فيما يلي إلى انعكاسات تطبيقات هذا المشروع على الطبيعة البشرية وفق منظور الأخلاق الإسلامية.

لقد أصبحت الطبيعة البشرية موضوع رهان بين العلم وتقاناته الحديثة، وأضحى التحول يتهددها، سيما بعد أن بلغت التقنية شأواً عظيماً في عصر الجينوم وتطبيقاته، حيث أن ثورة الجينوم تطرح الكثير من الإشكاليات المعقدة التي تمتد بتأثيراتها إلى عمق المسائل الدينية والأخلاقية للمجتمع البشري.

إن الخريطة الجينية اليوم يمكنها أن تساعدنا في علم اليوجينيا على توقع تلك الاختلالات البيولوجية، بما يمكن من «تحسين» تلك الطبيعة وتعديلها بمواصفات جينية أفضل، فالهندسة الوراثية قد قطعت وعدا بتصنيع جيل حسب الطلب «يكون خاليا من العيوب الوراثية، وسيحاط المشتري مقدما بلون عيني الطفل، وشعره، وجنسه وبالمعلومات الخاصة عن احتمالات حجمه ونضجه ودرجة ذكائه...» (الحفار 1984، 114)، في هذا الإطار تأتي قرارات مجمع الفقه الإسلامي لتؤكد عدم «جواز استخدام أي من أدوات علم الهندسة الوراثية ووسائله للعبث بشخصية الإنسان، ومسؤوليته الفردية أو التدخل في بنية المورثات (الجينات) بدعوى تحسين السلالة البشرية» (القره داغي 2006، 325)، لأن «في ذلك تغيير لخلق الله بالتدخل في التركيب الوراثي للإنسان» (الألفي 2012، 24)، فهذا الأمر من المنظور الفقهي مرفوض شرعا لما فيه من مخاطر تتعلق «بالكرامة والحقوق الإنسانية والفضائل الأخلاقية وذلك من خلال التلاعب بالرصيد الوراثي للإنسان وتحويل البشر إلى آلات وأجهزة يتلاعب بها حسب الأهواء والرغبات» (الخادمي 2013، 288)، ولما فيه من أضرار ومفاسد «قد تؤدي إلى تغير البنية الأساسية للبشر بشكل دائم إذا ما استخدمت مورثات من أعراق أخرى» (باشا، 72)، حيث أن التلاعب بشفرة الجينوم يمكن أن يوجد نسلا غامض الهوية ضائع النسب.

لكن ألا يمكن من وجه ما اعتبار تفصيل اليوجينيا لأطفال ذوي بنية قوية وطول فارع عَودٌ إلى الطبيعة البشرية في جبلتها الأولى؟ فقد ورد في الحديث النبوي الشريف قوله صلى الله عليه وسلم: «إن الله خلق آدم على صورته، طوله ستون ذراعا» (البخاري/6227)، كما ورد في القرآن الكريم أن قوم عاد كانوا أطول من الأجيال التي سبقتهم يقول الله تعالى: «أوعجبتم أن جاءكم ذكر من ربكم على رجل منكم لينذركم واذكروا إذ جعلكم خلفاء من بعد قوم نوح وزادكم في الخلق بسطة فاذكروا آلاء الله لعلكم تفلحون» (سورة الأعراف/الآية69) كما كان طالوت أطول وأضخم من جيله زمن سيدنا داوود عليه السلام فالله يقول عنه: «إن الله اصطفاه عليكم وزاده بسطة في العلم والجسم» (سورة البقرة/ الآية 24)، مما يدل على أن البشر في فترات معينة، عبر التاريخ،

كانت قاماتهم أطول مما هي عليه الآن، وأجسامهم أكبر حجما بما يوحي بأن تغيرات بيولوجية هائلة قد حدثت على الجنس البشري، مما يعني كذلك أن الجينات البشرية نفسها تكون قد «تلاعبت» بالطبيعة البشرية وغيرت مسارها نحو جهات بيولوجية غير متوقعة، وقد ذكر الله تعالى أن هذا الاختلاف في الألوان البشرية هو من آياته الكونية فقال سبحانه: «ومن آياته خلق السماوات والأرض واختلاف ألسنتكم وألوانكم إن في ذلك لآيات للعالمين» (سورة الروم/الآية22).

ثم ماذا لو أدى تحسين البشر وراثيا إلى تحسين سلوكهم اجتماعيا وأخلاقيا، بتنحية مثلا جينات (العنف والكراهية)، وتعزيز جينات (السلم والمحبة)، أتصور أن ذلك سيؤدي إلى تقليل العدوانية في مجتمعاتنا العربية المسلمة، والأمر نفسه ينبغي أن ينسحب على السلوكات الأخلاقية وغيرها، إن علم الجينوم يقدم أجوبة عن الأسئلة المتعلقة بفهم الطبيعة البشرية وتوقع شخصية الإنسان وفهم سلوكه المستقبلي في صورة المعلومات المخزنة في الجينات المرتبطة بتلك السلوكات وغيرها، وعلينا أن نستثمر ذلك بما يعود بالنفع على مجتمعاتنا وديننا الحنيف وعلى الإنسانية جمعاء.

من جهة أخرى أجاز الفقهاء العلاج الجيني إذا لم يترتب عليه الإضرار والمفاسد من تخليط للجينات ودمجها مع بعضها لتغيير الصفات الوراثية الخلقية ومن احتمالية أن يؤدي تصحيح المورثات المصابة إلى إحداث تكوين طفرات وراثية تمتد آثارها إلى الأجيال القادمة، ووضعوا له شروطا وأحكاما خاصة بحسب كل نوع من هذا العلاج، وتستثنى منه الجراحات التي تخرج بالجسم أو العضو عن خلقته السوية (القره داغي 2006، 326) من عمليات تغيير الجنس، والجراحات التجميلية كالوشم وتغيير اللون، فهي ممنوعة لأن في ذلك تغييرا لخلق الله تبارك وتعالى الذي يحرص الشيطان على إيقاع الناس فيه قال تعالى «ولآمرنهم فليغيرن خلق الله» (سورة النساء/الآية119). وروى البخاري (4507) عن عبد الله بن مسعود رضي الله عنه قال: «لعن الله الواشمات والموتشمات، والمتنمصات والمتفلجات للحسن، المغيرات خلق الله».

كما أجاز أكثر الفقهاء والعلماء العلاج الجيني المتعلق بنقل الجين إلى الخلية الجسدية، «لأنه يعيد العضو إلى أصل خلقته القويمة التي خلقه الله عليها»، فما كان لإزالة تشوه أو عيب ناتج عن حادث أو مرض فهو جائز، لما روى أبو داوود (4232) عن عبد الرحمن بن طرفة «أن جده عرفجة بن أسعد قطع أنفه يوم الكُلَاب، فأتخذ أنف أنفا من ورق (فضة) فأنتن عليه، فأمره النبي صلى الله عليه وسلم فاتخذ أنفا من ذهب»، مما يعني أن علم الجينوم لا يعدل الطبيعة البشرية

ولا يغيرها بل يعيدها إلى طبيعتها، إذا ما فهمنا أن تلك الطبيعة «مخلوقة» على «أحسن تقويم» ، «لقد خلقنا الإنسان في أحسن تقويم» (سورة التين/الآية 04).

أما تعديل الطبيعة البشرية جينياً فأكثر أهل العلم يرون عدم جوازه «للحصول على صفات أحسن لأن في ذلك عبثا بمكونات الإنسان الوراثية وفقا لشهوات الناس وأهوائهم دون حاجة إلى علاج أو وقاية من الأمراض أو غرض صحيح شرعا» (الألفي 2012، 24)، ولما يمكن أن يترتب عليه من عواقب وخيمة وأحد هذه المخاوف هو أن هناك احتمال ضئيل بوصول الدنا إلى الخلايا التناسلية مما يؤدي إلى إحداث تغييرات قابلة للتوريث (باشا، 73)، وبالتالي «لا يؤثر فقط على المادة الوراثية الشخصية، بل على المخزون الوراثي لذريته أيضا، ومن ثمة على مجموع الصفات الوراثية للبشرية جمعاء» (درويش، العلي 2008، 249)، ومن أخطر الأضرار في ذلك ضياع الحصيلة الإرثية التي تتضمن الصفات العامة التي يشترك فيها سائر المجتمع البشري، وتجب المحافظة عليها كما هي دون تعريضها لأي تعديل سواء كان تعديلا متعمدا أو عشوائيا (درويش، العلي 2008، 249)، وقد كانت مسألة حق الكائن البشري في أن يولد بالإرث الوراثي الخاص به دون تعديل من المسائل القوية التي قادت نقاشا كبيرا بين الفلاسفة والهيئة الدينية حول مسألة التدخل في الطبيعة البشرية، على أن السؤال الرئيس هنا هو من له الحق في تعديل أو تغيير مجين طفل لم يولد بعد؟ ومن له حق الموافقة؟

ولما كان التعديل الجيني، بل وأي صورة من صور التدخل الوراثي المباشر، يؤثر في طبيعتنا البشرية الحيوية (درويش، العلي 2008، 250)، ظهرت في هذا السياق مفاسد أخرى تتعلق بالتدخل في خلق وتغيير خليقته من حيث الطول والقصر ونحو ذلك، وهو ما أنتج سجالات كثيرة تتعلق بصعوبة الفصل بين ما هو عادي طبيعي فطري بذاته لا يتعدى حدود التدخل في خلق الله، وبين ما هو محور ومعدل وراثيا وبالتالي غير عادي يدخل في باب تغيير خلق الله.

وهو ما لم تفصل فيه الأهداف اليوجينية فيما يتعلق ببحوث الجينوم، حيث أنها لم تسفر بعد عن تحديد دقيق للتراكيب الوراثية لأفراد من البشر بين ما هو الطبيعي فيها وما هو غير الطبيعي، المقبول منها وغير المقبول (كيفلس، هود 1997، 42) ومن ثمة تعرضت برامجها إلى الرفض والتقييد.

كما أن احتمال استخدامها في تعديل أو تعزيز القدرات البشرية قد يفضي إلى إشكاليات معقدة حول حدود استخدام هذا العلاج في تصميم أشخاص لهم قدرات بدنية وعقلية مبرمجة كالذكاء

والقدرات الرياضية لدى العباقرة والعدائين، مما يؤدي إلى تفاوت غير طبيعي بين الذوات البشرية في التركيبة العضوية والخصائص البيولوجية وكذا في القدرات الذهنية والسيكولوجية، وبالتالي ظهور نوع جديد من الطبقية يصبح الامتياز فيه موفرا للغني وصاحب النفوذ فحسب.

ولكن السؤال الآن: هل يمكن وضع حد فاصل بين ما هو علاج وما هو تعديل ومن ثمة قبول الأول ورفض الآخر؟

ينبغي أن نعترف بأن التمييز بين العلاج الجيني والتعديل الجيني للطبيعة البشرية مهمة صعبة، فهذا المعيار سيتغير تبعا للتقدم العلمي والتقاني، فما نراه الآن تعديلا قد يعتبر غدا علاجا وما هو تعديل في مجتمع ما قد يكون بالنسبة إلى مجتمع آخر علاجا ومن هنا يمكن إدراج هذه الظاهرة ضمن الشواهد على التعددية الثقافية أو التنوع الثقافي (الألفي 2012، 25)، ثم ما المانع إذا تمكنا من جعل أطفالنا أكثر صحة وعافية أو موهوبين بشكل أفضل؟ وعلى هذا أجاز بعض الفقهاء التعديل الجيني «لأن تحصيل الصفات الحسنة من الأمور المحمودة شرعا، ولا مانع من طلبها بالطرق المباحة، فالمؤمن القوي خير وأحب إلى الله من المؤمن الضعيف والله تعالى جميل يحب الجمال» (درويش، العلي 2008، 257)، وقد وهب الطبيعة البشرية هذه القوة والمسحة الجمالية، وهذا ما يدفعنا إلى إعادة النظر مجددا في مسألة التطابق مع ما نفترض أنه تعديل طبيعي للنوع البشري. إن الرهان المعرفي هنا يكون في ربط علم الجينوم بما يمكن للفلسفة أن توضحه عن الطبيعة البشرية بدءا بالفصل بين ثبات النموذج البشري عبر التاريخ، والذي يمكن لعلوم أخرى كالأنثروبولوجيا أن تحلله (أعمال كلود ليفي شتراوس مثلا)، حيث أن بنية العقل البشري ثابتة منذ ما قبل التاريخ، وبين ما يذكره التاريخ الإسلامي من تحول للطبيعة البشرية.

أما بالنسبة للاستنساخ البشري، فتذهب الأحكام الفقهية إلى تحريمه لأنه «تغيير» لخلق الله يحمل مخاطر ذاتية للإنسان من حيث تمييع ذاته وتغيير هويته (القره داغي 2006، 381، 390)، حيث أن النسخ البشرية المستحدثة قد تحوي عيوبا مطمورة لا تظهر إلا بعد أجيال كما أن النسخ المنتقاة لاستيلاد السلالات قد تحمل مورثات العنف والإجرام، ولما فيه من أخطار على الأخلاق والإنسانية، حيث أنه يفضي إلى تغيير العلاقة القديمة بين الرجل والمرأة ومن ثمة تختلط الأنساب وتضيع القيم التي قامت عليها البشرية، كما أن نقل الجينات معمليا يؤدي إلى تدمير الروابط العائلية المقدسة بين الأبناء وآبائهم (خطاب (د.ت)، 73)، والتعدي على كرامة الإنسان من حيث بيع الأجنة وإنشاء بنوك خاصة لتخزين الحيوانات المنوية، إضافة إلى أنه يفتح

آمالا لتخليق كائنات خرافية وتصنيع عضويات جرثومية ويؤدي إلى تغيير الأطقم الوراثية، ومن ثمة هدم التنوع الإنساني الذي أراده الله سنة إلهية.

وهكذا إذن نجد أن علم الجينوم يتوافق في عمومه مع منظور الإسلام إلى الطبيعة البشرية من حيث أن ما يحدثه من تحسين وتعديل هو من صلب تلك الطبيعة نفسها، حيث أن الأخلاق الإسلامية تتعامل بقدر كبير من المرونة والانفتاح مع القضايا العلمية والمشاكل التقنية والأخلاقية التي يفرزها هذا العلم من خلال الأبعاد الإنسانية والأخلاقية التي تحراها الفتاوى الفقهية، بخلق توازن بين الأضرار والفوائد المحتملة من التقنيات المتوفرة لدينا، وهي إنما تسدد الاعوجاج وتصلح الخلل الطارئ وتعيد الفرع إلى أصله بالنظر الفقهي القويم.

ومن ثمة وجب إحاطة علم الجينوم البشري بالضوابط الشرعية الإسلامية التي تنص على منع الضرر ابتداء وعدم رفع الضرر بمثله أو أشد وتقديم درء المفاسد على المصالح وغير ذلك من القواعد الشرعية التي تحكم باب العلاج والتداوي والتدخل في البدني الإنساني جملة، وتنزيل قواعد الشرع العامة ومقاصده وكلياته بحسب مجالاته واستخداماته ومآله، فينظر في هذا كله لتنزل القواعد على الوقائع (الألفي 2012، 25)، مع «رعاية المآلات والغايات والنتائج والآثار المترتبة» (الخادمي 2013، 290)، والموازنة بين المصالح والمفاسد والضرورات والكماليات لتحدد منافعه وأضراره (القره داغي 2006، 323) من حيث اتصاله بكرامة الإنسان وبقيم الدين وثوابته، وبأمن الدول والشعوب لمنع البشرية من السقوط على مستوى أخلاقها (النشمي 2013، 174)، لكن هل يمكن للأخلاق الإسلامية أن تجيب عن كل الإشكاليات المتشعبة والمفصلية التي يطرحها هذا العلم ومن ثم فتح أفق جديد مبتكر فيما يخص معنى الإنسان؟

خاتمة:

إن الطبيعة البشرية ذات قيمة وجدارة غير أنها تواجه في سياقات التطبيقات التقنية لعلم الجينوم قضايا أخلاقية جديدة لم تكن مطروحة من قبل على منظومة القيم والأخلاق الإسلامية، ومنها الجدل حول ما يظل من الحق الطبيعي في الحرية لمن تكون حياته ومستقبله محددة جينيا سلفا؟ هل هو مخير أم مسير؟ ووفقا لأي قدر وراثي محتوم؟ وتعمد إنجاب اليتامى بالإنجاب الاصطناعي

في حالة تخصيب زوجة بمني زوجها بعد وفاته وما يترتب عن ذلك من قضايا مستجدة في تقسيم الميراث بين الأم والأبناء، وتغيير جوهر الطبيعة البشرية بتغيير الجنس واستبدال الأدوار الطبيعية.

كما تطرح مسائل الاستنساخ وإطالة الحياة والموت الرحيم وغيرها المزيد من الأسئلة الأخلاقية حول ما يمكن أن نسميه شذوذا في الطبيعة وتصرفا في الجسد، مما يعني ضرورة إعادة النظر في عوائد ومفاهيم درج عليها البشر لآلاف السنين مثل مفهوم الإنسان، الحياة، الموت، مفهوم العائلة نفسه في هذا النوع الجديد من التشكل البشري، بل ينبغي إعادة النظر في مفهوم الطبيعة البشرية عينه الممزق بين نزعة العلم الجامحة إلى الاستزادة والنزعة الشمولية لمبدأ الإتيقا، وهنا توفر الأخلاق الإسلامية قدرا كبيرا من الحلول المثلى لضبط هذا العلم وترشيده وتهذيبه وتنقيحه مما علق به من الأضرار والمفاسد، كما تضمن الشريعة الإسلامية في هذا الإطار الحماية الشرعية للجسم البشري مما يضمن له الحفظ والكرامة الآدمية، غير أن ثمة قضايا تتجاوز النظر الفقهي والمقاصدي ومن ثمة وجب ربط علم الجينوم بالقراءات المعاصرة للتراث الإسلامي واستثمار علومه المختلفة كالأدب والتاريخ والفلسفة، والاستعانة بالتخصصات المعرفية الأخرى.

إن الجينوم البشري في الواقع تراث مشترك للإنسانية، وترشيد أبحاثه وتقنياته يتطلب توافقا عادلا بين جميع أعضاء الأسرة البشرية لحفظ كرامة البشر الكاملة وتنوعهم، وأخلقة هذا الحقل العلمي تقتضي تظافر مختلف الميادين المعرفية كالبيوإيتيقا، الفلسفة فضلا عن الدور الكبير الذي ينبغي أن يؤديه رجال الدين وعلماء الاجتماع والنفس والأخلاق الدينية، الأخلاق الإسلامية بالخصوص للبث في إشكالياته وقضاياه بدلا من فرادة وجهات النظر.

إن الجينوم البشري يشكل حقيقة معرفية حضارية بشقها المادي والروحي والأخلاقي والإنساني وهو يشكل تحديا معاصرا ومطلبا عاليا من حيث استثماره والنفع به ومن ثمة وجب على المجتمعات العربية الإسلامية أن تسارع للانخراط في توجهاته الكبرى، على أن يحتكم مجال العمل إلى المرجعية والأخلاق الإسلامية، في سياق تتماشى فيه قوة الكينونة الإنسانية مع مستجدات ونوازل التكنولوجيا العلمية حتى يكون القادم العلمي أفضل للبشرية جمعاء.

المراجع:

إبراهيم، زكريا. 1971. الطبيعة البشرية في فلسفة كارل ماركس. ضمن مجلة عالم الفكر 2(1):256-259.

أبو جزر، إبتهال محمد رمضان. 2008. العلاج الجيني للخلايا البشرية (في الفقه الإسلامي). إشراف مازن إسماعيل هنية. غزة: الجامعة الإسلامية.

أبوريان، محمد علي. 1976. تاريخ الفكر الفلسفي. أرسطو والمدارس المتأخرة. لبنان: دار النهضة العربية للطباعة والنشر.

الدر، إبراهيم فريد. 1983. الأسس البيولوجية لسلوك الإنسان. بيروت: منشورات دار الآفاق الجديدة.

أفلاطون. 1973. فيدون (في خلود النفس)، تر: عزت قرني. مصر: دار النهضة العربية.

الألفي، محمد جبر. 2012. الوراثة والهندسة الوراثية والجينوم البشري الجيني من منظور إسلامي. جدة: منظمة المؤتمر الإسلامي.

أوسموس وآخرون. 1979. موجز تاريخ الفلسفة، تر: إبراهيم سلوم. ط3. موسكو: دار الفكر.

أوفراي، شارل. 2012. ما الجينات؟، تر: عبد الهادي الإدريسي. أبو ظبي: هيئة أبو ظبي للسياحة والثقافة.

باشا، حسان شمسي.2013. الهندسة الوراثية والبصمة الوراثية مفهومها وتطبيقاتها. ضمن بحوث وتوصيات الندوة العلمية حول الوراثة، والهندسة الوراثية والجينوم البشري من منظور إسلامي. الرياض: جامعة الإمام محمد بن سعود الإسلامية.

بوكاي، موريس. (د.ت). أصل الإنسان بين العلم والكتب السماوية، تر: فوزي شعبان. بيروت: المكتبة العلمية.

تشومسكي، نعوم، وميشيل فوكو. 2015. عن الطبيعة الإنسانية. تر: أمير زكي. مصر، تونس، لبنان: دار التنوير للطباعة والنشر.

التكريتي، الناجي. 1982. الفلسفة الأخلاقية الأفلاطونية عند مفكري الإسلام. بيروت: دار الأندلس.

حسنين، كريم. 2004. الخلق بين العنكبوتية الداروينية والحقيقة القرآنية. مصر: نهضة مصر للطباعة والنشر والتوزيع.

الحفار، سعيد محمد. 1984. البيولوجيا ومصير الإنسان. الكويت: المجلس الوطني للثقافة والفنون والآداب.

الخادمي، نور الدين. 2013. الجينوم البشري وضوابطه في الشرع الإسلامي. ضمن بحوث وتوصيات الندوة العلمية حول الوراثة والهندسة الوراثية والجينوم البشري من منظور إسلامي.

خطاب، عبد المعز. (د.ت)، الاستنساخ البشري هل هو ضد المشيئة الإلهية؟ القاهرة: الدار الذهبية.

درويش، بهاء، والعلي خالد. 2008. مشروعية وحدود العلاج الوراثي (المورثي). ضمن أخلاقيات التعامل مع التقانات الحديثة. تونس: المنظمة العربية للتربية والثقافة والعلوم.

دولوز، جيل. 1999. التجريبية والذاتية بحث في الطبيعة البشرية وفقا لهيوم. تر: أسامة الحاج. بيروت: المؤسسة الجامعية للدراسات والنشر والتوزيع.

ديران، غي. 2015. البيواتيقا الطبيعة، المبادئ، الرهانات، تر: محمد جديدي، بيروت: جداول للنشر.

روز، ستيفن وآخرون. 1990. علم الأحياء والأيديولوجيا والطبيعة البشرية. تر: مصطفى إبراهيم فهمي. الكويت: المجلس الوطني للثقافة والفنون والآداب.

ريدلي، مات. 2001. الجينوم والسيرة الذاتية للنوع البشري. تر: مصطفى إبراهيم فهمي. الكويت: المجلس الوطني للثقافة والفنون والآداب.

زكريا، فؤاد. 1962. نظرية المعرفة والموقف الطبيعي. القاهرة: مكتبة النهضة المصرية.

صبحي، أحمد محمود، وزيدان محمود فهمي. 1993. في فلسفة الطب. تقديم محمود مرسي عبد الله. بيروت: دار النهضة العربية للطباعة والنشر.

صحيح البخاري. 2002. بيروت: دار ابن كثير.

العجمي، أبو اليزيد. 1983.حقيقة الإنسان بين القرآن وتصور العلوم. الرياض: رابطة العالم الإسلامي.

فوكوياما، فرانسيس. 2006. مستقبلنا ما بعد البشري عواقب ثورة التقنية الحيوية. ترجمة إيهاب عبد الرحيم سعد. أبو ظبي: مركز الإمارات للدراسات والبحوث الاستراتيجية.

القره داغي، علي محي الدين، وعلي يوسف المحمدي. 2006. فقه القضايا الطبية المعاصرة، دراسة فقهية طبية مقارنة. بيروت: دار البشائر الإسلامية.

كردي، فوز بنت عبد اللطيف كامل. 2015. المؤثرات الغيبية في النفس الإنسانية بين الدين والفلسفة. الرياض: مركز التأصيل للدراسات والبحوث.

كيفلس، دانييل، ووليروي هود. 1997. الشفرة الوراثية للإنسان القضايا العلمية والاجتماعية لمشروع الجينوم البشري. ترجمة أحمد مستجير. الكويت: المجلس الوطني للثقافة والفنون والآداب.

مصباح، عبد الهادي. 1999. العلاج الجيني واستنساخ الأعضاء البشرية. بيروت: الدار المصرية اللبنانية.

مصطفى، إيمان مختار. 2012. الخلايا الجذعية وأثرها على الأعمال الطبية والجراحية من منظور إسلامي دراسة فقهية مقارنة. الإسكندرية: مكتبة الوفاء القانونية.

النشمي، عجيل جاسم. 2013. الوصف الشرعي للجينوم البشري والعلاج الجيني. ضمن بحوث الندوة العلمية حول الوراثة والهندسة الوراثية والجينوم البشري من منظور إسلامي.

هابرماس، يورغن. 2006. مستقبل الطبيعة الإنسانية نحو نسالة ليبيرالية. تر: جورج كتورة. بيروت: المكتبة الشرقية.

هيوم، دفيد. 2008. مبحث في الفاهمة البشرية. تر: موسى وهبة. بيروت: دار الفارابي.

يوسف، بوشي. 2013. الجسم البشري وأثر التطور الطبي على نطاق حمايته جنائيا. رسالة دكتوراه، جامعة أبو بكر بلقايد.

Moore, David. 2015. *The Developing Genome: An Introduction to Behavior Epigenetics.* New York: Oxford university press.

Bertand, Jordan. 2007. *Thérapie Génique: Espoir ou Illusion?.* Paris: éd. Odile Jacob.

Durozoi, Gerrard, et André Roussel. 1987. *Dictionnaire de Philosophie.* Paris: Nathin.

Watson, James, and Andrew Berry. 2003. *ADN: Le secret de la Vie. Trad. Barbara Hochstedlt.* Paris: Odile Jacob.

Pastermak, Jack. 2003. *Génétique Moléculaire Humaine.* Trad. Dominique charmot, Paris : éd. Deboeck.

Rabinow, Paul. 2000. *Le Déchiffrage du Génome.* Paris : éd. Odile Jacob.

Ricœur, Paul. 1967. *Philosophie de la Volonté : Le Volontaire et L'involontaire.* Paris: Aubier, Montaigne.

Watson, James. 2003. *Gènes, Génomes et Société. trad. Jean Mouchard.* Paris: éd. Odile Jacob.

الفصل 11

سؤال الجينوم بين الخِلْقة والأخلاق: مقاربة دلالية معرفية في أخلاقيات علم الجينوم من منظور إسلامي

عباس أمير[1]

أولاً: مقدمة منهجية في مشكلة البحث ومساراته:

الغاطس المعرفي الذي يشكّل القضية الرئيسة ومتغيرها المعرفي، في البحث، هو؛ الصلة الكائنة بين النمط الظاهري والنمط الجيني، لسؤال الجينوم، لا من حيث البعد المادي لتلك الصلة، بل من حيث البعد المعنوي أو الأخلاقي. والذي يفترضه البحث أصالة، ويسعى إلى التدليل عليه، هو أن ثمة علاقة وطيدة بين الخلقة والأخلاق، صلة تجعل أية تغيير نوعي ومهم، في الأخلاق يعني تغييرا في الخلقة -وإن على المدى البعيد-، والعكس صحيح.

ولكي يدلل البحث على صدق فرضيته الأولى، يلجأ إلى إثارة سؤال رئيس، مفاده؛ ترى هل يضبط علم الأخلاق علم الجينوم، وإن تم له ذلك، فأي نسخة من نسخ ذلك العلم-علم الأخلاق-، كافية لضبط الإجابة عن سؤال الجينوم؟ أو، لا، وإنما الذي يضبط علم الأخلاق بمرجعياته التاريخية السابقة لثورة البيولوجيا الجزيئية وتقانات الأحياء المعاصرة وكشوف الجينوم، والذي يدفع به إلى الأمام، هو ما رشح عن بحوث الجينوم وتجاربه من معطيات جديدة، فرضت نسخة أخرى لعلم الأخلاق ذاك وسمها المعنيون، هكذا؛ (البويوطيقا)؟ وإن كانت الأخيرة، فأي (بويوطيقا) هي؟ وهل بالإمكان صناعة بويوطيقا منضبطة بضابط ديني إسلامي؟

---

1 عباس أمير، أستاذ ورئيس قسم علوم القرآن، كلية التربية بجامعة القادسية، العراق
abbas.muariz@qu.edu.iq; abbasameir@gmail.com

ولتحقيق ما يمكن تحقيقه من إجابات عن تلك الأسئلة، اتخذ البحث لنفسه مسارا تطوريا، دعاه إلى التوقف ابتداء عند السياق الثقافي والحضاري لتلك الفرضية، ثم دعته الخلاصات المعرفية التي انتهى إليها من خلال توقفه عند العتبة الثقافية لسؤال الجينوم إلى التوقف عند البعد الأخلاقي للجينوم في ضوء البويطيقا الإسلامية المقترحة.

ومن أجل التدليل على تلك الصلة الكائنة بين البعدَين البروتيني والأخلاقي للكائن الجينومي أولا، ثم التدليل على إمكان ضبط بحوث الأخلاق الحيوية الجينومية بضابط الدين، لجأ إلى تأكيد هذا الدليل من خلال الممايزة بين المشتركات الجينية للكائنات الحية قبل تشكّل المفاهيم الأخلاقية لديها مجتمعة أو كلا على حدة، وبين الخصوصية الفردية لكل فصيلة من الكائنات ثم لكل كائن، ورحلة لكن، بعد تشكّل المنظومة الأخلاقية، وهذا ما جرى بسطه ضمن محور؛ الجينوم من المشتركات الكونية إلى الخصوصية الفردية. وهكذا ترشّح لدى البحث أهمية أن يتوقف عند علاقات التوازن والاعتدال بين النمط الجيني والنمط الظاهري، ليؤكد مرة اخرى وثاقة تلك العلاقات، سواء ضمن البعد البروتيني أم ضمن البعد الأخلاقي، وهذا ما يقوي فرضية البحث الأولى مرة أخرى.

ولقد كان البحث ضمن مساره ذاك يؤمن بمفهوم النص، وينتهج منهجا نقديا تحليليا، يستعين في أثنائه بحوث الأبستمولوجيا والسيمياء والتأويل، خاصة وهو يقارب موضوع بحثه مقاربة دلالية.

ثانيا؛ المسار التطوري لسؤال الجينوم، منظور ثقافي:

لعلنا لا نجافي الحقيقة إذا قلنا إن المنظور الثقافي للعلم يتأثر بالتصور الحضاري السائد في اللحظة التي يشكل فيها العلم ملامحه ورؤاه ومعطياته. وإن المتأثرين بهذا المنظور، العلماء أنفسهم، فضلا عن أولئك الذين ينظرون إلى العلم من خارج المؤسسات العلمية، ذلك أن «التصور السائد في حضارة ما هو الذي يحدّد معالمها، ويشكّل اللحمة بين عناصر معارفها، ويملي منهجيتها، ويوجه تربيتها. وهذا التصور يشكّل إطار الاستزادة من المعرفة والمقياس الذي تقاس به» (أوغروس، ستانسيو 1989، 15).

ومن هنا فإن وقفة منهجية أولى عند العتبة الثقافية والحضارية للحظة التاريخية التي يشكّل فيها سؤال الجينوم معالمه وملامحه الأولى هي وقفة فاعلة ومؤثرة في النظر إلى الرائز الذي تراز به

بحوث الجينوم وتساؤلاته المعرفية. فاللحظة تلك، بكل ما لها من مستندات معرفية، لحظة فاعلة إبان ترسيم معالم سؤال الجينوم ومقاصد بحوثه التي ما زالت تشكّل قلقا معرفيا وحضاريا، ولكن تلك اللحظة التاريخية-حقا-ستكون منفعلة بعد حين من الدهر يفقد في خلاله سؤال الجينوم قلقه الحضاري ويرضى أن يكون بديهيا. ولكننا الآن نعيش اللحظة التي تسبق، أو على الأكثر، نحن نعيش اللحظة البرزخية الكائنة بين كينونة الجينوم تساؤلا وبين كينونته إجابة. ومن هنا تأتي أهمية أن تملأ هذه اللحظة التاريخية ومراحلها الأولى بإجابات وافية، بما في ذلك الإجابة الأخلاقية. ويقينا إن هذه الإجابة لا تقف بالضد من بحوث الجينوم بقضّها وقضيضها، ولكنها تجهد في تفكيك حمولتها المعرفية والثقافية، وما يترتب على تلك الحمولة من تهديد براغماتي للنوع البشري وأنساقه الثقافية، مقصودا كان ذلك التهديد أو غير مقصود، إن أحسنّا الظنّ.

على كل حال، يعلمنا تاريخ العلوم أن إدراك العلاقة بين المعلوم تماما والمجهول تماما إدراك متقلّب، « وعندما يأخذ المرء بالتفكير في أنه استنفد كل ما هو ممكن أن يعرف، سرعان ما يدرك أن ما يمكن أن يعرف هو أوسع بكثير مما كان يتخيل. وينتج نوع من الانقلاب-وهي حالة علم الجينوم الذي أثار إحساسا بالانفتاح وبالإمكانات المتجددة. ويجب على هذا الإحساس بالانفتاح أن يدفعنا إلى التفكير، ليس فقط على المديين القريب والمتوسط، بل على المدى البعيد أيضا. إن التفكير في شيء، وليس التفاعل معه، هو نوع من الاستباق التأملي» (دوبرو 2007، 512). وعلى هذا يبدو أننا ملزمون بنوع من الاستباق التأملي الذي ننظر من خلاله إلى سؤال الجينوم، ولكن من على شرفة البحث الأخلاقي، حتى نشيّد فهْما آخر له يحقّق لنا مزيدا من الوعي به وبمضامينه الثقافية والحضارية، بما يعمل على تخطّي مشكل (اللبرلة) المادية للوعي إلى فضاء آخر يسمح لنا بالانخراط في فهم مشروع الجين، ولكن، دونما مجافاة المحددات التاريخية والأنساق القيمية للإنسان، بِعدِّهِ كائنا أخلاقيا.

وتأسيسا على ما سبق يبدو واضحا السبب الكامن وراء الاختلاف في النظرة الحضارية لعلم الجينوم من حيث القبول التام أو القبول المشروط. أما الاتفاق على قبوله فمردّه إلى الاتفاق على مقاصده المعلنة، متمثلة ببلوغ الإنسان كماله، والبلوغ به سعادته وتجنيبه ألمه وتعاسته. وبدليل قول المسلمين في موضوع علم الأخلاق، إنه: «الملكات النفسانية من حيث تعديلها بين الإفراط والتفريط (...) ومنفعته أن يكون الإنسان كاملا في أفعاله بحسب الإمكان ليكون في أولاه سعيدا وأخراه حميدا» (زاده 1998، 90-91). فإذا ثبت لنا أن غاية علم الأخلاق هي الوصول

بالإنسان إلى مرتبة الكمال، «لأنه لا يمكن في شيء من الأشياء أن يتشوّق ما ليس من طباعه وطبيعته، ولا أن ينصرف عما يكمّل ذاته ويقوّم جوهره» (ابن مسكويه 1985، 7)، محققا ذلك بلا روية ولا اختبار مرة، وبالرياضة والاجتهاد مرة أخرى[2]، إذا ثبت لنا ذلك، خلصنا، إلى أن محاولات تأسيس علم أخلاق حيوي هي محاولات مقصودة لذاتها ولغيرها، لأنها هي الأخرى تجهد، في أصلها الموضوعي السابق للأدلجة، من أجل إيصال الإنسان إلى كماله. والمشكل إذاً ليس في الأصل، وإنما في الذي تطوّر عن الأصل، وفي الكيفية التي ينتهي بموجبها الإنسان إلى كماله، ثم في الحدود الضابطة لمفهوم الكمال من وجهة نظر الباحثين في البيولوجيا المعاصرة وأخلاقياتها، ومن وجهة نظر أولئك الباحثين، ولكن ضمن الضابط الإسلامي لذلك النظر.

ولعلنا لا نبتعد عن الحقيقة كثيرا، إذا قلنا إن علم الأخلاق نفسه خاضع لهذا المنطق المعرفي، حينما ننظر إليه، على ضوء الفلسفات والأيديولوجيات التي تتشكّل فيها أدبياته، ذلك أن مائز الكمال الأسمى أو الخير الأسمى، مما يختلف فيه المختلفون، «فمنهم من يراه في تحصيل اللذة ومنهم من يراه في تكميل النفس بل إنهم ما يزالون مختلفين في تعريف الخير مطلقا، والشر من الأفعال، والصواب والخطأ من المناهج» (دني 2012، 10). ولهذا تعددت المذاهب الأخلاقية بناء على الأصل المعرفي الذي تنطلق منه في النظر إلى السلوك الإنساني، فثمة المذهب الفطري أو الوحياني، ومفاده أن الأفعال تكون حقا إذا هي طابقت معايير وقيم سابقة متفق على معياريتها دينيا واجتماعيا، وهناك المذهب اللذّيّ الذي يسم الفعل بالحقانية إذا أدى إلى اللذّة الفردية أو الغيرية، مع غض النظر عن معيارية الفعل ومجتمعيته أو وحيانيته، أما المذهب الثالث فهو المذهب النشوئي، ومفاده أن الفضيلة نتاج سياق تدريجي من النشوء والتطور، وهي لأجل هذا مرهونة بسياقها التاريخي (دني 2012، 12).

والذي لنا بناؤه، بلحاظ ما سبق، هو أن النظرة إلى بحوث الجينوم ومعطياته والآثار المستقبلية لتقانته لا تبتعد أخلاقيا عن تلك المذاهب. وأن كلا من علم الأخلاق وعلم الجينوم ينطوي على معايير أولى وأصول سابقة تؤسس لكل منهما التأسيس القابل لأن يتفرع منه علم أخلاق (جينومي). أما النظر في كل منهما بمعزل عن الآخر، أو النظر في رابطة واحدة تربط بينهما قانونية أو فقهية فهذا اختزال تام للحقيقة.

---

2 انظر: (حاتم، عديّ 1985، 47).

وهذا ما يقودنا إلى تساؤل مهم، مفاده: هل يضبط علم الأخلاق علم الجينوم، وإن تم له ذلك، فأي نسخة من نسخ ذلك العلم كافية لضبط الإجابة عن سؤال الجينوم؟ أو الذي يضبط علم الأخلاق بمرجعياته التاريخية السابقة لثورة البيولوجيا الجزيئية وتقانات الأحياء المعاصرة وكشوف الجينوم، والذي يدفع به إلى الأمام، هو ما رشح عن بحوث الجينوم وتجاربه من معطيات جديدة جاست خلال الديار التي لم تسكن من قبل إلا ما استطاعه العقل البشري ما قبل عصر الجينوم، متمثلا بإضاءتها تصوريا وتأمليا؟ ومن ثم هل صالح أن نقول إن البيولوجيا، وما له صلة بها ممثلا بكتاب الجينوم هو الأصل الذي تنبني عليه فصول كتاب الأخلاق، أو إن الأخلاق المنضبطة بضابط الدين والخير الأسمى هي البوصلة التي تضبط حراك بحوث الجينوم؟ أو إن كلا من الكتابين أصل وأن لا علاقة لفحوى كل منهما ومؤداه بفحوى الآخر ومؤداه؟

وهاهنا، ثمة الرؤية الكانتية للكمال الإنساني، وهي تدفع بعيدا كل إرادة خيرة خارج حدود كتلة البروتين المسمّاة: الإنسان، إذ يقول: «افعل الفعل بحيث يمكن لمسلمة سلوكك أن تصبح مبدأ تشريع عام» (كانت 2002، 11)، وهذا -وبتعبير استعاري مقصود- ما يجعل من كتاب الأخلاق وكتاب الجينوم ضمن مستودع واحد، وبفهرسة واحدة. والواضح أن هذه الرؤية ستشكّل الأس المعرفي والبنية التحتية لأيديولوجيا الجينوم الحاضرة، وقبل ذلك، تشكّل هذه الرؤية مستندا شرعيا، فضلا عن رؤى أخرى، للمذاهب الأخلاقية التي تنحّي بعيدا العلقة الوجودية والمعرفية الكائنة بين الدين والأخلاق. ولكن، لنتفق، على حقيقة مفادها: أنْ ليس ثمة كالدين يقصد البلوغ بالإنسان كماله الأسمى، وليس ثمة كالجنبة الأخلاقية من جنبات الدين ميدانا رحبا لتحقيق ذاك الكمال، ولكن بلحاظ أن الدين ينظر إلى ديمومة الحياة بعد الموت، على حين يقتصر الماديون المخالفون للدين والتدين على الكمال ما قبل الموت. ولعله من المفيد تماما أن نخلص في نهاية المطاف إلى أن الفرق بين أصلَي الأخلاق السابقين، فرق بين الموضوعي والذاتي معرفيا، وهو كائن بين تفسيرين: ميتافيزي وآخر وضعي. وبمعية هذا التمييز، وبموازاته يجري التمييز بين بين مدخل أخلاقي ديني ثم إسلامي، جينومي، ومدخل مادي، كما لا نحمّل الأدوات المادية والتقانات المعاصرة مسؤولية الشطط في وضع المقدمات واستخلاص النتائج.

والآن، لنخط خطوة قصيرة إلى الأمام، فنقول: حقا، هي مفارقة لافتة، أن يقترح سؤال الجينوم الغربي ضمن مفهوم (الحتمية الوراثية) إجابة تتنافى مع مسلّمات ذلك الغرب، بكل ما لتلك المسلّمات من بعد فلسفي وثقافي مادي. فالديباجة الفلسفية الغربية ومنهجها العلمي التجريبي

يذهب إلى أن البعد المعنوي للإنسان نتاج البيئة الثقافية واللحظة التاريخية. وهنا تبدو مضامين العصر الجينومي مفارقة لتاريخ العلم الحديث ومسلماته التي جهدت في تجريد الإنسان من بعده المعنوي الغيبي. وهذا ما يدعو إلى القول، إن مضامين العصر الجينومي، في جانب من جوانبها، تمثل نكوصا معرفيا وثقافيا غربيا، إن لم نقل، إنها تمثل خيانة عظمى للفضاء الثقافي للنُخَب الغربية...!

ولعل هذا النكوص المعرفي والثقافي الذي يرجع بمسلمات الحضارة الغربية المادية القهقرى، وهي تجهد اليوم في القول، إن البعد المعنوي للإنسان نتاج المادة- البروتين← القواعد النتروجينية، لعله لا يمكن تفسيره، في جانب من جوانبه، إلا ونحن نستحضر القولة الفلسفية المتداولة شعبيا؛ (كل شيء زاد عن حدّه انقلب إلى ضدّه). وطبعا، لا بد بعد ملامسة الضدّ الضدَّ من العود إلى (الوسط الذهبي الأرسطي) ثانية، فيصير البعد المعنوي للإنسان نتاج عالم الغيب وعالم المادة، تترجمه - مع غض النظر عن كفاءة الترجمة- الوراثة والثقافة إلى حقيقة معيشة، وهو ما خلص إليه المسلمون منذ قرون عديدة.

من جهة أخرى، تبدو امتدادات الفكرة النيتشوية القائمة على إرادة القوة و (موت الإله) وعدمية الأخلاق[3]، واضحة بارتداداتها وانعكاساتها المضمرة ومُصَاداتها المتعددة في عصر الجينوم. والذي ترتّب على ذلك ضمنا، تقليد الجينوم وسام الألوهية، وتتويجه إلها جديدا تبدأ حدود مملكته بهذه الكتلة البروتينية المسمّاة، الإنسان، وتنتهي عندها، ومن ثم، مطالبته بتدبير شؤون تلك المملكة...!

ولا نختلف بعد هذ وذاك، على أن سؤال الجينوم سؤال متشظّ فلسفيا وثقافيا، ولعلّنا لا نعدو الحقيقة إذا قلنا ؛ إنه متشظ علميا أيضا. وغير بعيد عن الواقع، قولنا ؛ إن تشظّيه، راجع في جانب منه، إلى تشظّي السياق الثقافي الذي تشكّلت فيه بحوث الجينوم، غير غافلين عن الأصول العلمية الموجهة لتلك البحوث، التي تشكّل البيولوجيا التطورية ضابطها. ومن هنا يبدو، أنه مهم جدا تفسير سؤال الجينوم على ضوء البعد الأخلاقي الذي له الأولوية في ترسيم فردية تلك الكتلة البروتينية، المسمّاة، الإنسان، مقارنة ببقية الكائنات الحية.

من هنا، نقول، إن التأسيس الأمريكي لمشروع الجينوم البشري، وما تلاه من تجارب شرق آسيوية وأسترالية وكندية وإسرائيلية مرة، وأوربية- محدودة- مرة أخرى، وما يُعدّ ثمة له متمثلا

3 انظر، (نيتشه 2011).

بمشروع (الجينوم الأعلى) عام 2008، وما يحيل إليه هذا التأسيس وما تبعه، من خضوع الحضارة الأمريكية لمذهب البراغماتية الأخلاقية وضابطها اللذّيّ أو النفعيّ، ضمن فضاء معولم ثقافيا، وحر اقتصاديا، يعني أننا بإزاء تفسير جديد لمفهوم الأخلاق سيلقي بظلاله الوارفة على مفهوم الإنسان الجديد. ومن هنا، ولكي نفهم محددات هذا الإنسان الجديد ال(مُجَيْنَن) بيولوجيا، و(المُعوْلَم) ثقافيا، لا بد لنا ابتداء من إعادة قراءة فهمنا لإنسان ما قبل (الجينوم). على أن لا ننسى أن الخطورة الكامنة في بحوث الجينوم تتجلى في قدرة هذه البحوث المستعينة بتقانات العصر على التأسيس لـ (براديغم) علمي جديد يؤسس هو الآخر لأفعال تتواتر حتى تصير عادات فتصير العادات سلوكا مرغوبا ومشروعا ثم يصير السلوك خلقا جديدا. إنه (براديغم) خلقي ناشئ في رحم (البراديغم) العلمي ابتداء ثم متناغم معه بما يشكّل فضاء جدليا لحاضر علمي و أخلاقي جديد يضطر منظومة الأخلاق القديمة إلى أن ترتدي طوعا أو كرها ثوب الجينوم الفضفاض تماما...!

ثالثا: البعد الأخلاقي للجينوم بين الأمن الديني والحرية الشخصية:

إن التلاقي الكائن بين الوراثة والتنشئة في النمط الظاهري للجينوم، وهو عينه التلاقي الكائن في النمط العميق، على الرغم من الكثير من المحاولات الغربية الحثيثة لتنحية ذلك التلاقي عند الإجابة عن سؤال الجينوم، يكشف عن تداخل عميق بين الخلْق (بسكون اللام)/ البعد البروتيني للجينوم، والخلُق (بضم اللام)/ البعد المعنوي. وهنا، لا بد من التنويه بأن اللسان العربي الذي يفارق بين صيغة (الخلْق)، بسكون اللام، وصيغة (الخلُق) بضم اللام، يجمع بينهما بصيغة (خليقة)، «فالخليقة تدل على ((المخلوق)) كما تدل على ما خلق به من سجايا؛ من هنا، يصح أن نقول بأن كل تشكّل في الخلق يقترن به تهيؤ في الخلُق]بضم اللام[، وأن كل زيادة في التشكّل الخلْقي]بسكون اللام[، تقارنه زيادة في التهيؤ الخلُقي]بضم اللام[؛ والعكس أيضا صحيح، فكل نقص في التشكل يقارنه نقص في التهيؤ» ، وهذا يعني أن (الخلق الآدمي) ليس ظاهرة متمتعة بالحياة فقط، وإنما هو ظاهرة قيمية أخلاقية (عبدالرحمن 2012، 273-274). ولا شك أن حرية الإرادة مظهر مهم من مظاهر تلك الظاهرة الأخلاقية، المسماة؛ الإنسان.

ولا شك أن المسؤولية الأخلاقية لذلك الكائن بشقّيه، الظاهري والعميق، مشروطة بقدرة الإرادة على التصرف، «إذ لا يصح طلب شيء من غير إرادة، إذ هي المحرك الباعث]على البحث[والتفتيش، والإرادة من خاصتك المصرفة لعامتك» (ابن عربي 2003، 46)، وإلا فكيف نجيز لأنفسنا أن نسأل شخصا عن فعله السيء مثلا، في الوقت الذي يفتقد فيه ذلك الشخص امتلاك قراره الشخصي وإرادته الحرة؟ ومن هنا نقول مع أحد الدارسين: «ما دامت حرية الإرادة شرطا أساسيا للمسؤولية الأخلاقية، فلا بد أن نكون على يقين من أن نظريتنا في حرية الإرادة تقدم لنا الأساس الكافي لهذه المسؤولية. وعندما نقول إن فلانا مسؤول عن أفعاله مسؤولية أخلاقية، فإن ذلك يعني أن فلانا هذا يمكن أن يُعاقب أو يُثاب، يُلام أو يُمتدح، بحق، على ما يقوم به من أفعال. لكن ليس من الإنصاف معاقبة انسان على فعل لم يكن في استطاعته أن يمتنع عنه» (ستيس 1989، 295-296). ولكن ، وبالمقابل، لا يمكن القول إن الحتمية تتناقض مع المسؤولية الأخلاقية، «فأنت لا تلتمس الأعذار لرجل لفعل خاطئ ارتكبه لأنك تعرف شخصيته. أو لأنك تشعر عن يقين أنه سوف يقوم به مقدّما. كما أنك لا تحرم انسانا من المكافأة أو الجائزة، لأنك تعرف جانبه الخيّر، أو قدراته، أو لأنك تشعر عن يقين مقدّما أنه سيفوز بهذه الجائزة» (ستيس 1989، 96). وهكذا يتضح أن الحرية شرط المسؤولية الأخلاقية، ويبدو أن الأخيرة تتطلّب الحتمية، «فالافتراض الذي يقوم العقاب على أساسه هو أن السلوك البشري محتوم سببيا. فإذا لم يستطع الألم أن يكون السبب في قول الصدق، فلن يكون هناك مبرر على الإطلاق لمعاقبة الكاذب» (ستيس 1989، 97).

من جهة أخرى، يبدو أن الميل التام إلى إحدى الجنبتين: الحرية أو الحتمية يعني، مما يعني، تفكك مبدأ التوازن الضامن ديمومة التنوّع والاختلاف مرة، وانهيار الصفة المجتمعية مرة أخرى. نعم إن ديمومة المنظومة الأخلاقية تستلزم الإعلاء من شأن الحتمية الأخلاقية المجتمعية، ولكن، وبالقدر نفسه، تستلزم المسؤولية الأخلاقية الإعلاء من قيمة الحرية الفردية. ولكن الذي يجري في عصر الجينوم هو إرادة عولمة الأخلاق، وتشكيل منظومة أخلاقية جديدة تحول معها المعايير الأخلاقية القارّة للشعوب إلى أعراف وتقاليد ليس إلا، وهذا ما يحكم عليها بالنسبية، ويسمح بعدم ثباتها، ما يعني اختلاف الحكم على طبيعة الفعل الأخلاقي الواحد باختلاف الزمان والمكان وطبيعة التركيبة الاجتماعية، وهكذا يصير تحديد النسل مستحبّا والإجهاض مباحا في عرف

المجتمعات التي تعلي من شأن الحرية الفردية غير المنضبطة بضابط سماوي، ويصير تحديد النسل عيبا والإجهاض جريمة في المجتمعات المحافظة أو القليلة الكثافة (ماير 2002، 284).

ولنستعن بأسلوب إثارة السؤال، ونحن نجهد في إضاءة القضية إضاءة جديدة، فنقول، ترى هل التعديل الاضطراري السابق للصفات الوراثية من (الكسب)؟ وكيف يحاسب من عدّلت صفاته الوراثية المتعلقة بالذكاء مثلا، فتصرّف بتلك الإمكانات بعد بلوغه سن التكليف تصرفا سالبا؟ وكيف يحاسب من عُدّلت صفاته الوراثية الجسمانية، فامتلك بذلك التعديل وسامة بالغة، ثم حينما بلغ سن التكليف انتهى به ذلك التعديل إلى فتنة الإغواء فأغوي بسبب وسامته؟ ترى هل من العدل أن يحاسب كما يحاسب دميم الخلقة ضمن هذه الحيثية؟

ترى كيف يحاسب من عُدّلت صفاته الوراثية فطال عمره، فارتكب ضمن هذه الزيادة العمرية المعاصي؟ ترى هل من العدل أن يحاسب كما يحاسب لو إنه بلغ من عمره ما كتب له قبل التعديل ثم قضى؟

لا شك أننا لا نقبل تلك المساواة على مستوى الحواسيب المبرمجة والأخرى غير المجهزة بعدّة برمجية، فكيف نقبلها للإنسان؟

فإن قيل، إن الانسان أصالة لا يمتلك الخيار في صفاته البيولوجية في كل الأحوال، قلنا، ولكن الذي منحه صفاته الأصيلة عادل يتجلى عدله في أن الجزاء عنده على قدر العطيّة.

فإن قيل، ولكن صفاته التي اكتسبها بعد تعديله ليست من كسبه، فلا يجازى عليها خيرا أو شرّا، قلنا، فيُجازى من تولى تعديله إذن، ومن رضي بتعديله ترتّبا. وهكذا نجدنا نقوّض مسلّمة الكسب الذاتي، ونصير غرضا للآية التي تقول، ﴿كُلُّ نَفْسٍ بِمَا كَسَبَتْ رَهِينَةٌ﴾ (المدّثر،38)، ونبتعد كثيرا عن مفهوم (الكدح) الإنساني الذي تؤكده الآية المباركة مشيرة إلى أن صناعتنا الشخصية لأنفسنا التي ترتفع بنا من مستوى البشرية إلى مستوى الإنسانية هي صناعة ذاتية مرهونة بمدى قدرتنا على المحافظة على معنوياتنا النفسية والسموّ بها من خلال بذل المزيد من الجهد في اكتشاف النفس وممارسة النشاط العقلي البناء وترجمة كل ذلك بالعمل الصالح، بدليل قوله تعالى، ﴿يَا أَيُّهَا الْإِنْسَانُ إِنَّكَ كَادِحٌ إِلَى رَبِّكَ كَدْحًا فَمُلَاقِيهِ﴾ (الانشقاق،6)...!

ولعلنا، ومن زاوية أخرى، نواجه هاهنا فكرة النسبية الأخلاقية، ونحن نحاول أن نجيب عن السؤال الآتي، ترى هل، حقا، أخلاقنا نحن- النوع البشري- نتاج سلوك النمط البشري الذي نشأ، كما يقول علماء الأحياء التطورية، خلال حقبة التكيّف التطوري قبل نحو100,000 سنة،

وهل نحن -النمط البشري- متكيفون أخلاقيا مع العالم الخارجي، التكيف الذي يحقق لنا كرامة، هي ببساطة نتاج تغيرات بيئية (فوكوياما 2006، 191-192)؟ وهل فعلا، يستطيع الإنسان أن يمنح نفسه معايير أخلاقية جديدة للحياة الكريمة في أوضاع جديدة؟ وهل لنا أن نخلق معايير كرامة إنسانية جديدة غير تلك التي انتهت إلينا عبر التاريخ المادي والمعنوي للبشرية؟

ورد عن رسول الله صلى الله عليه وآله وصحبه، قوله في محاسن الأخلاق ومكارمها: «إنَّ هذه الأَخلاقَ مَنَائِحُ يمنحُها الله عزّوجلّ مَن يشاءُ من عبادِهِ فإذا أَرادَ اللهُ بِعبدٍ خيراً مَنَحَهُ منها خُلُقاً صَالحاً» (ابن أبي الدنيا، 26)، وعلى هذا، وقبل الإجابة عن التساؤلات السابقة، لابد من تقرير حقيقة مهمة، بالإفادة من الحديث الشريف، مفادها بالنسبة لنا نحن المؤمنين بالغيب، أن الكرامة وأخلاقها أصالة، منحة وليست نتاج اجتهاد أوغنيمة، وهنا نجدنا نقول، مع من يقول: « يجب على طبيعة الأخلاق بدورها أن تحرّض على الفعل وتحرّكه من أجل خير الجميع، كما يجب عليها أن تحكمه، فالواحد منهما لا يلغي الآخر، بل يجب أن يسير الاثنان متوافقين. وتكمن المهمة السياسية، المهمة الصعبة إذا كانت كذلك، في تعريف نقطة التوازن» (دوبرو 2007، 515). ونقطة التوازن، هاهنا، تكمن في اللحظة التي يصير فيها الثابت الأخلاقي عِدْل الكرامة الإنسانية، فيكون المتغيّر منضبطا بضابطه ومُمَسّكا بعروته الوثقى، حتى لا تنفلت الموازين أخلاقية كانت أم كيميائية، وتحصل القطيعة بين المتحرك وثابته. فإذا كان ذلك كذلك، انسجم المتغير الأخلاقي مع ثابته، واتسقت خطوات الحرية الفردية على سطر الحتمية الأخلاقية، وامتلك الكسب البشري صفات حسنه أو قبحه على ضوء تلك الموازنة.

وتأسيسا على هذا، تبدو محاولة التدخل السلبي في البعد الجينومي للإنسان ومن ثم التحكم في سلوكه، محاولة مخادعة من محاولات سلبه إرادته، تحت عنوان تحريره من ميثولوجيا الدين، وإن أمكن إضافتها إلى تلك المحاولات ذات الصبغة الدينية الجبرية التي شهدتها حقب تاريخية سابقة. ولكن الأبعد غورا من سلب الحرية المترتب على التدخل الجينومي، هو عدم إمكان ضبط المستقبل البشري، وفقدان القدرة على التنبؤ بما سيترتب على محاولات (الكولجة) و (المونتاج) الوراثي الناشئ من عمليات التركيب الجيني الجامع بين الأنواع.

يبدو أننا ليس لنا أن ننحّي المستند الأخلاقي في مقاربة سؤال الجينوم، ذلك أن المستند المذكور يستمد مشروعيته من صلة ذلك السؤال بشخصية الإنسان وامتيازه وكرامته وإرادته، نعم إن التحوّلات الاجتماعية تنحو بالمجتمع البشري منحى جديدا يجعله غير آبه كثيرا بالمعايير الثابتة التي

ورثها من أسلافه المتفلسفين والمتدينين، لأنه ما عاد يؤمن بغير القيمة الذاتية للحياة، وما يترتب على تلك القيمة من تضحية بالمعايير الجمعية. ولكن مجرد اختيار معايير بعينها للكائن الجديد، كائن المستقبل الذي لم يشعر بذاته لحظة اختيار مواصفاته يعني التضحية التامة بالفردية التي يسعى إنسان المستقبل إلى عيشها. ومن هنا، نقول: ما زال سؤال الجينوم لم يستكمل إجابته، وما زالت فروضه غير مستوفية نصابها، وهذا يعني أن إمكان تقدير نتائجه تقديرا دقيقا مرهون بمدى التزامه بعديد الأبعاد التي تتشكّل منها الحقيقة الإنسانية، فضلا عن الحقيقة البشرية البيولوجية، وإلا فإن مجافاة تلك الحقيقة يعني الخلوص بنتائج تلك البحوث إلى مثابة الجور على الشخصية الإنسانية من خلال الانكفاء بها إلى حيث مستواها البشري الذي لا تمتاز فيه من غيرها من الكائنات ذات البعد البيولوجي، فضلا عن ظلمها الاجتماعي والحضاري الذي انضبطت مقاصد الشريعة بضابطها الكلي الأول بغية دفعه، « وذلك مؤذن بانقطاع النوع البشري، وهي الحكمة العامة المراعاة للشرع في جميع مقاصده الضرورية الخمسة: من حفظ الدين والنفس والعقل والنسل والمال» (ابن خلدون 2004، 479). ومن مظاهر ذلك الظلم –وما أكثر تلك المظاهر- الذي اتضح بعد مرور أقل من عقدين –ليس إلا- على الفراغ من مشروع الجينوم البشري، ما يترتب على الفحص الجيني وهو يتعامل مع حامل المرض بعدّه مريضا، وهو ما لا يتماشى مع مراد الشريعة أخلاقيا. ويتنافى مع فقه الأسرة وفقه العقود في أغلب الأحوال، ذلك أن الذي نفيده من منطق الآية المباركة التي تقول: ﴿يُرِيدُ اللَّهُ أَنْ يُخَفِّفَ عَنْكُمْ وَخُلِقَ الْإِنْسَانُ ضَعِيفًا﴾ (النساء: 28)، يخلص بنا إلى حقيقة مفادها، أن الإنسان –أيّ إنسان-، مستعد للإصابة بالأمراض كلها أو هو حامل للأمراض، ما ظهر منها وما بطن، وهذا يعني أن تحميل إنسان بعينه دون الآخرين تبعات كونه حاملا للمرض مجافاة لمنطق العدالة والمساواة، ودليل على غرور علمي، فاسد.

رابعا: رحلة الجينوم من المشتركات الكونية إلى الخصوصية الفردية:

ما بين القول بكمال الخلقة الأولى دينيا أو القول بتطورها عبر الدهور والأزمان وصولا إلى كمالها ماديا، ثمة مساحة للقول، مفادها: أن الخلقة الأولى الجامعة لكل الأنواع غير المتميز بعضها من بعض في حقبة (الرتق) وما يرتبط بها من حالات المادة السائلة، ﴿أَوَلَمْ يَرَ الَّذِينَ كَفَرُوا أَنَّ السَّمَاوَاتِ وَالْأَرْضَ كَانَتَا رَتْقًا فَفَتَقْنَاهُمَا وَجَعَلْنَا مِنَ الْمَاءِ كُلَّ شَيْءٍ حَيٍّ أَفَلَا يُؤْمِنُونَ﴾ (الأنبياء:30)، إن

هذه الحقبة امتلكت استعدادها التكويني لبلوغ غاياتها النوعية والمظهرية التي يتميز في خلالها كائن من كائن عبر الزمان والمكان وما يرتبط بهما من ظروف وأحوال بيئية نشأت بعد تميّز الأرضي من السماوي بعد تحقق (الفتق). والذي جرى ويجري بعد تمام حقبة (الفتق) هو حركة تطورية مقصودة، وذات معنى، وليست (صدفوية)، جرت لكثير من المخلوقات، وما زالت تجري لغيرها مما لم يخلق بعد. وهذا يعني أن الغايات كائنة بالخلقة وقبلها، وأن أي خروج عليها يعني خروجا على المقادير التكوينية ومقاصدها. ولكن الإيغال في سحر البحوث الجينية، بما يعمل على تجاوز الغايات التكوينية للمخلوقات يعني التمرّد على التقدير التكويني الأول وموازينه الضابطة وغاياته المقصودة. أما استغلال مرانة التكوين البيولوجي للخلق وليونته أمام البحث العلمي فلا يعني ضعفا في بناه التكوينية وفقدانا لوضوح مقاصده وغاياته المركوزة فيه، بل يعني تسخيره للبحث العلمي وتطويعه لإرادته، تحقيقا لقوله تعالى: ﴿قَالَ الَّذِي عِنْدَهُ عِلْمٌ مِنَ الْكِتَابِ أَنَا آتِيكَ بِهِ قَبْلَ أَنْ يَرْتَدَّ إِلَيْكَ طَرْفُكَ فَلَمَّا رَآهُ مُسْتَقِرًّا عِنْدَهُ قَالَ هَذَا مِنْ فَضْلِ رَبِّي لِيَبْلُوَنِي أَأَشْكُرُ أَمْ أَكْفُرُ وَمَنْ شَكَرَ فَإِنَّمَا يَشْكُرُ لِنَفْسِهِ وَمَنْ كَفَرَ فَإِنَّ رَبِّي غَنِيٌّ كَرِيمٌ﴾ (النمل:40).

ومن هنا، فإن التمرّد على الطبائع التكوينية للخلائق وغاياتها المركوزة فيها يعني تجريدها من خصوصيتها وتعميتها وسلبها تمايزها المادي والمعنوي، وصولا إلى كائنات مشوّهة، نعم تتطاول بناها الفوقية ولكنها بالمقابل تفقد متانة بناها التحتية والتكوينية، ما يعني انهيارها المريع وقد أُفقِدت منعتها وصلابتها الظاهرة بسبب تلك المداخلات الإكراهية التي تسوق الأنواع إلى غاياتها أو إلى غايات ليست لها سوقا شديدا عنيفا لا يتماشى مع صالح مقاصد الخلقة وغاياتها الموزونة بوازن العدل والقدرة والتأنّي. فالتدخل الجيني العلاجي أو التحسيني باستخدام وسائل التكنولوجيا الحيوية، وتطبيقات ذلك التدخل على الخلايا الجنسية، ومدى محافظة ذلك التدخل الجيني على الأنساب خاصة حينما تقحم مورثة خارجية-مع غض النظر عن مصدرها- أو تطبيقات ذلك التدخل على الخلايا الجسمية لا ما كان تصحيحا لتشوّه أو شذوذ بل ما كان داخلا في دائرة اتباع الهوى، كل ذلك يجري بعيدا عن الهمّ الأخلاقي غالبا، ما يعني مداخلة الشريط/ السطر الصبغي بما لا يتماشى مع الإرادة التكوينية وغاياتها، وبالمحصلة، انتزاع ما يجب أن لا يُنتزع من استعدادات الخلقة، ذلك أن انتزاعه كذلك من طريق شريط/سطر صبغي مختلف عن إمكاناته الجينية يعني الخلوص به إلى وجهة غير وجهته. نعم قد تبدو مداخلة بعض الكائنات البيولوجية ممكنة لبساطة تركيبها ومحدودية غاياتها، ولكن مداخلة الشريط/السطر الجيني البشري

مداخلة مختلفة، يرجع اختلافها إلى ارتباطها الوثيق بجهازه العصبي المتطور مرة، وببعده المعنوي الأخلاقي مرة أخرى، بل إن مداخلة الشريط/ السطر الصبغي البشري تختلف من شخصية انسانية إلى أخرى، خاصة ما يتعلق بالشخصية الأخلاقية التي تميز هذا الإنسان من ذاك، كأن يكون (س) معتقدا بالغيب موحّدا، و (ص) منكرا (لاأدريا)، وهذا يعني أن المنظومة الجينية لـ(س) تختلف كثيرا-على مستوى البنية التحتية- عن (ص)، وهو ما نأمل أن يتحقّق المعنيون من صدقيته...!

ولعلنا، بعد هذا وذاك، نجيز لأنفسنا القول؛ إن الانخراط التام في عملية الاحتيال على التركيبة الجينية لن ينتهي بنا إلا إلى خديعة كبيرة نُعبِّد بموجبها أنفسنا لمآلات ذاك الاحتيال الذي سيمكّن الخيال من الحقيقة، والسلطة المتفردة من العدالة، والآلة من الإنسان...الخ، وبما يجعل من البيولوجيا الإنسانية مصنعا ومختبرا لخيالات فاسدة هنا وهناك. وحينها، وبأثر من ذلك الجنون العلمي المأخوذ ببراءات الاختراع ومساندة رأس المال العملاق، ليس ثمة إلا أحد خيارين؛ استكانة الانسان وقبوله قدره الجديد، قدر الطواعية التامة وفقدان الإرادة، أو رفض الطبيعة الإنسانية محاولات التدخل في خصوصيتها وميزتها، الرفض الذي سينتهي بالبشرية إلى تدمير ذاتها بذاتها بعد غير قليل من الصراع عبر معدود من الأجيال، إن لم يكن ذلك بعد جيل على الأكثر.

وبهدْيٍ من آية الرتق والفتق، وما يتعلق بها من ارتباط الخلق بالتدبير، فلا فكاك بينهما، ذلك أن القدرة على إظهار الخلق تستبطن قدرة كمثلها على التدبير، ولهذا تستنكر الآية أن يجعل أولئك الكافرون أن الذي خلق إله، والذي دبّر إله غيره[4]. بهدْيٍ من ذلك، هاته وقفة من جنبتين؛

أولا؛ لا شك أن علماء البيولوجيا الغربيين لا يزعمون أنهم خلقوا الحياة بنمطيها الجيني والظاهر، وإنما هم يزعمون، بلحاظ النموذج المادي في تفسير ظاهرة الحياة عندهم، أن المادة خلقت نفسها بنفسها، وأن فكرة الكرامة أو حرية الإرادة محض وهم، « وأن جميع عمليات اتخاذ القرار يمكن عزوها في نهاية المطاف إلى أسباب مادية» (فوكوياما 2006، 190). وإذن، ومع غض النظر عمّن خلق، الله سبحانه أم المادة، فالذي لا ينكر هو أن الذي له الخلق، له التدبير أيضا، فليتركوا المادة تدبّر نفسها بنفسها، وإلا فإن إرادتهم القفز على ترميمها والتماشي مع إرادتها إلى تدبيرها، وإكراه مجسّاتها على الاستجابة لتدبيرهم يعني الانتهاء إلى تدميرها. ذلك أن الذي وصل بالمادة

4 انظر؛ (ابن كثير 1999، 5: 338-339).

إلى هذه الرتبة من الكمال في الخلق لا يباريه أحد في تدبير شؤون الخلق. أما نحن فلننظر في تدبير فهمنا لذلك الخلق، بضابط من بيان خالق الخلق...!

ثانيا، يؤكد بيان الآية أن الحالة التكوينية الأولى للسماوات والأرض كانت حالة لا يتمايز فيها أرض من سماء، وإنما الحالة حالة تضامّ والتصاق، وأن الجميع متصل بعضه ببعض، متلاصق متراكم بعضه فوق بعض، في ابتداء الأمر، ففتق هذه من هذه، فبان كل منهما من الأخرى، وامتازت أبعاض كل منهما من أبعاض الأخرى[5]. والذي نفيده من هذا البيان هو أن الأصل التكويني للموجودات الحية كان أصلا جامعا لا ميزة فيه لموجود من آخر، بل إن إمكان الوجود وشرائط الكينونة الحية موزعة بالتساوي في تلك الحقبة التي مازالت فيها الطبائع ملتئمة، وهذا يعني أن العدّة الجينومية هي فلا ميزة لعدّة موجود على عدّة غيره طالما أن التمايز النوعي والمظهري لم يتحقق بعد للموجودات. ومن جهة أخرى، وبعد تحقق الفتق وتشكّل الموجودات بأنواعها ومظاهرها عبر التاريخ، صار لكل موجود هويته الجينية، كل حسب استعداده وكفاءته ووظيفته، دونما تخلّ للموجودات جميعا عن عدّتها الأصل المشتركة الجامعة، وهذا ما يفسّر وجود ذلك المشترك الجيني-على اختلاف-بين الموجودات الحية.

أمر آخر مفاده، أن ذلك التمايز الذي جرى فتقا بين السماوات والأرض لم يعدم كمون الصفة الأرضية في السماوات كما لم يعدم الصفة السماوية في الأرض، وأن تبادل الصفات ذاك وتناوبها ظل جاريا في الموجودات جميعا، وظلت مظاهره تتسلسل ماديا ومعنويا، لتتمظهر مادة وطاقة، ليلا ونهارا، صلابة وسيولة، حياة وموتا، روحا وجسدا، ذكورة وأنوثة، صغيرا وكبيرا، وصمتا وكلاما...وهكذا، يتماهى الرتق بالفتق ظهورا لصفة وبطونا لأخرى. ولعل نظرة سريعة إلى الشريط الجيني تكشف عن تمثل تلك الحقيقة فلسفيا ومختبريا.

إن تلك الحزوز البيولوجية الكائنة بين الجين والآخر وهي تتوزع على السطر الجيني تعني، على ضوء مبدأ التوازن الزوجي للموجودات أن آلية الرتق والفتق هي هي على مستوى الصغير والخفي والعميق، وأن استقاء المعرفة الجينومية الحق كائن عند هذه الربيطة التي تصل الناطق بـ(الصامت)، والبروتين بالجينوم. ولكن الذي يجري في التجارب العلمية هو حزّ الجين وقطعه دون أصله والاكتفاء به ميدانا للمعرفة والتجريب، والحق أن الصمت الجيني الفاصل بين جين مشفّر وآخر يتلوه لا يعني صمت العدم، كما لا يعني، عدم تدخّل ذلك الصمت في تنظيم معنى

---

5 انظر: (ابن كثير 1999، 5: 339؛ ابن عجيبة 1999، 3: 457؛ البقاعي، 12: 412).

الكلام، وإنما هو صمت ضابط لتوسيع دلالة الجين وتغييرها، وفاعل في محوها إن استلزم الأمر، وما فاعليته تلك- ونحن نتذكر (الفتق والرتق) - إلا تمثل لفاعلية تلك الحالة الكونية الأولى، حالة المشتركات الجينية التي تحدث عنها آية الرتق والفتق.

خامسا: علاقات التوازن والاعتدال بين النمط الجيني والنمط الظاهري:

لعلنا لا نجافي الحقيقة إذا قلنا إن ثمة صلة كائنة بين الأخلاق والخِلْقة التي بموجبها يصير ظاهر الخلقة ونمطها البروتيني معادلا موضوعيا لباطنها ونمطها المعنوي، أي، التي يصير بموجبها النمط الظاهري صورة للنمط الجينومي، فالأخلاق (لصورة الإنسان الباطنة... بمنزلة الخلْق لصورته الظاهرة) (ابن منظور 1993). ومن جهة أخرى، الأخلاق (طبْع)، «والطَّبْع ابتداءَ صنْعةِ الشيء تقول طبعت اللَّبِنَ طبْعاً وطَبعَ الدرهم والسيف وغيرهما يطْبَعُه طبْعاً صاغَه والطَّبّاعُ الذي يأْخذ الحديدةَ المستطيلة فَيَطْبَعُ منها سيفاً أَو سِكّيناً أَو سِناناً أَو نحو ذلك وصنعتُه الطِّباعةُ وطَبَعْتُ من الطين جَرَّةً عَمِلْت» (ابن منظور 1993). والبادي من الاستعمالات الاجتماعية الواردة في معاني الخلْق والإيجاد هو معنى الاستعداد والتهيؤ والتقدير، والمرونة الأولى القابلة للتشكّل والإملاء والإنزال بتوفر الظروف والدواعي. وأن تطبيع الشيء بغية مناسبته الخلقة المبتدعة أصالة على مثال بعينه لم تسبق إليه يستلزم خلاقة (الخَلاقة وهي التَّمَرُّنُ من ذلك أن تقول للذي قد أَلِفَ شيئاً صار ذلك له خُلُقاً أي مَرَنَ عليه) (ابن منظور 1993). وهذا يعني أن الصورة الظاهرة امتداد مناسب للصورة الباطنة، وأن هذا الامتداد مناسب ومنسجم عند تحقق (التمرن)، وغير مناسب عند عدم تحققه. وأن مجافاته تعني قطع الامتداد بين النمطين أو تشويهه، وبالنتيجة إعاقته من بلوغ غاياته، على أن ذلك البلوغ مشروط ب(الخلاقة) أو التمرين أو التطبّع.

ولننظر في النصين الآتيين، بما يكشف عن بعيد معرفة بذلك الامتداد الذي توصّل إليه المسلمون، مؤكدين فكرة الامتداد الكائن بين الصورة الباطنة والصورة الظاهرة، وبين الأخلاق والخِلْقة. فمرة حُسْن الخُلُق كاشف عن نقاوة العرق؛ «حسن الأخلاق برهان كرم الأعراق» (الآمدي 1992، 189)، ومرةً، وُسوم أحسن الخلُق (بضم اللام) منكشفة بعلامات أحسن الخلْق (بسكون اللام). «فالفراسة أكرمك الله، نور من أنوار الله عزّ وجل، يهدي بها عباده، ولها دلائل في ظاهر الخلْق؛ جرت الحكمة الإلهية بارتباط مدلولاتها بها (...) فاعلم يا أخي،

وفقنا الله وإياك، أن أحسن الهيئات وأعدل النشآت الذي ينبغي لك أن تتخذه سفيرا وليلك سميرا وللملك وزيرا، من ليس بالطويل ولا بالقصير، ليّن اللحم رطبه، بين الغلظة والدقة، أبيض مشوب بحمرة وصفرة، معتدل الشعر طويله...» (ابن عربي 2003، 58)، أما تفسير ذلك كله، ف«اعلم رحمك الله ونوّر بصيرتك أن عالم الملكوت هو]المحرك لعالم[الشهادة، وهو تحت قهره وتسخيره حكمة من الله تعالى لا لنفسه استحق ذلك؛ فعالم الشهادة لا تصدر منه حركة ولا سكون ولا أكل ولا شرب ولا كلام ولا صمت إلا عن عالم الغيب. وذلك أن الحيوان لا يتحرك إلا عن قصد وإرادة وهما من عالم القلب وهو من عالم الغيب، والحركة وما شاكلها من عالم الشهادة» (ابن عربي 2003، 65).

والمهم في تلك المقابلة بين النشأة الطينية الجسمانية والنشأة المعنوية الروحانية، وصيرورة النشأة الجسمانية دليلا على النشأة الروحانية، وباعثا على النظر في ما بين النشأتين من تعالق منضبط بضابط العدل والاعتدال، المهم في هذا الترابط الكائن بين الأخلاق والخلقة، على الرغم من أنه قابل للاستثناء، هو أن النظر في النمط الظاهري للخلقة البشرية لا يتوقف عند الطبيعة البروتينية، والأبعد من ذلك، أن ذلك النمط لا ينسج في النمط الجيني المقروء، وإنما هو يتجاوز ذلك إلى غير المقروء. والمهم جدا هو النظر في منطقة ما بين الجين والجين، لأنها نقطة الوصل بين عالم الملكوت وعالم الشهادة، على مستوى الصغير (الكايوسي)، حيث لا يمكن التنبؤ بمسارات النظم الجينومية دونما نظر في ذلك البعد الخفي الذي يشكّل المعطى الأولي لحساب النتائج.

إن هذا الفضاء السيميائي المرئي من على كتلة البروتين تلك، المشروط باعتدال الهيئة المظهرية القائمة على أساس علاقات التناظر والتقابل والانسجام، بين المادة والصورة، وبين الطبع والتطبّع «فما في الطبع هو العدل» (الفارابي، 157)، وهو، من حيث كونه فعلا واعيا مرادا، نتاج ما في الطبع، فضلا عن الكثير من علاقات التفاعل والتواصل الكائنة بين الوراثة والتنشئة، وبين الحتمية والحرية، وبين الأخلاق والخلقة. ومن هنا تبدو الصلة الكائنة بين الجنبة الأخلاقية والجنبة التطورية البيولوجية للإنسان صلة بنيوية. ويبدو المدخل الأخلاقي للنظر في سؤال الجينوم ليس مدخلا افتراضيا مقحما كرها، وإنما هو مدخل تكويني تماما، يعنى بالطريقة التي تنسخ فيها الجينات معلوماتها الوراثية وهي تخضع لمبدأ الانحياز الانتخابي، مؤكدة ترابط الخُلق والخلْق ترابطا لا يقبل التفريق، ومن هنا يتكشّف لنا أهمية الربط بين بحوث الأخلاق وبحوث الجينوم، ومن ثم، يبدو لنا أهمية أن نطرح تفسيرا أخلاقيا لحركية الجينوم، لا أن نكتفي بطرح الحكم الشرعي

من هذه الفكرة الجينية أو ذلك المقترح الوراثي. ولعلنا بعد ذلك كله نجيز لأنفسنا أن نفترض أن تلك المنطقة الصحراوية والصامتة المتوسطة بين جينين ناطقين هي الجنبة الأخلاقية في سطر الجينوم، وهي الصغير الذي لا بد لانتظام الكبير على وفقه، من تعليق الكبير به، وتأخّره عنه في الرتبة، وانتظامه معه على سطر واحد، بدليل قوله تعالى: ﴿وَكُلُّ صَغِيرٍ وَكَبِيرٍ مُسْتَطَرٌ﴾ (القمر: 53). أما المنطقة المشفرة، والمقروءة، فهي الكبير الخلقي، على أن الذي هو باطن اليوم وارد فيه أن يكون ظاهرا غدا، وكمثله، الذي هو صغير اليوم بإمكانه أن يكون كبيرا غدا في حال تم تجاوز عتبة المجهول بوساطة التقانات العلمية. وفضلا عما سبق، تخلص بنا مقاربة الآية المباركة إلى ظنّ راجح مفاده: أن العمليات التأليفية التي تتوالى بموجبها انتظامات الخرائط الجينية سواء كانت على مستوى الجين أم على مستوى الخلية، ذات ترابط وثيق لا بد من مراعاته في كل مرة نحاول من خلالها أن نفكك او نركب تلك الجمل المتوالية حيويا، وإلا فإن عملية تقطيع أوصال تلك المنظومة بما لا ينسجم وإرادة الاستطار الكوني تعني نشأة اخرى فيها من الفوضى أكثر مما فيها من الانتظام، وهذا يعني في الأفق المستقبلي فناء مقصودا للكائن.

وتأسيسا على ما سبق كله، ينهض لدينا، أن العلاقة بين الصغير/ النمط الجينومي، والكبير/ النمط الظاهري هي علاقة استطار، تصير بموجبها الجنبة الخلقية والجنبة الأخلاقية للمخلوقات -بلحاظ أن الأخلاق بمعناها المقصدي سعي الكائنات إلى المحافظة على كمالها التكويني جينيا، أو استعادته- وجهان لحقيقة واحدة: وجه خفي وآخر ظاهر لـ (سطر) جينومي واحد، على اختلاف بين الأنواع والأفراد والأمم، في اللغة والمنهج والشرعة، بدليل قوله سبحانه: ﴿وَأَنْزَلْنَا إِلَيْكَ الْكِتَابَ بِالْحَقِّ مُصَدِّقًا لِمَا بَيْنَ يَدَيْهِ مِنَ الْكِتَابِ وَمُهَيْمِنًا عَلَيْهِ فَاحْكُمْ بَيْنَهُمْ بِمَا أَنْزَلَ اللَّهُ وَلَا تَتَّبِعْ أَهْوَاءَهُمْ عَمَّا جَاءَكَ مِنَ الْحَقِّ لِكُلٍّ جَعَلْنَا مِنْكُمْ شِرْعَةً وَمِنْهَاجًا وَلَوْ شَاءَ اللَّهُ لَجَعَلَكُمْ أُمَّةً وَاحِدَةً وَلَكِنْ لِيَبْلُوَكُمْ فِي مَا آتَاكُمْ فَاسْتَبِقُوا الْخَيْرَاتِ إِلَى اللَّهِ مَرْجِعُكُمْ جَمِيعًا فَيُنَبِّئُكُمْ بِمَا كُنْتُمْ فِيهِ تَخْتَلِفُونَ﴾ (المائدة: 48). ومن منظور الأممية البشرية الذي تتحدث عنه الآية المباركة، وبقرْن الآية بأختها، نقول: إن التكوين الأممي البشري ممتد في المخلوقات، فالأخيرة خاضعة لمثل ذلك التكوين، بدليل قوله تعالى: ﴿وَمَا مِنْ دَابَّةٍ فِي الْأَرْضِ وَلَا طَائِرٍ يَطِيرُ بِجَنَاحَيْهِ إِلَّا أُمَمٌ أَمْثَالُكُمْ مَا فَرَّطْنَا فِي الْكِتَابِ مِنْ شَيْءٍ ثُمَّ إِلَى رَبِّهِمْ يُحْشَرُونَ﴾ (الأنعام: 38).

ولعلنا، ونحن نستنطق قوله تعالى: ﴿إِنَّا كُلَّ شَيْءٍ خَلَقْنَاهُ بِقَدَرٍ﴾ (القمر: 49)، نملك مزيدا من الثقة بفرضيتنا العلمية التي تقول: إن جنبة الممكن والمحتمل في المستوى الجزيئي العميق للكائن

جنبة مرتبطة أصالة بجنبة الثابت غير القابل للتغيير، وإن محاولة تغيير ذلك الثابت تعني مجافاة حكمة القدر والتقدير وإلغاء عامل المقايسة والتقييس النوعيين، وبالمحصلة، إلغاء الذاكرة الجينية، وتعطيل عنصر التوقع الذي ينبني عليه أفق المستقبل الشخصي بالنسبة إلى الكائن. ومن هنا يبدو أن البعد الأخلاقي للممكنات الجينية مشروط بمراعاة التقدير الخلقي الأصيل للكائن، بلحاظ جنبتي الجبر والتفويض معا، فلا إمكان أخلاقيا للتحكم بالجهاز الوراثي البشري بعيدا عن تلك القدرة وذلك التقدير. ويتضح لنا، بناء على ما سبق، أن ثمة ثقلا اعتباريا للممكنات الجينية، لا بد أن يراعى فيه تعيير الخلق البايولوجي بالخُلق المعنوي تعييرا موزونا كيميائيا ضمن ثابت كيميائي رصين، وصبغة معنوية ثابتة، بدليل قوله تعالى: ﴿وَالْأَرْضَ مَدَدْنَاهَا وَأَلْقَيْنَا فِيهَا رَوَاسِيَ وَأَنْبَتْنَا فِيهَا مِنْ كُلِّ شَيْءٍ مَوْزُونٍ﴾ (الحجر: 19)، وهذا ما ينتهي بنا، ونحن نستدل بالنص الكريم، إلى أن ثمة ملازمة كبيرة بين الجنبتين لا بد من مراعاتها بغية تحقيق التوازن الجيني أخلاقيا، وما يترتب على تحقيق ذلك التوازن من أمن اعتباري إنساني، ممثلا بحرية الاختيار، وأمن بيولوجي، ممثلا بتعدد الهويات المظهرية والألسنية للكائن الحي، بدليل قوله: ﴿وَمِنْ آيَاتِهِ خَلْقُ السَّمَوَاتِ وَالأَرْضِ وَاخْتِلافُ أَلْسِنَتِكُمْ وَأَلْوَانِكُمْ إِنَّ فِي ذَلِكَ لَآيَاتٍ لِّلْعَالِمِينَ﴾ (الروم:22)، أما أن يجري التغيير دونما مجافاة ذلك التوازن الجزيئي الذي ينعكس في الواقع على شكل كيان متوازن، فذاك مما لا باس فيه.

وبنظرة أخرى إلى ما يؤكده اللسان العربي من فارق صرفي بسيط بين الجنبة الخلقية (بسكون اللام) والجنبة الخلقية (بضم اللام)، مفاده سكون لام الخلق أو ضمها، يتكشّف لنا، أن النظام التأليفي لمجموع الكتلة البروتينية للكائن لا يبعد كثيرا عن النظام التأليفي للغة الكائن. وعلى هذا، فإن تغييرا يحافظ للكائن على توازن بنيته البيولوجية تغيير وارد فلسفيا وأخلاقيا، أما أن يلجا البحث العلمي إلى عولمة الجين وفتح الحدود وضرب مبدأ السيادة الجينية، فهذا مما لا يمكن قبوله والاطمئنان إليه.

وبلحاظ ما سبق، وباستدعاء مقاصد هذا البحث ومقاصده، نقول: إن الباطن البيولوجي للإنسان متمثلا بنمطه الجيني ذي البعد الأخلاقي يحرك الظاهر متمثلا بالنمط الظاهري لذلك الكائن البايولوجي، نعم إن الظاهر حجة على الباطن، ولكن-إسلامياً-صيرورة الظاهر وتوسعته أمران مرهونان باستمداد الباطن. ولكن أن تسعى البيولوجيا المعاصرة إلى قطع صلة المادي/

البروتيني، بالمعنوي/ الأخلاقي، ومن ثم، السعي إلى إنفلات التضام البنيوي والجدلي بين ذينك البعدين، فهذا يعني صناعة بيوطيقا أو علم أخلاق حيوي فاقد لهويته الإنسانية المائزة.

المراجع:

القرآن الكريم

الآمدي، ناصح الدين. 1992. غُرر الحِكَم ودُرر الكلم المفهرس من كلام أمير المؤمنين علي بن أبي طالب عليه السلام. بيروت.

ابن أبي الدنيا. (د.ت). مكارم الأخلاق. تحقيق وتعليق مجدي السيد إبراهيم. بولاق: مكتبة القرآن للطباعة والنشر.

ابن خلدون، عبد الرحمن.2004. مقدمة ابن خلدون. تحقيق عبد الله محمد الدرويش. دمشق: توزيع دار يعرب .

ابن عجيبة. 1999. البحر المديد في تفسير القرآن المجيد. تحقيق وتعليق أحمد عبد الله القرشي رسلان. (3):457. القاهرة: الهيئة المصرية العامة للكتّاب.

ابن عربي. 2003. التدبيرات الإلهية في إصلاح المملكة الإنسانية. تحقيق الدكتور عاصم إبراهيم الكيالي. بيروت: دار الكتب العلمية.

ابن كثير. 1999. تفسير القرآن العظيم. تحقيق سامي بن محمد السلامة. (5):338-339. الرياض: دار طيبة للنشر والتوزيع.

ابن منظور، محمد. 1993. لسان العرب. بيروت: دار الصادر.

البقاعي. (د.ت). نظم الدرر في تناسب الآيات والسور. (12):412. القاهرة: دار الكتاب الإسلامي.

حاتم، جاد. 1985. يحيى بن عديّ و «تهذيب الأخلاق» دراسة ونص. بيروت: دار المشرق.

الخلف، موسى. 2005.العصر الجينومي- استراتيجيات المستقبل البشري. الكويت: المجلس الوطني للثقافة والفنون والآداب.

دوبرو، كلود. 2007. الممكن والتكنولوجيات الحيوية. ترجمة د. ميشال يوسف. بيروت: المنظمة العربية للترجمة.

ستيس، ولتر. 1989. الدين والعقل الحديث. ترجمة أ.د. امام عبد الفتاح امام. القاهرة: مكتبة مدبولي.

دني، ي. 2012. أصول الأخلاق. ترجمة إبراهيم رمزي. القاهرة: مؤسسة هنداوي للتعليم والثقافة.

عبد الرحمن، طه. 2012. سؤال العمل بحث عن الأصول العملية في الفكر والعلم. الدار البيضاء: المركز الثقافي العربي.

الفارابي، أبو نصر. (د.ت). كتاب آراء أهل المدينة الفاضلة. تحقيق الدكتور البير نصري نادر. بيروت: دار المشرق.

فوكوياما، فرانسيس.2006. مستقبلنا بعد البشري- عواقب ثورة التقنية الحيوية. ترجمة إيهاب عبد الرحيم محمد. أبو ظبي: مركز الإمارات للدراسات والبحوث الاستراتيجية.

كانت، إمانويل. 2002. تأسيس ميتافيزيقا الأخلاق. ترجمة الدكتور عبد الغفار مكاوي، مراجعة الدكتور عبد الرحمن بدوي. كولونيا: منشورات الجمل.

كبرى زاده، طاش. 1989. موسوعة مصطلحات مفتاح السعادة ومصباح الزيادة في موضوع العلوم. تقديم وإشراف ومراجعة د. رفيق العجم، تحقيق د. علي دحروج. بيروت: مكتبة لبنان ناشرون.

م. أوغروس، روبرت، و جورج ن. ستانسيو. 1989. العلم في منظوره الجديد. ترجمة د. كمال خلايلي. الكويت: المجلس الوطني للثقافة والفنون والآداب.

ماير، ارنست. 2002. هذا هو علم البيولوجيا- دراسة في ماهية الحياة والأحياء. ترجمة د. عفيفي محمود عفيفي. الكويت: المجلس الوطني للثقافة والفنون والآداب.

نيتشه، فريدريك. 2011. إرادة القوة. ترجمة محمد الناجي. الدار البيضاء: أفريقيا الشرق.

الفصل 12

الجينوم والحياة: تمديد الحياة وأثره الأخلاقي على المجتمعات الإسلامية

عمارة الناصر [1]

مقدمة:

إن رغبة البشر في حياة مديدة وشباب دائم هي رغبة نابعة من أمل فطري في البقاء والتشبّث ببقايا الوجود الذي يبدأ في التصدّع على أعتاب الشيخوخة، وهو الأمل الذي حوّلته الكيمياء القديمة إلى مشروعٍ لصناعة «ترياق الحياة» المُعبَّر عنه في ذاك العقار الأسطوري «إكسير الحياة» (Elixir of life)، في سعيٍ لشفاء جميع الأمراض وإطالة الحياة إلى أبعد مدى، ولم يتوقف هذا السعي بل ظلّ حاضراً في تاريخ العلم وفي إرادة بشرية عميقة في مقاومة آثار التقدّم في العمر وتجاوز لحظة الانهيار البيولوجي الذي يسحب معه كينونة نخَرَها الزمن من الداخل؛ فالشيخوخة تجربةٌ إنسانية للعيش بجسدٍ لا يستجيب لتطلّعات صاحبه بل يصبح عبئاً ثقيلاً عليه، وهنا تصبح مشاهدة الحياة من داخل جسد واهن وضعيف أمراً مرهقاً.

إن مَلْءَ تجربةِ الشيخوخة بمزيدٍ من الحياة وضخّها بزمنٍ أطول هو مشروعٌ علميٌّ يُبشّر اليوم بآمالٍ كبيرة، غير أنه ينطوي على مخاوف حقيقية تُنذر بحدوث انفلات «للكائن الأخلاقي» من حدود الطبيعة البشرية المحروسة بمفاهيم «الطبيعي»، «العادي» و«الأصلي»...، ويُراهن هذا المشروع على فكرة أنه يمكن للجسد البشري أن يُصلح نفسه بنفسه ليستمر في الحياة، بما يملكه من مَلَكات بيولوجية داخلية تحمل شيفرات البقاء وهذا عبر وسيط التقنية الحيوية، غير أن هذا يعتمد على تصورٍ عميق لمفهوم الحياة ذاتها وهو التصوّر الذي يضمن تماسك ماهية «الإنسان» وتناغم

1 أستاذ فلسفة التأويل والفلسفة الغربية المعاصرة، جامعة مستغانم، الجزائر،
naceur.amara@univ-mosta.dz

تركيبته البيولوجية مع هويته الأخلاقية، وهذا ما يضع العلم (البيولوجيا خصوصاً) أمام المسؤولية الأخلاقية لحماية الرأسمال الرمزي للإنسان الذي يُمثّله تراثٌ من القيّم الكونية التي يصطبغُ بها تاريخ الجسد البشري المتحرّك داخل بيئة مصطنَعة حضارياً ومزوّدة بجهاز ثقافي يمنع تسرّب الحياة خارج مجال العيش المشترك للمجموعة البشرية.

يُقدِّم علم الجينوم، اليوم، مَطلبَ تمديد الحياة على أنه «حقٌّ» إنساني «طبيعي ومشروع»، إذ توجد إمكانية تقنية لـ«تحرير» رسالة جينية تستطيع أن تُعدّل بعض ما كُتب فيها لتكتبَ من جديد ما يُفيد بأنه يوجد في جسد الكائن الحي ما يسمحُ له بالتمدّد في الزمن البيولوجي متخطياً حواجز المرض، وأنه يمكن مَنح الإنسان زمناً إضافياً «للعيش»، وأنَّ ما كنّا نعتقد أنه حياة ليس إلا جزءاً منها، بما أن مفهومنا للحياة نفسها بدأ يتغيّر بعد أن أدركنا أن ما يُحدّد شكل الحياة في الكائن الحي هو المعلومة الجينية وليس المادة الحاملة لها.

ولكن ما شكلُ الحياة التي يهدفُ علم الجينوم إلى تمديدها؟ ما الذي يُبرّر شغف الناس بمزيدٍ من الحياة؟ هل هو طول الحياة أم جودتها؟ وكيف ستتفاعل المجتمعات الإسلامية أخلاقياً مع مشروع تمديد الحياة باستخدام التقنية الحيوية للجينوم؟

أولاً: شكل الحياة: من الجينوم إلى ما بعد- الجينوم.

يبدأ فهم الحياة بوصفها «حدثاً بيولوجياً» بفهم ما يجري في هذا الحدث وتوقّع ما يمكن أن يؤول إليه خلال تطوّره المورفولوجي في الزمن، أي بالاقتراب من «ظاهرة» الحياة في مادتها الحيوية الأولى التي تُمثّل مصدر تَشكّلها، وفي هذا السياق يمكننا العودة إلى الفيزيائي النمساوي الشهير «إرفين شرودنغر» (Erwin Schrödinger) لطرح السؤال الذي صاغه في كتابه «ما الحياة؟»: «كيف يمكن للأحداث المكانية والزمانية التي تجري في الحدود الفضائية للكائن الحيّ أن تتمثّل بواسطة الفيزياء والكيمياء؟» (Schrödinger 2006, 03)، وتقوم البيولوجيا المعاصرة بمَهمّة الإجابة عن هذا السؤال الجوهري بالربط بين شكل الحياة وبنيتها الفيزيائية والكيميائية، حيث وضَع «اكتشافنا لـDNA نقطة نهاية لنقاش قديم حول الجنس البشري: هل للحياة ماهية سحرية أو لغزية، أم أنها، مثل أيّة ردّة فعل كيميائية درَسها تلاميذٌ في حصة الكمياء، من نتاج عمليات فيزيائية- كيميائية عادية جداً؟ وهل توجد في قلب الخلية شرارة إلهية تنفخ فيها الحياة؟» (Watson 2003, 14) حيث ليس لاكتشاف DNAهذا أهمية في فهم الترابط الجيني للكائن

البشري فقط بل له أهمية في فهمٍ أكثر عقلانية لماهية الحياة ومصدر تشكّلها ومن ثمة أمكن تصور «التاريخ» المُدمج في الخلية والذي يعبّر عنه بيولوجياً بالقدرة على البقاء على قيد الحياة.

إن فهم ما يدفع الحياة ويُطوّرها في الكائن الحيّ يسمح أيضاً بفهم ما يُوقفها ويكبح قواها الطبيعية في التطور، وهو ما يعني أن مفهوم «الطبيعة» نفسه قد تغيّر ولم يَعد يشير إلى الثبات تحت صفة «العادي»، «فعلماء البيولوجيا الجزئية يقولون أن نشاطاً بسيطاً كحديث أو لعبة الكلمات المتقاطعة يُعدّل الحالة البيوكيميائية للدماغ» (Rogers 1995, 106)، وبهذا تكون مَهمّة البحث الجيني في محاولة إيجاد الأشكال المشتركة للحياة عبر ضبط القاعدة الفيزيائية والكيميائية التي تشكل وسيط الحياة بين الكائن الحي والنشاط المتغيّر لوجوده.

لقد أشار أحد مكتشفي بنية الـ DNA جيمس واطسون (J.D.Watson) إلى أنه «منذ سنة 1944 ظهرت المؤشرات القاطعة التي تُظهر أن DNA هو الجهاز الجيني الذي يحمل مخطط الحياة العضوية كاملة، لكن لم تكن لأحدٍ الإجابة عن سؤال كيف يمكن لهذه المعلومة كلها أن تكون مخزّنة في مُركب كيميائي (Polymère)، أي في سلسلة وحدات كيميائية مرتبطة ببعضها البعض (les nucléotides)، لم يعلم أحدٌ كيف يمكن للمعلومة أن تنتقل إلى الأحفاد عندما تنقسم الخلايا» (Watson 2003, 15)، وبهذا بدأت مرحلة البحث في قراءة فيزيائية وكيميائية للترميز الجيني الذي يختزل تصميم الكائن البشري في سلسلة طويلة من القواعد التي يؤدي التبادل الوظيفي فيما بينها إلى «صناعة» الصفات والخصائص الخَلقية وحتى النفسية والسلوكية، حيث «إن المعلومة الوراثية التي انتقلت إلينا بواسطة آبائنا يتمّ تخزينها في جزيئات DNA، وفي عمل الكروموزومات التي تمّ استقبالها من أبوينا، فإن هذه المعلومة مشفرة بمعدل 3,5 مليار من الأزواج القاعدية لـDNA، ويمكن اعتبار هذه المعلومة كمجموع التعليمات الخاصة بتصنيع الكائن البشري» (Edelstein 2002, 59)، وهذا ما يفسر نسبياً الطبيعة الفيزيائية-الكيميائية للجينات وطريقة تخزينها للمعلومة الوراثية ضمن مرحلة مُهمّة من مراحل تطور البيولوجيا الجزئية (Molecular Biology) والتي تضع من بين أهدافها التجريبية مسألة البحث في الأشكال الجزئية المشتركة لجينات الكائن الحي بما يسمح بوضع المقاربات الوظيفية والتحويلية بينها.

يُمكن للجينوم، مَسْكن الجينات (gene's home)، أن يُجيبنا عن أسئلة مثل: «مَن نحن؟ من أين حصلنا على ما نحن عليه؟ أيُّ قَدَر محتوم لنا قبل حتّى أن نولد؟ أسئلة طُرحت قبل أن نتمكّن من ترجمتها جينياً (..) فأصبح لدينا جينات للأمراض، جينات للميول، جينات للسلوك،

وحديثاً علمنا باكتشاف جين الوفاء الزوجي (لدى الحيوان)» (Pouteau 2007, 08)، وقد مكّنت عملية السَلسَلة (Sequencing) الكاملة للجينوم البشري من وضعه على عتبة مشروع أكثر تطوراً في مدخل القرن الواحد والعشرين، في ما يُسمّى عصر»ما بعد الجينوم» (post-génome) حيث أمكننا التمييز بين مرحلتين من مشروع الجينوم البشري:

أ. المرحلة الجينومية:

وهي المرحلة التي «بدأت سنوات 1980 واكتملت في نهاية سنوات 1990 والتي تميّزت بتطور النظرية الداروينية الجديدة (neodarwinism) للتطور والبيولوجيا الجزيئية، حيث تحقق علماء البيولوجيا من أن وراء التنوع الهائل لأشكال الحياة تختبئ هوية جزيئية فائقة» (Perbal 2011, 35)، وهي المرحلة التي فتحت الباب البيولوجي أمام سؤال شرودينغر «ما الحياة؟» عبر الإجابة عن أسئلة أساسية: «ما طبيعة الشيفرة الوراثية؟ وبأية آلية يمكن للمعلومة الواردة في سلسلة DNA أن تنطبع على سلسلة بروتينية؟ ما دور RNA (الحمض النووي الوسيط في تركيب البروتينات) هذا المركب الكيميائي اللغز والذي يتم استقلابه بشكل سريع جداً في الخلايا النشطة» (Watson 2003, 19)، وأدّت هذه الأسئلة إلى توسيع دائرة فهمنا للعالم المتشكّل بواسطة البنية الجزيئية لخصائص الكائن الحي المنطبعة على «بروتين الحياة» والتي تترجم إلى خرائط جينية «تُستخدم في تحسين الأنواع الحيوانية والنباتية وكذا في مجالات مختلفة من البحث البيولوجي (..) عبر تحديد وضعية الجينات في الجينوم سواء من خلال المسافة الفيزيائية أو من خلال وضع مبني نسبياً على تعالقات فيما بينها» (Gibson 2004, 02)، فمعرفة عدد الجينات وأبعادها ومواضعها والمسافات الرابطة بينها تعني الحصول على معلومات مهمّة حولها كأن نعرف خطط إنتاج البروتينات المخزّنة للنظام التطوري للكائن البشري والذي يُمكّن من قراءة تاريخ المرض ومستقبله أو تحديد شكل النمو والطفرات التي تطرأ عليه.

لقد مرّت المرحلة الجينومية بطورين: الأول هو تحديد سلسلة DNA كاملة والتي تشكّل التراث الجيني للكائن البشري، أمّا الطور الثاني فهو التعرّف على أجزاء سلسلة DNA التي توافق الجينات «لأن الجزء الأكبر من سلسلة DNA الكاملة للجينوم البشري لا يحتوي على جينات؛ فهي مبعثرة على طول سلسلة DNA الكاملة، ولا تحتلّ إلا 3% من قواعد DNA (Edelstein 2002, 60) فتكون مَهمَّة علماء البيولوجيا الجزيئية في البحث عبر سلسلة DNA عن

المواضع الجينية ومسافاتها، وبالفعل فقد «قام جون دوسيه (Jean Dausset) ودانييل كوهن (Daniel Cohen) بتحديد تموضع الجينات جيناً بجين وكذا تحديد سلسلة الجينوم من نفس العائلات» (Rabinow 2000, 59)، وهذا يعني التوصّل إلى معرفة النظام التسلسلي الكامل للجينوم والذي يقدّم المؤشرات البيولوجية الكافية للتعرّف على الطريقة التي تُنتج بها المُركّبات العضوية بين الإنسان والحيوان البروتينات نفسها، حيث تُطوّر التقنيات البيوتكنولوجية هذه المعلومة وتحوّلها إلى معلومة علاجية.

ب. المرحلة ما بعد- جينومية (post-génomique):

وهي المرحلة التي بدأت «بفكّ شيفرة الجينوم بشكل شبه كامل (99%) في شهر جوان من سنة 2000، بفضل آلات قوية وجديدة واستراتيجيات مبتكرة» (Edelstein 2002, 59) ومع اعتبار هذه النتيجة حدثاً مهمّاً إلا أنها لا تمثّل إلا البداية لعصر جديد من مشروع الجينوم الذي بقيّ الكثير من ألغازه غامضاً وبحاجة إلى المزيد من الأبحاث المخبرية «فما تمّ فكّ شيفرته في الجينوم البشري هي «الحروف» «الأبجدية» الجزيئية للتراث الجيني للإنسان، لكن معنى تلك الحروف واستخدامها يبقى أمراً غامضاً» (Hervé 2006, 50)، لأن التفسير العلمي الذي تُقدّمه البيولوجيا الجزيئية حول الطبيعة الفيزيائية والكميائية للجينوم البشري لا يُبرّر الكثير من العلاقات الوظيفية بين مكوناته، فبقدر ما «مثّل نشر التسلسل الكامل للجينوم البشري منعطفاً كبيراً للبيولوجيا، إلا أن هذا «النص» الطويل بثلاث مليارات حرف، المستخدم لتهجئة من أربع جزيئات، مرمّزة بالحروف A,T,G,C، لا يكشف شيئاً كبيراً عن الكائن البشري وهو يُضاعف المشكلات أكثر ممّا يحلّها» (Pouteau 2007, 07)، إذ يظهر أن ما يدفع الحياة في بنية الكائن الحي أعمق من أن يُختزل في رموز حرفية في صلب الجينوم وهذا يعني المرور إلى تفسيرات ما بعد- جينومية لما يحدث في التركيبة الجينية ضمن حركة تطوّرية لشكل الحياة الذي تستندُ إليه البيولوجيا الجزئية لتفسير الروابط الأكثر عمقاً بين التراث الجيني للإنسان وتكوينه النفسي والسلوكي، إذ «إن اشتغال العلماء على الأشكال التي تُنتج الكائن الحي باستمرار (life-forms) يرتدُّ إلى أشكال حياتها (forms of life) أي إلى طريقة سَكَنها للعالم» (Rabinow 2000, 12) فالكشف عن حياة الجينوم لا يعني الكشف عن جينوم الحياة بشكل كامل أي أن فك شيفرة سلسلة DNA، لا يُفسر شيفرة الحياة كاملة لأن أشكال الحياة لا تخضع لذلك النظام الداخلي للجينات ولهذا طرح بول رابينو (Paul Rabinow) فكرة أن «تحديد هوية الـDNA مع «الشخص الإنساني» هو تحديدٌ

«روحي»» (Rabinow 2000, 38)، كما أن العديد من علماء البيولوجيا ينسبون تكيّف أنواع الكائنات الحية إلى قوة روحية كونية.

في المرحلة ما بعد- جينومية تتجاوز التفسيرات البيولوجية البنية الفيزيائية- الكميائية إلى البنية السلوكية لتطور الكائن الحي، كما أنه «في عصر ما بعد- الجينوم تتطور المفاهيم المفتاحية للبيولوجيا الجزيئية فيما يتعلق بمفهوم الجين، المعلومة أو البرنامج الجيني» (Perbal 2011, 15)، وهو تطورٌ لبراديغم العلم نفسه ليتكيّف مع التحدّيات الإبستيمولوجية التي تتعلق بمعرفة ما لا يستجيب للطرائق المنهجية أو التجارب العلمية المعروفة ولا يخضع لمنطق عقلاني معين، فسؤالٌ مثل: «كيف تتحوّل المعلومة المتواصلة في DNA جينٍ ما إلى سلسلة أحماض أمينية لأنزيم ما أو لسلسلة بروتينية أخرى، يبقى سؤالاً دون إجابة» (Pastermark 2003, 15) ممّا يستدعي البحث في أُطر أخرى يُنتج داخلها النظام الجيني الكائنَ الحي مثل البيئة والتي تمثّل مصدراً لأشكال الحياة من خلال السلوكات التي تعكس ارتباط الشيفرة الجينية بتلك الأشكال وهو ما يسمح بتفسير بعض الظواهر البيولوجية التي عجز عن تفسيرها مشروع الجينوم في مرحلته الأولى وعلى رأسها طبيعة المعلومة أو الرسالة المنتقلة عبر الجينات بما يرسم خيطاً للحياة، إذ قد بدأنا نُدرك أن «خيط الحياة ليس خيطاً مادياً، إنه رسالة، رسالةٌ وراثية متشكّلة عبر الزمن. كل جيل يُرسل رسالة إلى الجيل الذي يليه، إنّه يُكلّمه، بطريقة ما، لكي ينقل إليه طريقته في العيش. إذا أردنا البحث عن هذه الرسالة فيمكن أن نجد آثاراً لها في DNA كروموزوماتنا، في بنية أنظمتنا البيئية أو في المحادثات بين البشر، لكن هذه الرسالة لا تُختزل إلى آثارٍ مادية سريعة الزوال» (Dessalles 2016, 02)، فتفسير حياة الكائن الحي، إذن، يتجاوز المعطيات المادية التي يُقدّمها مشروع الجينوم إلى المعطى اللاّمادي الذي تُعبِّر عنه الأبعاد والمنظورات العقلانية للإنسان والمندرجة ضمن تاريخٍ من اللاّمُبرمَج واللاّمتوقَّع.

ثانياً: جين الشباب: تمديد الحياة ضد الشيخوخة.

لقد تسارعت التطورات العلمية في العلوم الطبية والبيولوجية متجاوزةً الأهداف العلاجية والعمليات التجميلية وعمليات زرع الأعضاء والتعديلات الجينية التقليدية كالاستنساخ وغيرها إلى عمليات ذات «رفاهية» عالية تتعلق بالبحث عن الجينات المسؤولة عن إطالة العمر لدى الإنسان وتحريرها بيولوجياً أو تطعيم الجينوم البشري بجينات تمديد الحياة الموجودة لدى كائنات

أخرى أو التنبؤ الجيني بالأمراض وتعديلها علاجياً قبل وقوعها..»، وقد بدأ «الدليل العلمي للتلاعب الجيني لتمديد الحياة، مع أندريه بارتك (Andrzej Bartke) وزملاؤه في مقالهما سنة 2001 المعنون: (Genes that prolong life and relationships of growth hormone and growth to ageing and life span) [الجينات المُطيلة للعمر والعلاقات بين هرمون النمو والنمو مع الشيخوخة ومدة الحياة] والذي صرّحا فيه بأن الدراسات المقامة على الخميرة، الديدان والذباب قد وفّرت أدلة كثيرة على وجود جينات تتحكم بالشيخوخة وطول الحياة» (Struckelberger 2008, 101)، ويُعيد هذا الدليل النظر في تصورنا لمفهوم الشيخوخة نفسه وكذا مفهوم الحياة ككل، تلك الشيخوخة التي يتمّ انتظارها في أفق نفسي يجمع بين المرض والضعف والانهيار العضوي والعقلي والنفسي كمقدمة سريعة ورهيبة للموت. إن «الشيخوخة ظاهرة بيولوجية تختلف بطريقة مدهشة من نوع إلى آخر؛ لنلاحظ مثلاً، عند الثدييات، الفرق الموجود بين الفأر والإنسان. معدّل مدة الحياة عند فأر المخبر العادي تمّ توقّعه، في دراسة حديثة حول الشيخوخة بـ761 يوما، وعند الإنسان، في البلدان التي تتمّع بمعايير عالية للصحة، فإن معدل الحياة هو حوالي 75 سنة» (Edelstein 2002, 85)، وعليه فإن الطبيعة البيولوجية لظاهرة الشيخوخة تسمح بإمكانية التدخل الجيني في مسارها والقيام بعمليات «تلاعب» جيني بالكائن البشري في هذه المرحلة العمرية «لمكافحة» الشيخوخة أو تأخيرها.

تعتمد مسألة تمديد الحياة وإطالتها على مؤشرات ديموغرافية تتعلق أساساً بـ «تضاعف أمل الحياة عند الولادة على مدى القرنين الماضيين، لدى الشعوب الأكثر حظاً، مدعوماً بالتناقص الكبير في الوفاة المتعلقة بالولادة وتطور التلقيحات. حالياً في فرنسا يستمر أمل الحياة عند الولادة في الارتفاع، ليتجاوز 74 سنة عند الرجال و82 سنة عند النساء» (Klarsfeld 2000, 51)؛ وتدلّ هذه المؤشرات أيضاً على أن الشيخوخة ظاهرة متغيّرة عبر التاريخ بفعل العوامل الوراثية والبيئية المختلفة أي أنه بالإمكان تمديدها وتأخير «أعراضها» بفضل التدخل في تلك العوامل وتغيير طريقة تفاعلها داخل أشكال الحياة التي يمكنها أن تتطور خطياً دون أن تفقد القدرة على ثباتها البيولوجي.

تواجه فرضية إطالة الحياة البشرية وتمديدها بتأخير الشيخوخة سؤالاً مُهمّاً: هل يمكن إيجاد جينات مسؤولة عن إطالة العمر؟ وهل بالإمكان التلاعب بتلك الجينات دون حدوث انتكاسات بيولوجية ونفسية وأخلاقية؟ تعتمد الإجابة عن هذه الأسئلة على مدى التحوّل في

تناول ظاهرة الشيخوخة والتحكم فيها إبستيمولوجياً، ولذلك «نحن نعرف الآن أن الشيخوخة توجد بوصفها ظاهرة منفصلة، وبدأنا نفهم ذلك، ونحن نريد بطبيعة الحال أن نرى ما إذا كنا نستطيع العبث بتلك العملية، فقد قال عالم شيخوخة بارز هو «ليونارد هيفليك» (Leonard Hayflik): «الشيخوخة منتوجٌ من صُنع الحضارة» (Hall 2003, 04)، ومن ثمّة أمكن إدخال ظاهرة الشيخوخة إلى المخبر عبر العمل على التجارب الجينية على الحيوانات، فقد تمّ «القيام بخمس عشرة عملية تلاعب جيني مختلفة لحث إطالة أمد الحياة في الكائنات الحية مثل ذباب الفاكهة والديدان الخيطية والفئران، وهو ما يعني أن الزيادة في متوسط العمر المتوقع في النماذج الحيوانية يمكن أن يتحقق جينياً» (Struckelberger 2008, 102) وهذا يعني أن هذه النماذج تمتلك القابلية الجينية لطول العمر وتأخير الشيخوخة، وبهذا نتوقف عملية التعديل الجيني لطول الحياة على إيجاد النظائر الجينية لدى الإنسان، وقد كان «أول جين لطول العمر للخميرة تمّ وصفه هوLAG1 (D'mello et at (LAG1) سنة 1994، وتمتلك الخميرة نظيراً لهذا الجين يسمى LAC1 (Jazwinski)) سنة 1996، (Jang et at) (سنة 1998)، وقد تمّ استنساخ نظائر هذا الجين لدى الإنسان، إذ يمكن للنظير البشري أن يُحسّن من وظيفة جين طول العمر في الخميرة» (Hekimi 2000, 26)؛ كما تحدّثت سينثيا كينيون (Cynthia Kenyon) عالمة بيولوجيا جزيئية كبيرة في جامعة كاليفورنيا، عن «عملها في تحديد «الجين الحاصد» (grim-reaper gene) و»جين نافورة الشباب» (Fountain of youth gene) في الديدان الخيطية، حيث توقعت أن يكون تمديد العمر الافتراضي (life-span extension) حقيقة في القرن الواحد والعشرين» (Hall 2003, 08)؛ إذ يمكن إيجاد نظائر جينية لدى الإنسان لإطالة الحياة عبر التلاعب الجيني بخريطة الجينوم البشري لتصبح الخلايا قادرة على تجديد نفسها لمدة أطول ممّا هي عليه في العادة، وهذا يؤخّر مظاهر العجز في الأعضاء ويؤجّل بعض الأمراض الأكثر شيوعاً والمرتبطة بمرحلة الشيخوخة.

ثالثاً: التمديد الجيني للحياة: حدوده وآثاره.

يُبشّر التعديل الجيني بتمديد عمر البشر من خلال إسكان روح الشباب في وظائفهم الحيوية لتتوهّج شرارة الحياة في خلاياهم فتتأجل آثار الأمراض المتوقعة، إذ «حتّى عملية الشيخوخة يمكنها أن تستمر بمساعدة التلاعب الجيني كما يمكن إلغاء الموت نفسه» (Hervé 2006, 51) ولكنها لا

تستمر دون مرافقة علاجية لما يُخلّفه تقدّم العمر من تآكل في البنية البيولوجية للجسد وانهيار للوظائف العضوية وانحسارٍ للقوة المناعية للجسم، ولهذا «تبدو فكرة الشفاء بواسطة الجينات فكرة مغرية وطبيعية (...) ولذلك تستخدم كثيراً عبارة «DNA الدواء» والتي تعبّر جيداً عن هذا الأمل، في فكرة أنه بمجرد إدخال الجين، فإن الميكانزمات الخلوية «تقوم بعملها» في تعبيرها وتعديلها، منتجة البروتين العلاجي» (Jordan 2007, 10)؛ ويتدخّل في هذه العملية علمان آخران الأول قبْلي والثاني بَعدي هما:

أ. اليوجينا (Eugenics):

علم تحسين النسل، وهو المصطلح المطوّر سنة 1883 من طرف فرانسيس غالتون (Francis Galton) والذي يستند إلى فكرة أنه يجب على العملية التطورية أن تساهم في تحسين النوع الإنساني(Perbal 2011, 22)، وذلك من خلال «إمكانية إزالة كل «ضعف» وكل مرض، وتقليص «الجينات المضرّة بالناس» وتحسين بنائها البيولوجي بواسطة اختيار أفضل الصفات» (Hervé 2006, 50)، وتُجرى هذه العملية قبل الولادة أي قبل اكتمال التشكيل الخَلقي للجهاز البيولوجي للإنسان الذي يخضع لطبيعة النمو حيث تشارك الجينات في عملية بنائية معقدة من التفاعلات الخلوية؛ ولهذا يتمّ مثلاً «استباق ولادة طفل حاملٍ لمرض وراثي بـ «تصحيح» الجين الذي به عيب قبل الولادة» (Beland 2006, 42) وهذا يؤدي إلى استباق لتطور المرض، في مرحلة الشيخوخة، إلى أمراض أكثر فتكاً وهو ما يعني نظرياً تأخير عملية الشيخوخة وتمديد العمر بتجاوز الأمراض التي تُعطّله.

ب. الإيبيجينا (Epigenetics):

وهو العلم الذي يدرس التغيّرات الحاصلة في الكائنات والناجمة عن تعديل الجينات بدلا من تغيير الشيفرة الوراثية نفسها (Moore 2015, 08)، وهو من علوم مرحلة ما بعد الجينوم أي فيما بعد التعديل الجيني وما يرافقه من تغيّرات فيزيولوجية على مستوى الوظائف الحيوية وآدائها والتفاعلات بين الجينات.

يندرج تحت علم الإيبيجينا علم آخر هو علم التخلّق السلوكي (Behavioral Epigenetics) «والذي يدرس تأثير علم التخلّق على العمليات السيكولوجية مثل الانفعالات الشعورية، الذاكرة

والتعلم، الصحة العقلية والسلوك» (Moore 2015, 08) لأن الأمر لا يتعلق لطول الحياة فقط بل بجودتها كذلك، فالأمل في تمديد العمر يمكنه أن يتوقف أو يتراجع أمام التهديدات الصحية الأكثر قساوة والتي تصاحب فترة التمديد الجيني للحياة البشرية.

على الرغم من التحقق الفعلي للتجارب العيادية لطب الشيخوخة بالعلاج الجيني، إذ «قد بدأت بعض العيادات المتخصصة في محاربة الشيخوخة في توفير جواز سفر جيني لمرضاها لزيادة نسبة نجاح التدخّلات العلاجية عبر ما يسمى «عيادة التنبؤ الجيني» (Gene Predict Clinic): «نحن ندعوك للحصول على جواز سفرك الجيني- الملف الشخصي- والذي من شأنه أن يؤدي إلى علاجات مصمّمة خصيصاً لك، وأدوية ذات فعالية (دون آثار جانبية) وإلى شيخوخة متباطئة في إحدى آأمن الدول وأكثرها سرية (سويسرا)» (Struckelberger 2008, 105)، إلا أن هذا المشروع ما بعد الجينومي يواجه مشكلات تقنية متعلقة بالأبحاث الجينية وعوائقها الإبستيمولوجية ومشكلات أخلاقية واجتماعية متعلقة بالانزلاقات الإيتيقية لهذا المشروع والتي تؤدي إلى انثقاب كينونة الإنسان وتسرّب أشكاله الجوهرية المتماهية مع الحياة في نواتها الأنطولوجية:

أ. المشكلات التقنية:

تعترض عمليات التعديل الجيني المستهدفة للعلاج الجيني وتأخير الشيخوخة، عوائق مرتبطة ببنية الجينات نفسها وعلاقتها ببعضها، إذ «في الواقع، على الرغم من أن الجين يمكنه أن يندرج فعلياً في DNA الخاص بالخلايا المستهدفة، إلا أن مجرد وجوده في الكروموزوم لا يضمن أن يتمّ التعبير عنه آلياً (أي ترجمته إلى شكل من أشكال البروتين): حيث في الحقيقة، حسب المكان الذي يحتله على الكروموزوم، فإن التعبير عنه في شكل بروتين سيكون أكثر أو أقل كثافة، فيكون من الضروري إذن إجراء أبحاث جديدة لمعرفة وضع الجينات المحوّلة على أجزاء الكروموزومات والتي تؤدي إلى التعبير عنها بأكثر فعالية» (Edelstein 2002, 26)، لأن عملية إنتاج البروتين المعبّر عن الكروموزوم المعدّل بجين جديد هي عملية غير متوقعة النتائج وليست دقيقة مخبرياً حيث إن المشكلة قائمة في التعارض الذي يحدث بين البروتينات «الأصلية» الناتجة عن التكوين البيولوجي لحظة الولادة والبروتينات الناتجة عن جينات معدّلة وهذا ما يؤدي إلى تحفيز الجهاز المناعي ضد الجينات المعدّلة (كما في بعض حالات زرع الأعضاء) لأن «النظام المناعي مؤسس على مفاهيم «الذات» و»اللاّ- ذات» وبالتالي فإنه في ما يأتي من الحياة، إذا دخل عامل غريب إلى الجسم

أو إذا بدأت خلايا الجسم في إنتاج بروتين شاذ، فإن النظام المناعي يحثّ على مهاجمة بروتينات «اللاّ- ذات»» (Edelstein 2002, 35)، فترجمة الجين إلى بروتين ليست مرتبطة بالتموضع داخل الكروموزوم الحامل له والمراد علاجه أو تحسين الصفات التي يحملها فقط بل هي مرتبطة بالعلاقات بين الجينات ببعضها، ولهذا فإنه «في الكثير من الحالات العيادية البشرية، لا تخبر المعرفة بالسلاسل الجينية إلا باحتمال المرض دون يقين يتعلق بحدوثه أو خطورته» (Pouteau 2007, 08)، إذ لا يمكن التنبؤ بما يمكن أن يحدث في الكروموزوم الحامل للمرض الذي يمكن أن يختفي في المستقبل أو يبقى في حالة كمون أو وجود بالقوة، وهذا ما نستطيع تسميته بـ «الجين الحامل» والذي قد يؤدي تعديله علاجياً إلى إنتاج أمراض أخرى غير متوقعة وغير مُسيطَر عليها، ولهذا «تبقى الطبيعة الحقيقية للعلاقة بين الجين المتضرر ومرضٍ ما لغزاً كبيراً» (Pastermark 2003, 14)، وهذا يدعو إلى البحث في الجزيئات المتحكمة في إنتاج البروتين انطلاقاً من الجينات وفهم آلية التعبير عنها بالميكانزمات غير التقليدية للبيولوجيا الجزيئية.

تُمثّل عملية التدخّل الجيني المتعلق بإطالة العمر، إذن، تحدّياً أمام العلاج الجيني للشيخوخة لأنه «عكس الخلايا الدموية، فإن أغلب الأنواع الأخرى من الخلايا يصعب عزلها أو انتقاؤها بواسطة قوة موجِّهة معينة، وأكثر من ذلك، في بعض الحالات، تصدر ردّة فعل مناعية ضد الفيروس، ممّا يحدّ من إمكانيات الاستخدام، ففي بعض المحاولات العيادية الحديثة للأمراض الجينية في الولايات المتحدة الأمريكية، فإن بعض المرضى كانوا قد توفّوا» (Edelstein 2002, 26)، ممّا يدل على وجود علاقات وظيفية بين الجينات أكثر دقة ممّا هي عليه التوصيفات البيوتكنولوجية الحديثة كما أن درجة التشفير في الجينات هي أعلى ممّا تمّ فكه إلى غاية الزمن الحاضر بالإضافة إلى صعوبة تطبيق ما تمّ التوصل إليه في هذا المجال لدى الحيوان على الإنسان، إذ «على الرغم من أنه تمّ تحديد الجينات [المسؤولة عن طول العمر]، إلاّ أنه ليس من الواضح كيف تشارك البروتينات المشفرة بواسطة هذه الجينات في تحديد طول العمر» (Struckelberger 2008, 102) لأن أعراض تمديد طول العمر لا تختلف عن أعراض الشيخوخة نفسها ومن ثمّة لا يمكن تمييز مسؤولية الجينات المعدّلة بغرض العلاج في إطالة العمر لأن ذلك لا يُعرف إلا من خلال عملية واسعة على عدد معتبر من الأفراد الذين يملكون المعطيات الصحية والعمرية والجينية نفسها. ومع ذلك توجد «اليوم، آمالٌ في تلاعب جيني مُراقب ومخصوص والذي ينشأ مع إمكانية استخدام الخلايا الجذعية الجنينية البشرية، حيث يمكن التلاعب جينياً بهذه الخلايا لإدخال الجينات

العلاجية» (Struckelberger 2008, 102) لأن هذا النوع من الخلايا له القابلية للانقسام المتجدد بالجينات العلاجية المدمجة فيه بما يساعد على إنتاج بروتينات من الأصل.

إن توجّه البيولوجيا الجزئية إلى الخلايا الجذعية الجنينية البشرية يأتي بعد الإخفاقات الناتجة عن الحدود المخبرية التي تحدّ الخلايا العادية، «ففي سنة 1961 اكتشف هيفليك (Hayflik) أن الخلايا البشرية العادية المزروعة في المخبر لها مدة حياة محدودة أي أنها مبرمجة لتنقسم إلى عدد ثابت من المرات ثمّ يتوقف الانقسام المتكرر، (وهو ما يُعرف بـ «حدّ هيفليك»)» (Hall 2003, 04) حيث يقلّ عدد الانقسامات الخلوية كلما تقدّمت الخلية في العمر أي أن هناك حدّاً لحياة هذه الخلايا، «ويوجد عدد من النظريات التي تفسّر سبب ما يُدعى «حدّ هيفليك»، حيث ترى أهم تلك النظريات بوجود تراكم لتلف جيني عشوائي أثناء نسخ الخلايا، إذ مع كل انقسام خلوي يمكن للعوامل البيئية مثل الدخان والإشعاع ومواد كميائية تدعى جذور الهيدروكسيل الحرة والنفايات الخلوية، أن تعطّل النسخ الدقيق لـ DNA من جيل خلوي إلى جيل خلوي يليه. ولدى الجسم عدد من الأنزيمات التي تقوم بإصلاح DNA والتي تشرف على عملية النسخ والتدخل لإصلاح مشكلات النسخ عند بروزها، ولكنها لا تنجح في اكتشاف جميع الأخطاء وتحديدها. ومع استمرار نسخ الخلايا، يتراكم عطب الـ DNAبداخلها، ممّا يؤدي إلى إنتاج بروتيني خاطئ وخلل وظيفي، وتمثّل هذه الاختلالات والأعطاب سبب الأمراض الخاصة بالشيخوخة مثل تصلّب الشرايين وأمراض القلب والسرطان» (Fukuyama 2002, 58 - 59)، ويظهر أن هذا الحدّ غير قابل للاختراق الجيني كلياً لأنه يمثّل التوازن الحيوي للكائن البشري كما يعبّر عن تاريخ طويل من التشكّل الجيني؛ إذ إن «أكثر من 200 جين بشري تمّ التعرف على تسلسلها ظهرت ناشئة بشكل مباشر عن بكتيريا معادلة لها، وعليه فقد استمّد الجينوم البشري جزءاً كبيراً من معلومته القاعدية للـDNA من بكتيريا عمرها آلاف السنين» (Mattei 2001, 26) ممّا يجعل جينات إطالة العمر وتأخير الشيخوخة في مواجهة هذا التاريخ من المعلومات التي حافظت على الطابع الجوهري للبشر وخصائصهم وصفاتهم الجسدية والنفسية؛ وما تحريك الجهاز المناعي لمهاجمة هذه «الجينات الغريبة» إلا حشداً لهذا التاريخ ضد المؤقت والطارئ؛ وعليه فإن هذه الحدود التقنية لمشروع إدخال الشيخوخة إلى المخبر وعلاجها جينياً تمثّل حدود الطبيعة البشرية في حدّ ذاتها وكل تجاوز لها يؤدي إلى انتكاسات بيولوجية للجسد البشري ممثلة في الأمراض المؤثرة على سلامة الدماغ والأعصاب والقوى الحركية والقلب..وغيرها.

ب. المشكلات الأخلاقية والاجتماعية:

لا يقتصر التلاعب بالجينوم البشري بغرض إطالة العمر ومحاربة الشيخوخة على تغييرات بيولوجية تطرأ على الكائن البشري والتي يمكن مراقبتها مخبرياً بل يتعدّاه إلى التلاعب بالبنية الأخلاقية والاجتماعية لهذا الكائن لأنه سيندمج بهذه التعديلات الجينية في مجموعته الاجتماعية التي ستتعرض لانزلاقات أخلاقية غير معتادة حيث «تقتصر قدرة التقنيات الطبية الراهنة على إبقاء أجساد البشر على قيد الحياة، ولكن بجودة حياة رديئة كثيراً، ولهذا برزت في السنوات الأخيرة- في الولايات المتحدة الأمريكية وغيرها- قضايا مثل المساعدة على الانتحار والقتل الرحيم، وفي المستقبل، من المحتمل أن تعرض علينا البيوتكنولوجيا صفقات تُفاضل فيها بين طول العمر وجودة الحياة، فإذا ما تمّ قبول تلك المفاضلات، فستكون الآثار الاجتماعية مأساوية» (Fukuyama 2002, 69) لأن إطالة العمر باستخدام تقنيات التعديل الجيني تعني مزيداً من «الأعباء» الإضافية على الحياة اليومية، إذ سيرافق تمديد الحياة الفردية للبشر تمديداً لمشاكلهم اليومية ولروتين عملهم ومسؤولياتهم الاجتماعية والمهنية وما يتبع ذلك من تراكم للمشاكل النفسية.

إن النظر إلى المشكلات الأخلاقية التي تعيشها المجتمعات المعاصرة اليوم يشير إلى أن إطالة العمر في متوسط معدلات الحياة فيها لا يعني إلا تفاقم تلك المشكلات وليس حلّها، حيث سيزداد العمر الافتراضي لارتكاب الجريمة أكثر من المعتاد بسبب تمديد الحالة الصحية للمجرمين سنوات إضافية مما يعني مزيداً من الجرائم وتكرارها، كما أن تزايد الأمل في الحياة يعني تزايد فرص اقتراف المزيد من الجرائم لأن مرحلة الشيخوخة، والتي هي المرحلة التي يرتكب فيها الناس جرائم أقل بسبب ضعف أجسادهم وكثرة أمراضهم، في الأعمار العادية، ستوافق مرحلة الكهولة في الأعمار الممدّدة؛ كما أنه لا يمكن التنبؤ بالحالات النفسية المصاحبة لذوي الأعمار الممدّدة جينياً إذ يمكن أن يقع انفلاتٌ أخلاقي بسبب الفترة الممدّدة من حياتهم والتي تُضاعف من حالات القلق والتوتر وحينها يكون التفكير «فيما إذا كان الناس سيحتفظون بنشاطهم البدني والذهني طيلة فترات الحياة الممدّدة، وما إذا كان المجتمع سيتحول تدريجياً إلى ما يشبه داراً ضخمة للعجزة» (Fukuyama 2002, 67) بما ينذر بتصدّع العلاقات الأخلاقية بين الآباء والأبناء، خصوصاً في المجتمعات الإسلامية التي تعتبر المرافقة الصحية والاجتماعية للآباء في فترة الشيخوخة واجباً أخلاقياً مقدساً بل عبادة من أعظم العبادات.

أما على المستوى الاجتماعي فإنه «ستكون لإطالة العمر بالبيوتكنولوجيا آثاراً مأساوية على البنيات الداخلية للمجتمعات، وأهمها تلك المرتبطة بتدبير السلاسل الهرمية الاجتماعية (..) إذ سيُلحق تمديد العمر أضراراً كبيرة بتلك السلاسل المنتظمة حالياً حسب العمر، وتفترض هذه السلاسل بنية هرمية بشكل تقليدي، لأن الموت يُصفّي مجموعة المتنافسين على المناصب العليا ويغربلها» (Fukuyama 2002, 64 - 65) لأن الموت يمثل بالنسبة للأعمار الحالية «عتبة طبيعية» لتطورات اجتماعية وتنظيما «عادلاً» لإعادة «تدوير» البنى الاجتماعية ودمجها في مسارات متجددة من العمل والثقافة والحضارة، حيث «إن أشكال الاختفاء بالموت «الطبيعي» تقدم تنوّعاً هائلاً، بإيجاد أشكال لإعادة الإنتاج» (Klarsfeld 2000, 59) وهو ما يعني أن الموت في الأعمار الممدّدة سيكون ظاهرة «غير طبيعية» لأنه سيُعتبر «حدثاً اجتماعياً» ينتهي معه شكل من أشكال الهيمنة الهرمية غير العادية، فمع نهاية الجسد تنتهي الحمولة الاجتماعية الرمزية التي يمثّلها، ولهذا يقول عالم الاجتماع لوبرتون (D. Le Breton): «يبقى الجسد مادة متاحة بصعوبة لأنه محميٌّ بمقاومات شعورية (وكذلك لا شعورية) لجزء كبير من الجماعة الاجتماعية» (Cadré 2001, 104)، فما إن يدخل جسد الفرد في السلسلة الهرمية لمجتمع ما فإنه لا يصبح مِلكاً لصاحبه وستكون الآثار الاجتماعية جزءاً من الآثار المَرَضية التي تلحق به مع تقدم العمر.

يؤدي تمديد الأعمار إلى تمديد سن التقاعد في الوظائف العامة وإلى تمديد مدة البقاء في المناصب والهيئات الاجتماعية ممّا يؤدي إلى تأخير سن العمالة وإلى تراجع الدور الاجتماعي للفئات الشابة حيث «إن فكرة بقاء الكفاءات مناسبة على امتداد حياة العمل متواصلة لخمسين أو ستين سنة، هي فكرة مناقضة للعقل بشكل كبير (..) وستُضاف الحرب بين الأجيال إلى الصراع الطبقي والعرقي، كعامل فاصل في المجتمع. وستتحول عملية إزاحة الشباب لمسنين إلى صراع أساسي، وقد تضطر المجتمعات إلى طرق غير شخصية مبنية على أنماط من المؤسسات العمرية في عالم مستقبلي يكون فيه العمر المتوقع أطول» (Fukuyama 2002, 67)، فيظهر أن عملية إطالة العمر تُخلّ بالتوازن الاجتماعي الذي يقتضي تداول فئات عمرية معينة على هرم النظام الاجتماعي في العمل والهيئات النظامية المختلفة بما يدفع الصراع ويحافظ على استمرارية البنيات التقليدية للمجتمع.

رابعاً: الإسلام والتمديد الجيني للحياة:

أ. القرآن وإمكانية تمديد الحياة:

يفتحُ القرآن مجالاً للإمكانات الإنسانية والطبيعية التي قد تتعارض مع التصورات المألوفة ولكنها لا تتعارض مع طبيعة العلم نفسه لأن العلم نفسه يندرج ضمن إمكان الخلق الإلهي، وبهذا المعنى فإن ما نعتبره معجزة لا يعني أنه غير قابل للفهم أو أنه لا يمكن التعبير عنه بالدقة العلمية التي تتطور عبر التاريخ ليظهر أن ما يوصف بـ «غير الطبيعي» متضمّن في «الطبيعي» نفسه، وفي هذا السياق يمكن الحديث عن إمكانية «تمديد الحياة» في القرآن والتفسير الذي يرافقها، حيث «إن الإمكانية «غير الطبيعية» لحياة مديدة ليست مقصورة في القرآن على الأنبياء كنوح، فقد أشار القرآن إلى قصة أهل الكهف الذين هربوا إلى كهف» (Maher 2009, 125)، ممّا يعني أن تهيئة الأسباب لتمديد الحياة تتجلّى في الطبيعة البشرية كتخطيط مسبق لملَكات الجسم التي يمكن من خلالها أن يتجاوز حدود الطبيعة التي تعني الثبات ولكنه ثبات في زمن محدود غير أن طول هذا الزمن يوحي بعدم إمكانية تغيّر المعطيات البيولوجية.

لقد أورد ابن قيم الجوزية في تفسير قوله تعالي «إنه على رجعه لقادر» (سورة الطارق، الآية 8): «قال مقاتل: «إنْ شئتُ رَددته من الكبر إلى الشباب ومن الشباب إلى الصبا، ومن الصبا إلى النطفة» (الجوزية (د.ت)، 164)، وهذا يعني أن الكائن البشري يتوفر على القابلية للعودة إلى الشباب بفضل تخليق ذي دافعية مسبقة «ولو كُشف لك الغطاء لرأيت التخطيط والتصوير يظهر في النطفة شيئاً بعد شيء من غير أن ترى المُصوّر ولا آلته ولا قلمه، فهل رأيتَ مصوّراً لا تمسّ آلته الصورة ولا تلاقيها؟» (الجوزية (د.ت)، 608)، وهذا يشير تماماً إلى البرمجة الجينية للإمكانات الطبيعية وغير الطبيعية التي يمكن أن تكون عليها بيولوجيا الجسد البشري ومنها إمكانية تمديد الحياة، وبهذا «يقدّم القرآن للمسلمين معاني لفهم إمكانية تمديد الحياة، ويرى المسلمون أن العلاقة بين العلم والدين تدعم قبول هذه الإمكانية» (Maher 2009, 125)، ما دام تمديد الحياة أمراً ممكنًا في القرآن فإن ذلك يفتح الباب واسعاً للعلم ليُجسّد هذه التجربة دون مخاطر على التركيبة الفيزيولوجية للوظائف الحيوية للكائن البشري.

ب. طول العمر ومسوّغاته في الإسلام:

إن الحياة في الإسلام ذات شأن بما يُعمَّر فيها وبما يتمّ إصلاحه من أخلاق وعمل، ولأن الإسلام دين فطرة فإن كراهية الموت، بما هي فطرة بشرية، لا تُخالف المقاصد الكبرى للدين ولا تعني فساد نفس المسلم، فحُب الحياة ليس هو نفسه حُب الدنيا المذموم، فالحياة تجربة إنسانية محايثة لكينونة الإنسان بينما الدنيا هي الحمولة المادية التي تُثقل تلك التجربة بإغراءاتها وغوايتها؛ فقد جاء في الحديث النبوي «عن أبي هريرة قال: قال رسول الله صلى الله عليه وسلم: إن الله قال: مَن عادى لي وليّاً فقد آذنته بالحرب. وما تقرب إليّ عبدي بشيء أحب إليّ ممّا افترضته عليه. وما يزال عبدي يتقرب إليّ بالنوافل حتى أُحبه، فإذا أحببته كنتُ سمعَه الذي يسمع به وبصره الذي يُبصر به ويده التي يبطش بها ورجله التي يمشي بها، وإن سألني لأعطينّه، ولئن استعاذ بي لأعيذنّه. وما تردّدتُ عن شيء أنا فاعلهُ تردّدي عن نفس المؤمن يكرَه الموت وأنا أكره مَساءته» (صحيح البخاري 2002، 1617: 6502)؛ فقد أثبت هذا الحديث أن كراهية الموت فطرة حتى في نفس المؤمن الذي يُحبّه الله وقد أتى بما يتقرّب به إليه، ويستلزم عن هذا حب الحياة وطولها إلا إذا جاء ما يُفسدها ويُفقدها طابعها الجوهري ويَلبس عليها كالفتن فتتحوّل كراهية الموت إلى محبّته، فـ«عن أبي هريرة عن النبيّ صلى الله عليه وسلم قال: «لا تقوم الساعة حتى يمرّ الرجل بقبر الرجل فيقول: يا ليتني مكانه» (صحيح البخاري 2002، 1759: 7115)، وليس هذا حبّاً في الموت بل كُرهاً للفتن وكثرتها وتأثيرها على إيمان المؤمن الذي يكون كالقابض على جمر من شدّة ما يُلاقي منها.

لمّا عُلِم من محبّة الإنسان لطول الحياة فقد جُعلت كمكافأة على القيام بواجب اجتماعي هو صلة الرحم فـ«عن أبي هريرة رضي الله عنه قال: سمعتُ رسول الله صلى الله عليه وسلم يقول: مَن سَرّهُ أن يُبسط له في رزقه، وأن يُنسأ له في أثره فليْصل رَحِمه» (صحيح البخاري 2002، 1503: 5985)، وهذا يعني أن الإنسان يُحب أن يطول عمره ونتأخر شيخوخته بما يمثّل خاصية بشرية لم يُكرّهها الدين الإسلامي بل ذهب، على العكس من ذلك، إلى النهي عن تمنّي الموت (مجرّد التمني)، فقد قال رسول الله صلى الله عليه وسلم: « لا يتمنى أحدكم الموت إمّا مُحسناً فلعلّه يزداد، وإمّا مُسيئاً فلعلّه يستعتب» (صحيح البخاري 2002، 1788: 7235)؛ فطول العمر مرغوب إذن بشرط ربطه بالعمل الصالح والخُلق الحسن، لأن هذا العمل وهذا الخُلق متى ما طال بهما الزمان كان ذلك خيراً لصاحبه وللناس من حوله وأثراً لما بعده ومَدَداً لأهل الأخلاق الحسنة

في مواجهة أهل السوء، فقد قال رسول الله صلى الله عليه وسلم: «خيرُ الناس من طال عمره وحسُن عمله» (الإمام النووي 45: 108)، ولهذا فإن طول العمر هو من الخيرية إذا ارتبط بحسن العمل فيكون مطلوباً لهذا الغرض وتزداد معه كراهية الموت لأنه يقطع ديمومة ذاك العمل فيحبسه عن الناس.

يتبيّن من خلال ما سبق أن طلب تمديد الحياة والأعمار ليس مذموماً، كما أنه طلبٌ قديم منذ آدم عليه السلام، فعن أبي هريرة أن رسول الله صلى الله عليه وسلم قال: «لما خلق الله آدم مسح ظهره، فسقط من ظهره كلّ نَسَمة هو خالقها إلى يوم القيامة أمثال الذرّ، ثم جعل بين عيني كل إنسان منهم وبيصاً من نور، ثم عرضهم على آدم، فقال: يا رب من هذا؟ فقال: هذا ابنك داود، يكون في آخر الأمم. قال: كم جعلتَ له من العمر؟ قال ستين سنة، قال: يا رب زِده من عمري أربعين سنة، فقال الله تعالى: إذاً يُكتب ويُختم فلا يُبدّل.

فلمّا انقضى عُمر آدم جاءه ملَك الموت. قال: أو لم يبق من عمري أربعون سنة؟ فقال: أوَ لم تجعلها لابنك داود؟ قال: فجحدَ، فجحدت ذريته، ونسي فنسيت ذريته، وخطئ فخطئت ذريته». (..) وزاد محمد بن سعد: «ثم أكمل الله لآدم ألف سنة، ولداود مائة سنة» (الجوزية (د.ت)، 455-456)، فطلبُ الزيادة في العمر لا يتناقض مع الأقدار والآجال والسنن الإلهية في تحديد الأعمار، لقوله تعالى «ولكل أمة أجل فإذا جاء أجلهم لا يستأخرون ساعة ولا يستقدمون» (الأعراف، الآية 34)، أو لقوله تعالى: «وما يُعمّر من معمّر ولا يُنقص من عمره إلا في كتاب» (فاطر، الآية 11)، لأنه إذا كان تحديد الأعمار أمراً غيبياً فهذا يعني إمكانية زيادة بعض الأعمار أو نقصانها دون أن يكون للبشر علمٌ بذلك؛ وعليه فإنه إذا كان طلب تمديد العمر وزيادته بالدعاء لا يتعارض مع القدر أو الشرع فإن طلبه بالطرق التقنية الطبية والعلاجات المختلفة يدخل ضمن اللاتعارض نفسه.

لقد حدّد الرسول صلى الله عليه وسلم متوسط أعمار أمته، فعن أبي هريرة رضي الله عنه، قال رسول الله صلى الله عليه وسلم: «أعمار أمتي ما بين الستين إلى السبعين، وأقلّهم من يجوز ذلك» (الترمذي (3550)، وابن ماجه (4236))، وتتفق أغلب شروح هذا الحديث على أن هذا التحديد العمري هو تحديد أخلاقي بالأساس أي دفعاً للبطر والاستكبار، فقد جاء في «فيض القدير» للمناوي: «أعمار أمتي ما بين الستين إلى السبعين وأقلهم من يجوز ذلك»، قال الطيبي هذا محمول على غالب بدليل شهادة الحال فإن منهم من لم يبلغ ستين وهذا من رحمة الله بهذه الأمة

ورفقه بهم أخّرهم في الأصلاب حتى أخرجهم إلى الأرحام بعد نفاد الدنيا ثم قصّر أعمارهم لئلا يلتبسوا بالدنيا إلا قليلا فإن القرون السالفة كانت أعمارهم وأبدانهم وأرزاقهم أضعاف ذلك كان أحدهم يعمّر ألف سنة وطوله ثمانون ذراعاً وأكثر وأقل وحبّة القمح ككلوة البقرة والرمانة يحملها عشرة فكانوا يتناولون الدنيا بمثل تلك الأجساد وفي تلك الأعمار فبطروا واستكبروا وأعرضوا عن الله «فصبّ عليهم ربك سوط عذاب» فلم يزل الخلق ينقصون خلقاً ورزقاً وأجلاً إلى أن صارت هذه الأمة آخر الأمم يأخذون أرزاقاً قليلة بأبدان ضعيفة في مدة قصيرة كيلا يبطروا فذلك رحمة بهم» (المناوي 1972، 11)، فإطالة العمر أو تقصيره مرتبطان بالعمل المنجز فيهما وليست إطالة العمر لذاتها بل لما يتحقق بها من أخلاق وإصلاح في الأرض ولما تكون لها من فعالية على المجتمعات.

أما ذمّ طلب تمديد العمر وتمنّيه فيكون في من أفسد وأراد المزيد من الحياة هروباً من عقابه، فقد جاء قوله تعالى: «ولتجدنهم أحرص الناس على حياة ومن الذين أشركوا يودّ أحدهم لو يُعمّر ألف سنة وما هو بمزحزحه من العذاب أن يُعمّر» (البقرة، الآية 96)، وقد جاء في تفسير هذه الآية قول الطبري: «وإنما وصف الله جل ثناؤه اليهود بأنهم أحرص الناس على الحياة، لعلهم بما قد أعدّ لهم في الآخرة على كفرهم» (تفسير الطبري 2001، 276)، وهو على قول ابن كثير: «أي أحرص الخلق على حياة أي: على طول عمر، لما يعلمون من مآلهم السيئ وعاقبتهم عند الله خاسرة، (..) فهم يودون لو تأخروا عن مقام الآخرة بكل ما أمكنهم» (تفسير ابن كثير1999، 334)، ففي هذا فصلٌ بين طول العمر (بالمعنى البيولوجي) والغرض منه (المعنى الأخلاقي) ولذلك ذكر الطبري قول أهل التأويل في «»أن» التي في قوله «أن يعمّر» بمعنى: وإن عُمّر» (تفسير الطبري 2001، 280) أي في إمكانية التعمير البيولوجي دون التعمير الأخلاقي.

وعليه فإن «الخوف من الموت هو أحد الانفعالات الإنسانية الأكثر عمقاً وثباتاً، ولذلك يكون مفهوماً الاحتفاء بكل تطور في التقنيات الطبية والذي من شأنه أن يُبعد شبح الموت عنّا» (Fukuyama 2002, 67)، ولأن هذه التقنيات المتعلقة بالعلاج الجيني للشيخوخة وإطالة العمر ستكون متوفرة في العالم الإسلامي مثلها مثل العمليات التجميلية أو زراعة الأعضاء والتخصيب الاصطناعي.. فإنها ستكون في صُلب النقاش الديني والأخلاقي، لأن المسألة، كما تقدم، مرتبطة بالقيم الأخلاقية المرافقة لكل تطور طبّي والتي هي من معالم الدين وركائزه.

ج- النظر الفقهي والأخلاقي في التعديل الجيني:

تعتبر الشيخوخة وما يؤول إليه صاحبها من عجز بدني وتراجع في القدرات العقلية والنفسية سنة إلهية ثابتة لقوله تعالى: «الله الذي خلقكم من ضعف ثم جعل من بعد ضعف قوة ثم جعل من بعد قوة ضعفاً وشيبة يخلق ما يشاء وهو العليم القدير» (الروم، الآية 54) أما بالنسبة للعلم فإن علماء البيولوجيا «يميلون للاعتقاد بأن الشيخوخة هي نتيجة للتفاعل بين عدد كبير من الجينات، وعليه فإنه لا توجد طرق جينية مختصرة لتأجيل الموت» (Fukuyama 2002, 58)، ومنه فإن العلم لا يُعارض السنن الإلهية بل يتحرك داخل مجالها المحدد ويوافق ما هو متاح فيها من إمكانيات، لأنه لا شيء في الطبيعة البشرية يخضع للعلم إلا إذا كان مخلوقاً بهذه الإمكانية ومجبولاً على إظهار ما يمتلكه من معلومات جينية مشفرة، غير أن الخوف المشترك بين العلم والدين هو أن «يؤدي التلاعب بالجينوم البشري إلى آثار كارثية هي مقدمة سيناريو نهاية الإنسان» (Hervé 2006, 49) أي نهاية الطبيعة العادية للإنسان واختفاء ملامحه التقليدية ليتحول إلى مخلوق ممسوخ مشوه وهذا ما يدعو إلى البحث في الطابع الأخلاقي للتعديل الجينومي المؤدي إلى إطالة العمر وتأجيل الشيخوخة من منظور الأخلاق الإسلامية.

من حيث الأحكام الفقهية المتصلة بجواز نقل الجينات أو عدمه فإن «منهم من منع نقل الجين إلى الخلايا الجسدية، لأن في ذلك تغييراً لخلق الله بالتدخل في التركيب الوراثي للإنسان، ولأن هذه العمليات لا تخلو من أضرار ومفاسد، ودرء المفاسد مقدم على جلب المصالح؛ وأكثر العلماء والباحثين أجازوا عمليات نقل الجين إلى الخلية الجسدية، على ألاّ يؤدي ذلك إلى ضرر أعظم من الضرر الموجود فعلاً وألاّ تكون هناك وسيلة أخرى لعلاج المرض» (الألفي 2012، 24)، وهنا يكون التقدير الفقهي لمسألة إطالة العمر جينياً من زاوية المفاسد التي تترتب على ذلك وعلى رأسها التمديد في عمر الظالمين فيزدادوا ظلماً أو ما يكون من طمع في الحياة بمزيد من الشهوات وكثير من الخطايا التي يجلبها طول العمر والتأخر في الشيخوخة بما يوهم ببُعد الموت وطول الحياة الدال على الخلود «الرمزي»؛ أما من زاوية المصالح فيكون على سبيل العلاج المفيد بحيث «أن التحكم في صفة وراثية تشتمل على خلل أو مرض وراثي أو مرض ناتج عن خلل في المورثات (..) أمرٌ جائز شرعاً لأنه من باب العلاج الذي شرعه الله تعالى يدل عليه قوله صلى الله عليه وسلم: «تداووا عباد الله، فإن الله تعالى لم يضع داء إلا وضع له دواء غير داء واحد: الهرم» (أبو يحيى 2011، 31)، إلا أن هناك مشكلة علمية وأخلاقية هي: هل إطالة العمر وتأجيل

الشيخوخة يدخلان في باب العلاج أم لا؟ لأنه يظهر من معنى الحديث السابق أن الشيخوخة المُعبّر عنها بلفظ «الهرم» تعتبر «داءً»، وأمّا أنه ليس لها دواء فمعناه ليس لها دواء يقضي عليها أمّا ما دون ذلك أي في إمكانية تأجيلها فأمر ممكن.

إذا فُهم من التعديل الجينومي، المؤدي إلى إطالة العمر، تحسين النسل، فقد صدر قرار من المجمع الفقهي التابع لرابطة العالم الإسلامي، يرى بأنه «لا يجوز استخدام أي من أدوات علم الهندسة الوراثية ووسائله للعبث بشخصية الإنسان، ومسؤوليته الفردية، أو للتدخل في بنية المورثات (الجينات) بدعوى تحسين السلالة البشرية» (القرة داغي 2006، 325) لأن هدف العلاج غير متوفر وهذا ما يجعل التعديل الجينومي بغرض تحسين النسل أمراً غير مبرر طبياً، ولكن ماذا لو أدّى تحسين النسل جينياً إلى تحسينه أخلاقياً أو تغيير بعض الطبائع المشينة التي اكتسبها وراثياً أو بيئياً وهذا إذا نظرنا إلى هذه المسألة بناءً على قاعدة المآلات، لأن «النظر في مآلات الأفعال معتبر مقصود شرعاً سواء كانت الأفعال موافقة أو مخالفة، وذلك أن المجتهد لا يحكم على فعل من الأفعال الصادرة على المكلفين بالإقدام أو بالإحجام إلا بعد نظره إلى ما يؤول إليه ذلك الفعل مشروعاً لمصلحة فيه تستجلب أو مفسدة تدرأ، ولكن له مآل على خلاف ذلك، فإذا أطلق القول في الاول بالمشروعية فربما أدّى استجلاب المصلحة فيه إلى مفسدة تساوي المصلحة أو تزيد عليها. فيكون هذا مانعاً من إطلاق القول بالمشروعية، وكذلك إذا أطلق القول في الثاني بعدم المشروعية ربما أدى استدفاع المفسدة إلى مفسدة تساوي أو تزيد فلا يصح القول بعدم المشروعية» (القرة داغي 2006، 323)، إذ يمكننا أن نبني على هذه القاعدة ما نستطيع تسميته بـ «أخلاق المآلات» أي تلك الأحكام الأخلاقية المتعلقة بأغراض التعديل الجينومي والتي تنظر في ما يؤول إليه هذا التعديل وليس لما يكون عليه في الحاضر.

وجبَ النظر إذن إلى ما سيتغيّر في الإنسان عند إطالة عمره جينياً، هل يتغيّر خَلقه أم يتمدد فقط؟ هل هو من باب تحسين الصفات أم مجرد تأجيل لمرحلة محتومة من حياة الإنسان وليس تغييرها؟، هناك «من أجاز نقل الجين إلى الخلية الجسدية للحصول على صفات مرغوبة؛ لأن تحصيل الصفات الحسنة من الأمور المحمودة شرعاً، ولا مانع من طلبها بالطرق المباحة، فالمؤمن القوي خير وأحب إلى الله من المؤمن الضعيف، والله تعالى جميل يحب الجمال» (الألفي 25، 2012)، وعندئذ يكون طول العمر من قوة المؤمن وجماله وهو مَطلب غير محصّل بتغيير خَلق

ظاهر ولا تشويه لخُلق فطري لأن طول العمر لا يُغيّر الصفات أو يُحسّنها بل هو تمديد للحياة في الزمن بما لا يتناقض مع أجَل الموت، أي أنه تمديدٌ لقدَر إلى قدَر آخر.

كيف يكون التأثير الأخلاقي لإطالة العمر جينياً على المجتمعات الإسلامية؟ إن الأعمار الحالية متأقلمة مع تطور المجتمعات المعاصرة لأنها من صناعة الحضارة أي أنها مرتبطة بتطور النظام الصحي وتزايد الرعاية الاجتماعية وتحسّن ظروف العمل ووفرة الغذاء وجودته وتطور الدواء وتقنيات العلاج الجسدي والنفسي وكذا نمو أشكال الرفاهية، وقد استفادت المجتمعات الإسلامية من هذه المظاهر الحضارية، غير أن كل تغيير سريع في معدل الأعمار ستكون له عواقب أخلاقية متسارعة، فإذا كنا نشهد اليوم حالات إبعاد الوالدين إلى دور العجزة والذي يُعدّ عملاً غير أخلاقي لأنه من العقوق فإن تمديد الأعمار سيزيد من هذه الحالات بما يُشكّل ظاهرة دراماتيكية في المجتمعات الإسلامية.

ستفتح إطالة الأعمار في المجتمعات الإسلامية الباب أمام أطماع غير إنسانية في السيطرة والتحكم في السلسلة الاجتماعية للأسرة مما سيحرم الأعمار الدنيا من تسلّم المسؤوليات والقيام بالوظائف الأخلاقية المتداولة ضمن حدود الأعمار التقليدية ممّا سيؤجّج الصراع (الصراع على الميراث مثلاً) كما أن الموت المرتبط بالأعمار الحالية يُنهي بشكل آلي بعض المشكلات الاجتماعية والأخلاقية والتي ستطول مع طول الأعمار.

يؤدي تمديد الحياة وإطالتها إلى مزيد من طول الأمل المذموم لقوله تعالى: «فذرهم يأكلوا ويتمتّعوا ويلههم الأمل فسوف يعلمون» (سورة الحجر، الآية 3)، وقد جاء في رياض الصالحين الحديث الشريف: «عن ابن مسعود رضي الله عنه قال: «خطّ النبيّ صلى الله عليه وسلم خطّاً مربعاً، وخطّاً في الوسط خارجاً منه، وخطّ خُططاً صغاراً إلى هذا الذي في الوسط من جانبه الذي في الوسط، فقال: «هذا الإنسان، وهذا أجله محيطاً به- أو قد أحاط به- وهذا الذي هو خارج أمله، وهذه الخُطط الصغار الأعراض، فإن أخطأه هذا، نهشه هذا» رواه البخاري وهذه صورته (الإمام النووي 184):

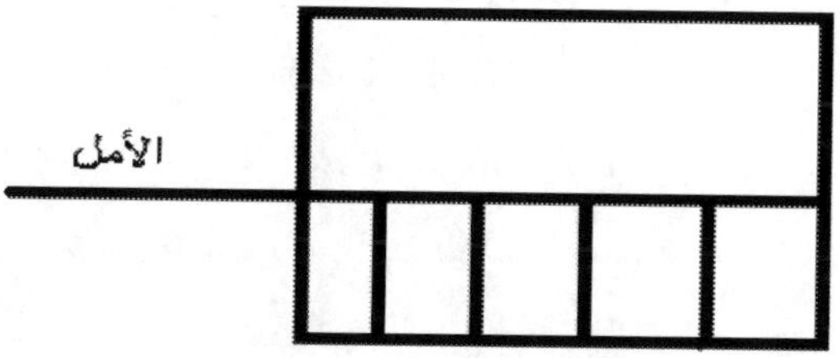

أي أن تمديد الشيخوخة وتأجيل أعراضها يدفع إلى الوهم بالخلود ويترتب على هذا الوهم تأجيل الواجبات الأخلاقية الضرورية بسبب وهم القيام بها في المستقبل كما يختل التوازن الاجتماعي باختلال التوازن الطبيعي للبيولوجيا البشرية حيث إن «توازن الطبيعة يقتضي أن تختفي كائناتٌ حية لتظهر كائنات حية أخرى» (Klarsfeld 2000, 59) لأن الجسد ليس مادة بيولوجية خالصة بل هو مادة اجتماعية أيضاً أي أنه خاضع للتمثّلات الاجتماعية (كسلطة كبار السن، أو التقدير الاجتماعي لهم، أو الأولوية التي يكتسبونها في المؤسسات السياسية...) كما أنه فاعلٌ في حركة المجتمع (مثل الفرق بين تأثير صورة أجساد الشباب وصورة أجساد الشيوخ في المخيال الاجتماعي..)، لِيُشكّل الجسد «ثقافة» خاصة، «ثقافة يكون فيها الجسد في التقاطع بين الشعور الحميم بهوية الأشخاص وتحقيق مشاريع عقلانية، أي بين الموضوعية والبُعد الرمزي للجسد» (Cadré 2001, 105) ومنه فإن تمديد الأعمار يعني تراجع البُعد الرمزي للجسد وخُفوته في كثافة الحضور المُطوَّل لصاحبه في الحياة اليومية والفضاء العام ممّا يشكّل ثقلاً إضافياً على العلاقات الاجتماعية التي تبدأ في التصدّع والتفكك.

خامساً: نكسة الجينوم: من الإنسان إلى ما بعد الإنسان.

يُمثّل البُعد الأخلاقي لثورة الجينوم بصفة عامة، وللتعديل الجينومي المتعلق بتمديد الحياة بصفة خاصة، تحدّياً علمياً وإنسانياً لأن مسألة الجينوم وتعديلاته تتعلق بماهية الإنسان بما هو كائنُ الأبعاد، ولذلك فإن السؤال الذي يُطرح في سياق تمديد العمر وتأخير الشيخوخة هو: أيُّ إنسان هو ذاك الذي يبقى حيّاً بعد عمره التقليدي؟ أيّة حياة تكون في الأعمار الممدّدة؟ ما الذي يتغيّر في القيم الأخلاقية الحالية وما مصيرها في المستقبل؟

يمكننا أن نمثّل إخفاق التعديل الجيني المؤدي إلى تمديد الحياة بـ«إخفاق تيثونوس»[2] والذي يعني فقدان السيطرة على آثار الشيخوخة في المرحلة الممدّدة جينياً من حياة الناس والتي ستظهر عليهم بشكل أكثر حدّة ممّا يجعل تمديد الحياة مشكلة أخرى، «لأن الناس ينتابهم القلق أيضاً حول جودة حياتهم وليس حول طولها فقط» (Fukuyama 2002, 67)، وتنشأ هذه المشكلة من الصناعة الجينومية لإنسانٍ يشترك في خصائصه البيولوجية مع بقية الناس وهذا ما يهدم خصوصية كينونته المبنية على التنوع، لقوله تعالى: «ومن آياته خلق السموات والأرض واختلاف ألسنتكم وألوانكم إن في ذلك لآيات للعالمين» (الروم، الآية 22)، وحينها نعتبر أن الإنسان الموحّد بيولوجياً (بفعل التوحيد الجيني) هو ضد الإنسان المتعدّد في جوهره والمتنوع بيولوجياً حيث إن «التطور هو خاصية للحياة وليست هذه العملية ممكنة إلا لأن كل فرد يمكنه أن يكون مختلفاً عن الأفراد الآخرين من جنسه. لكن مع اكتشاف جزيء DNA والشيفرة الجينية الكونية (تقريباً) أصبح مفهوم اللاّتنوع مفهوماً مركزياً» (Perbal 2011, 43)، فاللاتنوع يعني صناعة منمذجة للإنسان بواسطة المخبر العلمي الذي تغطّي سلطته المعرفية على الأبعاد الأخلاقية للإنسان والتي تتجاوز حدوده البيولوجية وتُحدّد وضعه البشري.

تواجه عملية تأخير الشيخوخة سؤالاً مهمّاً هو: ماذا نريد أن نكون؟ هل نريد نسخة عن تيثونوس هي الأكثر تشوهاً؟ أم إنساناً هو ما بعد إنسان؟ أو «هل من المشروعية «في ذاتها» أن تُقام التجارب العلمية التي تشوّه «الممتلكات» الفيزيائية والكيميائية والنفسية المكونة للإنسان وتقوّض إنسانية الإنسان؟ إن فرضية السؤال هي أنه توجد طبيعة إنسانية لها طابع دائم، «نواة أنطولوجية صلبة»، وتشويه هذه النواة يتضمن خطراً جذرياً، هو التراجع نحو شكل أدنى من الحياة أو الوجود، التطور نحو أشكال الحياة غير الإنسانية، الخروج من الإنسانية، «القفز خارج ماهية الإنسان»، فقدان الملَكات التي تؤسس خصوصية الإنسانية وكرامته» (Beland 2006, 61)، فالشيخوخة جزء من النواة الأنطولوجية الصلبة للإنسان التي تتضمن التطور البيولوجي

2 - نسبة إلى الأسطورة اليونانية التي تتحدث عن تيثونوس (Tithonus) ابن ملك طروادة الذي أحبّته إيوس (Eos) (آلهة الفجر) فطلبت من زوس (Zeus) أن يحقق لها أمنية واحدة هي أن يمنح تيثونوس حياة خالدة، فاستجاب لها زوس (الغيور) مبتسماً، لكنها نسيت أن تطلب له الشباب الخالد، فبدأ تيثونوس يشيخ ويضعف ويهرم، فوضعته إيوس في حجرة، ثم حوّلته إلى حشرة هي جندب الحقل. أنظر: (Turner 2009, 27)

«الطبيعي» للجسد البشري في الزمن عكس الثبات البيولوجي الذي يُحدّده التوجيه الجينومي للصفات الإنسانية أي تلك «الآلة الثابتة التي تربط النمط الجيني (génotype) بالنمط الظاهري (phénotype) (..) ضمن الصيغة (النمط الظاهري= النمط الجيني+بيئة)» (Pouteau 2007, 08) وتؤدي الآلة الثابتة هذه إلى الدخول إلى المرحلة ما بعد إنسانية للإنسان وهي المرحلة التي يميّزها التحوّل الأخلاقي للحياة اليومية التي تحدُّ أنطولوجياً الإنسان من الخارج وتحافظ على نواته الأنطولوجية الأصيلة ولهذا فإن «المرور من الإنساني إلى ما بعد إنساني ليس قراراً خاصاً بالعلم، ولكنه خاصٌ بالأخلاق (ethics) وبالسياسة» (Beland 2006, 90)، وعلى رأس هذه الأخلاق مبدأ المسؤولية الذي يرتكز على الطبيعة الإنسانية المعطاة في بنيتها البيولوجية المتنوعة وكلّ تشويه لهذه الطبيعة بالتلاعب الجيني هو تخريبٌ لمبدأ المسؤولية وإفقادٌ لطابعها الأخلاقي.

ما «فضائل» تمديد الحياة جينياً؟ أيّة «سعادة» أو «لذّة» يمكن الحصول عليها لحياة ممدّدة؟ هل لو عاش بعض البشر زمناً أطول سيزدادوا خيراً أكثر مما فعلوا أو ازدادوا شراً أكثر ممّا اقترفوا؟ سيكون تمديد الحياة تمرّداً على الحياة نفسها وخرقاً «للالتزام» البيولوجي بعمر معيّن، ضمن حدود التناهي الإنساني، الحياة كتجربة إنسانية تَعتبر الموتَ جزءاً مهمّاً من تلك التجربة وبهذا فقط تُعتبر إنسانية، كما أن الإنسان خالدٌ ضمن حدود حياته والزمن المستنفد فيها ومصادر تلك الحياة العميقة والتي تستمدّها من الفناء، لأن الإنسان عندما يولد في الحياة يولد معه موته، لقد قال شرودينغر (Schrödinger): «الكائنُ خالدٌ، بالنسبة للقوانين التي تحتفظ بكنوز الحياة التي يرتسم بها الكون جمالاً» (Schrödinger 2006, 19)، فالخلود يتدفق من نواة الحياة ذاتها ويعود إلى منبعها، والحياة لا تتمدد ولا تطول وإنما توجد وفقط وتلك غواية الحياة ذاتها ومصدر الفضائل فيها.

في المستقبل القريب، سيعود إلى ساحة النقاش الفلسفي والعلمي سؤال: ما الإنسان؟ بأكثر حدّة، لأنه السؤال- المفتاح إلى إعادة تحديد وضبط مفهوم الطبيعة البشرية، وإعادة طرح هذا السؤال تعكس الهاجس الوجودي لكينونة كائن تتمفصل حياته بين منظور علمي ينظر إلى تلك الحياة كجسد معزول يمكن مراقبته عن قرب وتثبيته في إطار فيزيائي- كيميائي ومنظور اجتماعي أخلاقي ينظر إلى الحياة بوصفها حياة يومية أي ككينونة مندرجة في الصيرورة التاريخية بحيث لا يمكن التعالي على هذه الصيرورة كما لا يمكنها أن تكون معزولة خارج نفسها.

يمثّل الجينوم جزءاً من الحياة الداخلية للجسد وليس آخر حصون تلك الحياة بحيث يمكن جعله مُفسّراً وحيداً لما «يحدث» في الإنسان أو لما سيحدث فيه وله، فدَرسُ «التناهي» الإنساني

يجعل كل طرح بيولوجي بخصوص الحياة البشرية أو تمديدها وإطالة شيخوختها هو طرحٌ محدودٌ أنطولوجياً أي أنه لا يقترب من قاعدة الوجود التي يُبنى عليها تصوّر خالص لمفهوم «الإنسان»، وبقدر ما يمثّل العلم، على المستوى الإبستيمولوجي، تفسيراً مقنعاً لماهية الكائن البشري ولبنية الحياة ذاتها، إلا أنه، على المستوى الأنطولوجي، يُعبّر عن تآكل البناء التاريخي والثقافي للإنسان في صورة الانهيار القيمي- الأخلاقي لمنظومة عيشه.

إن تجربة «العيش» في الحياة اليومية ليست تجربة علمية محضة، فحتّى العلوم تجد حدودها في الحياة اليومية، كما يرى الفيلسوف الألماني هيدغر، وهو ما يجعلنا نطرح السؤال: ما الذي تبقّى من الإنسان؟ كيف تتفوّق بيولوجيا الجسد المحدود على الأبعاد غير المحدودة لكائن إيتيقي تستمرّ «حياته» الأخلاقية بعد موته أي بعد النهاية البيولوجية لجسده؟ ولسنا هنا بصدد مواجهة بين العلم والأخلاق ولكننا بصدد مواجهة بين الإنسان واللاّإنسان، بين الجوهر والعَرض.

هل يمثّل تمديد الحياة جينياً انتصاراً للإنسان على الحياة؟ أمْ إنها غواية جديدة بعد تلك التي عاشها آدم في الجنة؟ هل انتصر «الشيطان» من جديد (الشيطان بالمعنى الديني وبالمعنى الأخلاقي للكلمة) وها هو يريد أن يُخرج بني آدم من الأرض كما أخرج أباهم من الجنة؟ تمثّل هذه الأسئلة ما بعد علمية مقدمةً لفهم الإنسان في أبعاده غير التقليدية أي لفهم صورة كائن ما بعد «عودة الديني» إلى صُلب الجدال العلمي والفلسفي، بعد أن انتهى هذا الجدال إلى طرق مسدودة، وهي عودة تمّت عبر نافذة الأخلاق التي تشكّل اليوم المدخل إلى مستقبل الإنسان المهدّد بـ «الانقراض الرمزي» أي بالتحوّل إلى كائن «يشبه الإنسان» ولكنه غريب عنه.

المراجع:

أبو يحيى، محمد حسن.2011. حكم التحكم في صفات الجنين في الإسلام. الأردن: دار يافا العلمية للنشر والتوزيع.

الألفي، محمد جبر. 2012. الوراثة والهندسة الوراثية والجينوم البشري الجيني من منظور إسلامي. جدة: منظمة المؤتمر الإسلامي، مجمع الفقه الإسلامي.

الإمام النووي. (د.ت). رياض الصالحين. مصر: دار الريان للتراث.

تفسير ابن كثير. 1999. تحقيق: سامي بن محمد السلامة. (1):334. الرياض: دار طيبة للنشر والتوزيع.

تفسير الطبري.2001. جامع البيان عن تأويل آي القرآن. تحقيق: عبد الله بن عبد المحسن التركي. (2): 276-280. مصر: هجر للطباعة والنشر والتوزيع والإعلان.

الجوزية، ابن قيم. (د.ت). كتاب الروح. تحقيق: محمد أجمل أيوب الإصلاحي. خرّج أحاديثه: كمال بن محمد قالمي. مكة: دار عالم الفوائد.

الجوزية، ابن قيم. (د.ت).التبيان في أيمان القرآن. جدة: دار عالم الفوائد للنشر والتوزيع.

صحيح البخاري. 2002. رقم [6502]. دمشق- بيروت: دار ابن كثير.

القرة داغي، علي محي الدين، و علي يوسف المحمدي. 2006. فقه القضايا الطبية المعاصرة: دراسة فقهية طبية مقارنة مزودة بقرارات المجامع الفقهية والندوات العلمية. بيروت: دار البشائر الإسلامية.

المناوي. 1972. فيض القدير، شرح الجامع الصغير. (2):11. بيروت: دار المعرفة للطباعة والنشر.

Beland, Jean-Pièrre et autres. 2006. *L'homme Biotech: Humain ou Posthumain?*. Canada: Les presses de l'université Laval.

Cadoré, B. 2001. «Réflexions sur le corps». In *Plus tôt de la Vie Plutôt que la Mort : Actes des 2e Journées Pédagogiques D'éthique Médicale d'Amiens*, edited by J.C. Boulanger, M. Laude, J. Petit and A. Safavian, 104-113. Paris : John Libbey Eurotext.

Dessalles, Jean-Louis, and Cédric Gaucherel, and Pierre-Henri Gouyon. 2016. *Le Fil de la Vie, La Face Immatérielle du Vivant*. Paris: Odile Jacob.

Edelstein, Stuart J. 2002. *Des Gènes aux Génomes*. Traduit de l'anglais par Marcel Blanc. Paris: Odile Jacob.

Fukuyama, Francis. 2002. *Our Post-Human Future, Consequences of the Biotechnology Revolution*. New York: Farrar, Straus and Ciroux.

Gibson, Greg, and Spencer V. Muse. 2004. *Précis de Génomique*. Traduit de l'américaine par Lionel Domenjoud. Bruxelles: De Boeck et Larcier s.a.

Hall, Stephen S. 2003. *Merchants of Immortality: Chasing the Dream of Human Life Extension*. Boston, New York: Houghton Mifflin Company.

Hekimi, Siegfried. 2000. *The Molecular Genetics of Aging*. Berlin Heidelberg: Springer Verlag.

Hervé, Christain, and Jaques J. Rozenberg. 2006. *Vers la Fin de L'homme*? . Bruxelles: De Boeck, et Larcier s.a.

Jordan, Bertrand. 2007. *Thérapie Génique: Espoir ou Illusion*?. Paris: Odile Jacob.

Klarsfeld, André, and Frédéric Revah. 2000. *Biologie de la Mort*. Paris: Odile Jacob.Maher, Derek F., and Calvin Mercer. 2009. *Religion and the Implications of Radical Life Extension*. New York: Palgrave Macmillan.

Mattei, J.F. et autres. 2001. *Regard Ethique: Le Génome Humain*. Strasbourg: éditions Conseil de l'Europe.

Moore, David S. 2015. *The Developing Genome: An Introduction to Behavioral Epi-*

genetics. UK: Oxford University Press.

Pastermark, Jack J. 2003. *Génétique Moléculaire Humaine, une Introduction aux Mécanismes des Maladies Héréditaires*. Traduction de 1re édition américaine par Dominique Charmot. Paris: Deboeck.

Perbal, Laurence. 2011. *Gènes et Comportements à L'ère Post-Génomique*. Paris: Librairie philosophique J. Vrin.

Pouteau, Sylvie et autres. 2007. *Génétiquement Indéterminé : Le Vivant Auto-Organisé*. Paris: Quae.

Rabinow, Paul. 2000. *Le Déchiffrage du Génome : L'aventure Française*. Traduit de l'anglais (États-Unis) par Frédéric Keck. Paris: Odile Jacob.

Rogers, Arthur, and Denis Durand de Bousingen. 1995. *Une Bioéthique pour L'Europe*. Strasbourg: Les éditions du Conseil de l'Europe.

Schrödinger, Erwin. 2006. *What is Life? The Physical Aspect of the Living Cell*. UK: Cambridge University Press.

Struckelberger, Astrid. 2008. *Anti-Ageing Medicine: Myths and Chances*. Zurich, Suisse: Vdf. Hochschulverlag AG mdr ETH.

Turner, Bryan S. 2009. *Can we Live Forever? A Sociological and Moral Inquiry*. UK and USA: Anthen Press.

Watson, James, and Andrew Berry. 2003. *ADN: Le Secret de la Vie*. Traduit de l'anglais (États-Unis) par Barbara Hochstedt. Paris: Odile Jacob.

Watson, James. 2003. *Gènes, Génomes et Société*. Traduit de l'anglais (États-Unis) par Jean Monchard. Paris: Odile Jacob.

فهرس

آدم 253، 273، 275، 320، 328

الأبحاث البيوطبية 259

أبي هريرة 319–320

أخلاق المآلات 323

أخلاق عملية 254، 272

أخلاقيات الطب 254

الأدوية 270

أرسطو 255–256، 281

الأشكال الجزيئية المشتركة 306

أفلاطون 255، 267، 281

الأمن الديني 290

الأنساب 278، 295

الأنسولين 270

إخفاق تيثونوس 326

إرادة القوة 289، 303

إطالة العمر 309–310، 314–317، 321–323

الإيبيجينا 312

الاخلال الوراثية 268

الاستنساخ 270، 280–281

استنساخ الإنسان 270

الأخلاق 254، 257، 267، 272، 274، 278–280، 284–291، 293، 298–300، 302–303، 319، 322، 327–328

الأخلاق الإسلامية 254، 272، 274، 279–280، 322

انتقال الصفات الوراثية 261

انفلاتُ أخلاقي 316

انقسام خلوي 315

بارتك، أندريه 310

براديغيم 290

البرمجة الوراثية 272

البرنامج الجيني 309

البروتينات 265، 307–308، 313–314

البروتين العلاجي 312

البنية السلوكية 309

البويوطيقا 284، 302

البيوإتيقا 271

تحسين النسل 267، 270، 312، 323

تداعيات التجريب 264

ترميز الجيني 306

التطور 253–254، 258–260، 266، 270–271، 274، 283، 306، 326

تعديل الجينات 264، 312

التعديل الجيني 269، 277–278

تعديل الطبيعة البشرية 264، 277

التفاعل الخارجي 259

تفسير الديني الإسلامي 253

التقاعد 317

التقنية الحيوية 261، 264، 266–267، 282، 303–305

تكنولوجيا النسل 270

تمديد الحياة 304–305، 309–310، 316، 318، 320، 324، 326–328

التناهي 327

التنبؤ الجيني 310، 313
التوازن الاجتماعي 325
التوجه التجريبي 257
تيموس 255
جالتون، فرانسيس 267
الجريمة 316
الجسد 256، 272، 280
جمهورية أفلاطون 255
الجنينية البشرية 269، 314–315
جواز سفر جيني 313
الجوزية، ابن قيم 318
جوهر الإنسان 273
الجينات الغريبة 315
الجين الخامل 314
جين الشباب 309
الحتمية 262–263، 274، 288، 291، 293، 299
الحتمية الاجتماعية 262–263، 274
الحتمية البيولوجية 262
الحتمية الوراثية 288
حدثاً بيولوجياً 305
حدّ هيفليك 315
حرية الإرادة 290–291، 296
الحرية الشخصية 290
الحرية الفردية 291، 293
حقوق الإنسان 255
الحياة البيولوجية للإنسان 253
الحيوية 255، 261، 264، 266–267، 271، 277، 282، 285، 295، 302–305، 311–312،
318
الخرائط الجينية 266، 300
الخرطنة الجينية 272
خريطة وراثية 266
الخلايا الجذعية 282، 314–315
الخلايا الجنسية 269، 295
الخُلق 290، 298–299
الخلْق 290، 298
الخلقة 284، 292، 294–295، 298
الخِلْقة 284
الخلقة الأولى 294
الخلود 255، 322
داروين 258، 260
البعد الأخلاقي للجينوم 285، 290
داود 320
الدنا 261–262، 264–265، 277
ديكارت 256
الرموز البيولوجية 254
رنا 265
الروح 272–273
سقراط 255–256
سلسلة بروتينية 307
السلوك الإنساني 254، 287
السلوك البشري 259–260، 262–263، 267، 273، 291
شخصية الإنسان 255، 262، 276
شرودنغر، إرفين 305
شرودينغر 327
الشفرة الوراثية 261، 282

الشيخوخة 263، 304، 309–316، 322–323، 325–326
شيمكوف، إيليا 261
صبدجيكت 257
الصورة الباطنة 298
الصورة الظاهرة 298
الطبائع التكوينية 295
طب الشيخوخة 313
طبيعة إشكالية 259
الطبيعة البشرية 253–268، 270، 272–282، 304، 315، 318، 322، 327
الطبيعة الفيزيائية والكميائية للجينوم البشري 308
الطبيعة المادية 255
عبد الرحمن بن طرفة 276
العدالة الاجتماعية 255
العدوانية 276
العقلان 257
العلاج الجيني 268–269، 276، 278، 281–282، 314
العلاقات الأخلاقية بين الآباء 316
علاقات وظيفية بين الجينات 314
علم الأخلاق 284، 286–288
علم التخلّق السلوكي 312
علم الجينوم 254، 268، 271–272، 274، 276، 278–280، 284، 286–288، 305
علوم الحياة 260، 263
عملية السَلسَلة 307
عن عبد الله بن مسعود 276
العنف 254، 276، 278
عيادة التنبؤ الجيني 313
الفتق 295، 297–298
فرويد، سيغموند 258
فطرة 274، 319
الفيزيولوجيا 259
قانون التطور 258
القتل الرحيم 316
القرطبي 274
الكائن الحي 260–261، 305–309
الكرامة 266، 273–274، 293، 296
الكراهية 254
كروموزوم 261
كريك، فرانسيس 261، 265
كوجيتو الديكارتي 256
كوهن، دانيال 261
كينيون، سينثيا 311
اللاّتنوع 326
لاستنساخ البشري 278
اللاشعور 258
اللسان العربي 290، 301
جون لوك 256–257
لوك، جون 256
لولب مزدوج 261
مادة اجتماعية 325
المادية الجدلية 258
ماركس، كارل 258
ماهية الإنسان 255، 326
مجمع الفقه الإسلامي 275، 328
المجمع الفقهي التابع لرابطة العالم الإسلامي 323

المذاهب الأخلاقية 287
مرتبة الكمال 287
المرحلة الجينومية 307
المسؤولية الأخلاقية 291، 305
المساعدة على الانتحار 316
مشروع الجينوم البشري 264–267، 289، 294، 307
معجزة 318
المعلومات الوراثية 261–262، 265
معمّر 320
المقاربة العلمية البيولوجية 260
المناوي 320
مندل، جريجور 260–262
المواضع الجينية 308
الموت 317، 327
موت الإله 289
مورغان، توماس 262
النسبية الأخلاقية 292
النظرة الحضارية لعلم الجينوم 286
النظرية الداروينية 253، 307
النفس 254–255، 257–258، 281–282، 287، 292
نقل المعلومات 265
النمط البشري 292
النمط الجينومي 300
النمط الجيني 284–285، 298–299، 327
النمط الظاهري 284–285، 290، 298–300، 327
هابرماس، يورغن 271–272، 283
هريرة، أبي 319
الهندسة الاجتماعية 258
الهندسة الوراثية 264–267، 275، 281
هيدغر 328
هيفليك، ليونارد 311، 315
هيوم، دافيد 257، 283
واطسن، جيمس 261، 265
الوراثة 260–263، 266–267، 271، 281–282، 289–290، 299، 328
اليوجينا 312
اليوجينيا 267–268، 275

Index

'Abbās, 'Abd Allāh ibn 11, 156, 194
'Ammār al-Ṭālibī 101
'Amr, Ḍirār b. 154
'Aql 152–160
'ilm 5–11, 51–61, 130, 159
'Uthmān, Muḥammad Ra'fat 31, 51, 63–67, 88
abortion 1–8, 35, 51, 84–94, 142–147, 169, 183–191, 245–247
active intellect 160
Adam 102, 129, 156–158 227–231
Adoration of the Mystic Lamb 205
aḥkām fiqhiyya 36
AIDS 51, 90
Akhenaten 218
al-aḥkām al-khamsa 36
Alahmad, Ghiath 71
al-Ash'arī 154
al-Attas 155
al-Azhar 18–23, 37
al-Bār, Muḥammad 'Alī 27–41, 53–56
al-Baṣal, 'Abd al-Nāṣir Abū 25
al-Baṣrī, Al-Ḥasan 156
al-ḍarūriyyāt 67
al-faqīh 49
al-Ghazālī, Abū Hāmid 160–162
al-Ghumārī, 'Abd Allāh Ibn al-Ṣiddīq 37
algorithmic governance 219
al-Ḥaqq, Shaykh Jād al-Ḥaqq 'Alī Jād 37
al-ijtihād al-jamā'ī 49–51, 63–73
al-Iṣfahānī, Al-Rāghib 154
al-Jundī, Aḥmad Rajā'ī 27–37, 55–68
al-Khādimī, Nūr al-Dīn 31–40, 69
al-Kurdī, Aḥmad 57–67
'Abd Allāh, 'Abd Allāh Muḥammad 37, 61–67, 87–92, 156
al-Marāghī, Shaykh Muṣṭafā 18–21
al-Mu'tamir, Bishr b. 154
al-Nashmī, 'Ajīl 59–68
al-Qaraḍāwī, Yūsuf 21–25, 37–38, 95
al-Qurṭubī, Abū 'Abd Allah 152–156
al-Qushayrī, 'Abd al-Karīm 152
al-Raḥmān, Ṭāha 'Abd 20–31, 54, 102
al-Rāzī, Abū Bakr 21
al-Rūkī, Muḥammad 65–66
al-Salām, al-'Izz Ibn 'Abd 67
al-Salāmī, Muḥammad Mukhtār 57, 101, 185
al-Sayyid, Ḥamdī 63
al-Sa'dī, Shaykh 'Abd al-Raḥmān 20–32
al-Shawkānī, Muḥammad 41
al-Shirbīnī, 'Iṣām 42, 186
al-Suddī, Ismā'īl ibn 'Abd al-Raḥmān 156
al-Sulamī, Muhammad b. al-Ḥusayn 153
al-Tabarī, Ibn Jarīr 156
al-Ṭālibī, 'Ammār 102
Alzheimer's 83, 206
al-Zuḥaylī, Muḥammad 89
al-'ādamiyya 188
al-'Ammārī, 'Abd al-Qādir 42
al-'Awaḍī, 'Abd al-Raḥmān 27–29, 54–61, 101
al-'Uthaymīn, Ibn 188
anti-female bias 96
appreciation of life 181–192
Aquinas, Thomas 211
Arabic language 5–10, 22, 41
Arḍ 151
Aristotle 11, 128, 159, 172, 203–214
ART 75, 205, 219–230, 248
artificial reproduction 8
ATCG 65
Augustine of Hippo 9, 230
autonomy 10, 58, 83, 124, 177, 197–198, 226, 245
Averroes 211
Avicenna 154–160, 211
Bachelard, Gaston 218
badan 153
Badrān, Ibrāhīm 61–63, 103
Bāsalāma, 'Abd Allāh 67
bashar 151
Bayya, 'Abd Allāh Bin 61–63
Bāz, Shaykh 'Abd al-'Azīz Ibn 37–41, 88, 185
beginning of life 184
behavior 40–41, 71, 144–160, 181, 246
behavioral traits 247
benefit-risk assessment 56, 73–74
biobank 39, 71
bioethical discourse IX, 2–18, 36–47, 69–75, 169
bioethical standards 245
bioethics 2–24, 40–54, 68–75, 113–119, 131–133, 203, 224–228, 243

biomoral modification 128
biopower 219
biotechnology 169–172, 203
blood money 190
blueprint 82, 122, 206–207
bodily identity 145
Bonhoeffer, Dietrich 9, 224–231
book of life 26, 82, 205–206
Boorse 228–229
breathing of the soul 184–188
CAGS 52, 71
Caliph ʿUmar 87, 101
carrier testing 82
CDA 118
Celera Corporation 244
CFTR 143
changing God's creation 97–99, 194
Chardin, Pierre Teilhard de 204–209
Christian theology 225–237
Christian tradition 8
chromosomes 82, 122, 242
CILE 1–28, 47–53, 71–75, 113, 169
Clark, Angus 9
clinging substance 184–190
Clinton, Bill 37, 204
cloning 9, 25, 37, 51–55, 86, 99–103, 120–131, 174, 242–248
co-creation 9, 210–215, 231
co-explorers 231
cognitive abilities 146
coitus interruptus 95–96, 186–191
Cole-Turner, Ronald 9, 231–236
collective duty 5, 32–39, 69
collective *ijtihād* 5–6, 23–51
collective unconscious 216
Collins, Francis 26–33, 68–72, 130, 203–210, 224
commodification of fetuses 180
commodification of genetic data 244
common morality 18
communitarian ethics 10, 245
concept of the family 6, 81
confidentiality 91–92, 191–193, 245–247
consanguineous marriage 87–91, 105, 246
Copernicus 141, 169
cosmic intelligence 160
creation 8–9, 57–132, 148–247
Crick, Francis 11, 65, 81, 141–145, 208, 224–234
criminal behavior 146, 246
CRISPR 127, 208
cybernetics 233
cystic fibrosis 84, 143
Dār al-Iftāʾ 38, 175
Darwin 11, 141, 225, 242
data-store 123
ḍawābiṭ Sharʿiyya 6, 27–36, 67–72
Dawkins, Richard 142–145
De Anima 159, 204–214
Delbrück, Max 204–213
designer babies 85
dharr 156
diagnostic testing 82
Dierickx, Kris 71
disability 9, 85–91, 103, 126, 232–246
disclosure 119–129, 172
discursivity 207–214
distinctiveness of life 192
divine creation 81, 100–104, 189–198
DNA 1, 26–35, 52–81, 98, 122–145, 204–242, 306–309, 312–313, 315, 326
DNA fingerprinting 52–57, 72, 98
Dolly the sheep 86
dominion 9, 132
double helix 81–82, 224
Down Syndrome 82–84
Dubai Declaration 52, 71
duty to warn 123–130
egg donation 125
ELSI 33, 47–48
embodiment 9
Embracement-Inclined Approach 5, 68–74
embryo 86–94, 125–131, 183–194
embryo donation 125
embryonic development 7
Encyclopedia of the Human Genome 9, 241
enforced sterilization 246
entelechy 154
Epigenetics 312, 330
equality 177–179
era of genomics 25–28, 40, 53
ethical being 179, 196
ethical harms 62
ethic of care 123

eugenics 9, 103, 144, 170–196, 242–246, 312
European Council for Fatwa and Research 25, 90
euthanasia 245–246
evolution 22, 127, 141–142, 169–180, 192–196, 231–242
experimental sciences 10, 172
Eyck, Jan van 205–206
fahm 159
family institution 6
family relationships 81–84, 104, 246
Fārābī, Abu Nasr 159
farḍ kifāya 5, 32–39, 69
farḍ ʿayn 69
Farīd, Muḥammad 21
female infanticide 96
Fenton, Elizabeth 178–180
fertilized egg 181–195
fetal reification 180
fetal sex selection 6, 85–105, 248
fetus 84–98, 127, 142, 174–198
fiqh 1–60, 75–103, 175–176, 191–195, 245
first covenant 155–164
fitrah 101
free will 40, 75, 145–147, 177, 224–232
Freud, Sigmund 11, 146, 207–215
fuqahāʾ 75
gamete donation 98–103, 243–248
gay couples 126
gender discrimination 85
gene mutation 82–84
gene therapy 22–28, 52–72, 85, 117, 143, 244
genethics 6–7, 115–133
genetic alterations 82
genetic counseling 54, 72, 91–93, 245–246
genetic determinism 11, 41, 130–147, 224–234, 247
genetic discrimination 62–67, 83
genetic disease 9, 90, 193, 242
genetic disorder 82–94, 143
genetic engineering 9, 27–35, 52–85, 101–103, 126–127, 170–176, 192–196, 231, 244
genetic enhancement 101, 120–132, 194
genetic essentialism 104, 145
genetic explanations 9, 242–247
genetic intervention 7–12, 98–100, 128–144, 170–196
genetic predisposition 146
genetics 6–11, 27–74, 99–103, 115–144, 172–175, 233–249
genetic structure 25, 65, 80–85, 99–103
genetic technology 81–86, 102–104, 125–128, 174–198, 244
genetic tests 82–92, 126
genome 1–13, 25–82, 103–179, 191–245
Genome Question 1–11, 53, 113, 140–147, 203
genomic data 123–130, 244
genomic metaphysics 236
genomics 1–83, 115–148, 162–169, 203–227, 242
genomic sequencing 6, 32
genomic technologies 2–7, 31, 122–132
genotype 12, 25, 132, 144–145
germline gene therapy 85
germline genetic modification 6, 87, 99–105
germline manipulation 127
Gestalt 218
gestational surrogacy 125–131
Ghudda, ʿAbd al-Sattār Abū 65–68, 87–103
GQ 1–5
guardianship 102, 191
Habermas, Jürgen 11, 169–184, 196–197
Ḥadīth commentaries 41
Hamer 209
Hare 229
Ḥathūt, Ḥassān 27, 55–69, 94–103
hawā 60
Ḥawāʾ 156–158
Ḥawwa, Saʿīd 152
Hayʾat Kibār al-ʿUlamāʾ 38
heart 24, 40, 57, 129, 159–163, 177, 197
Hegel 8, 181, 204–219
Heidegger, Martin 171–172
heliocentrism 141
hereditary diseases 82
HGP 26–33, 47–74, 203–247
ḥifẓ al-nafs 62
ḥifẓ al-nasl 62
ḥijr 159
ḥikma 159
ḥukm sharʿī 4, 59
human agency 40, 64–66, 145–153, 247
human being as an evolving entity 129–132
human genetic structure 25, 80–84, 99–103

Human Genome Project 9–11, 25–82, 117, 158–164, 203–204, 226–244
human identity 2
human milk banks 25, 51
human nature 6–12, 101–104, 116, 129, 170–180, 195–197, 236–237
humanness 12, 188–189, 236
human person 7, 139–165
human potential 237
human reproduction 97, 119–132
human reproduction related ethics 120
human transgression 157
ḥurma 158
hylemorphic 211–213
Ibn al-Jawzī 187
Ibn Rushd 211
Ibn Sīnā 21, 211
Ibn Taymiyya 8, 159
Ibn ʿAbbās 156, 194
Ibn ʿābidīn 187–191
ibtilāʾ 59
iconoclasm 218–219
Idrīs, Jaʿfar Shaykh 31, 51, 101
IFA 28, 48–52, 245
Iglesias, Teresa 182
IIFA 24–34, 50–72, 91, 191, 245
ijtihād 4–6, 23–73
image of God 9, 226–238
imago dei 9, 226–227
inborn disposition 101
inbreeding 246
individual freedom 10, 197, 245
individual *ijtihād* 24–32, 50
individual privacy 247
insān 64, 151–154
intellect 59–65, 81, 102, 152–163
interdisciplinarity 2–8
interdisciplinary communication 54
International Islamic Fiqh Academy 24–32, 50–55, 81–91, 103, 175–176, 191–195, 245
intervening observer 172–175
in vitro fertilization 37, 98
IOMS 24–70, 97–103, 244
Irenaeus 9, 230–231
ISESCO 27, 51–58, 70
Islamic bioethical framework 243
Islamic bioethics 2–24, 131–133
Islamic ethics 1–11, 28–33, 51–53, 73, 113, 132, 169, 203
Islamic family regulations 84
Islamic Fiqh Academy 24–32, 48–55, 81–91, 103, 175–176, 191–195, 245
Islamic jurisprudence 1–18, 36–48, 60, 75, 177
Islamic legal discourse 243
Islamic legal theory 66
Islamic model of the family 81
Islamic scholastic theology 171
Islamic theology 7, 75, 116, 132
Islamic tradition 1–10, 49–52, 67–75, 87–88, 116, 129–164
Islamic values 71–72
Islamic worldview 131
Islamization of Knowledge 8
Ismāʿīl, ʿAbd al-ʿAzīz 18–21, 156
IVF 22, 82–93, 126–131
jāh 158
Jasad 153, 194
Jewish tradition 128
Jonsen, Albert 18
Junker-Kenny, Maureen 228
justified intervention 172, 184
kalām 171–176, 198
kalimāt dīniyya 8
kalimāt kawniyya 8
Kant 179–181, 196–197
Karnein, Anja J. 181–183
karyotype 146
kasb 40, 59
Kaʿb, Muhammad ibn 156
Kennedy Institute of Ethics 37, 117
khalq 96–100, 194
Khāṭir, Shaykh Muḥammad 38
Kierkegaard 173–179
Kuwait Document 39
Lacan, Jacques 8, 204–219
language of life 65
legal capacity 40
lesbian couples 125–126
liberal eugenics 170–178
like God 9, 227–236
Locke, John 11, 180
lubb 159
mafāsid 56

Makkah Council 176
Manhattan Project 61
mankind 32, 69–70, 139, 151, 226–236
MAOA 235
maqāṣid al-Sharīʿā 59
marriage 62, 81–93, 105, 131, 174, 196, 246
maṣāliḥ 20, 56
maṣlaḥa maḥḍa 67
Mauron, Alex 210, 236
Maʾmūn, Shaykh Ḥasan 38
medical model 228–229
medical profiles 246
medical treatment 35, 62–69, 83–89, 101, 196
MEDLINE 117–119
monopolization 244
moral reasoning 116, 147
moral status of embryos 85
morphogenesis 182
MRT 126–131
muftī 49, 101
mujtahid 23, 50
Muslim biomedical scientists 48
Muslim World League 24–28, 48–50, 81–103, 191, 245
nafs 7, 62–66, 151–164, 191, 247
Nasim, Anwar 57–58
nasl 60–66
natural gender balance 85, 97
nature of man 139
neodarwinism 307
neural circuitry 162
neutral observer 172–175
newborn screening 82
Noble, Denis 59–65, 143–145
normativity 86, 180
nuclear bomb 61
nuclear energy 61
nuhā 159
objective morality 179
occidental perspective 8
ontology 115–133
Oppenheimer 61
organ transplant 20–21
original permissibility 60–66
Osler, William 75
ownership of data 120–124
parenthood 125–131, 248
parthenogenesis 86
particular morality 18
paternity testing 8, 98
paternity verification 84, 98
Peacocke, Arthur 231
personality 103, 177–195, 219
personalized medicine 26, 80
PGD 82–93
PGS 85–94
phenotype 12, 25, 132–144, 237
Plotinus 154
post-genomic era 25, 308, 330
precaution-inclined approach 5, 64–74
predictive testing 82
pre-implantation stage 7
premarital genetic testing 6, 87–91, 105
prenatal testing 82–85
pre-personal life 177–183
preservation of lineage 131, 193–196
preservation of progeny 60
preservation of self 184–196
preventive medicine 75
procreation 90, 169–191
Prophet Zakariyya 95
prospective offspring 81–85
provision 99
psychosocial risks 62
qalb 151–163
Qalʿajī, Muḥammad Rawwās 55, 67
Qatar Biobank 39, 71
Qaṭma, ʿAbd al-Majīd 56–68
QGP 39, 72
qiyam 102
quality of life 143–144, 181–192
Qurʾān 2–20, 40–41, 88–99, 148–161, 198
Qurʾān exegesis 2–3, 41
rational animals 203
rationality 146
realisation 207–213
regulation of knowledge 124
religious revelations 216
reproduction 8–9, 35, 91–98, 119–132, 169–181, 197, 242
reproductive cloning 86, 99–103, 120–131, 248
reproductive function of marriage 89–90
reproductive organism 6, 125–130

Ricci, Matteo 18
Riḍā, Rashīd 19–32
rizq 27, 54–55, 99
Rizq, Hānī 27, 54–55, 99
RNA 265, 307
Rovelli 210
ṣalṣāl 151
sanctified life 182
sanctity of human life 85
Savulescu, Julian 128–146
saw'a 158
scientific information 54–73
scientific revolution 58–74, 210
SCNT 86
selective abortion 1–8, 85
self-consciousness 146, 209–214
self-knowledge 65, 155, 178, 206–207
self-understanding 173–178, 196–203, 220
sensory reception 163
shakhṣiyya 188
Sharia 4–6, 18–72, 86–102
sicut deus 9, 227–228
Ṣidqī, Muḥammad Tawfīq 21
Singer 142–146
social dependency 178
social justice 103, 247
somatic gene therapy 85
soul 2–11, 75–80, 94, 140–164, 184–213, 225–247
speaking animals 217
speciation 127
speciesism 140–147, 227
spirituality 164, 209–218
stem cell research 22, 85
stem cells 85–86
structure of DNA 141, 224
Sufism 2–4, 41
surrogate motherhood 98
tadabbur 159
tadāwī 62–69
tafakkur 159
taḥqīq al-manāṭ 21
takhayyur 88
taklīf 40
taṣawwur ṣaḥīḥ 21, 49–56
Taylor, Charles 123, 141–145
Tay-Sachs 84
technical harms 62
techno-sphere 215
technotope 207
Teilhard 8, 204–217
telos 9, 224–238
textuality 8, 208–220
Thalassemia 84
the Fall 151–158, 224–232
theological framing 59–70
Ticehurst, Flo 9
ṭīn 151
transhumanism 103, 144
turāb 151
Tutu, Desmond 141
twins 144
two readings 8
umma 69
uniqueness 101, 156, 183, 207–208
unjustified intervention 172, 184
Venter, Craig 204–213
Watson, James 11, 33, 65, 81, 141–142, 207–208, 224–232
Weiner, Norbert 233
Wyatt, John 9, 230–231
XYY defence 146
yaqīn 63, 130
Yashū, Ḥasan 31–35
Zahra, Muḥammad al-Mursī 57
ẓann ghālib 63

Printed in the United States
By Bookmasters